Product Design and
Technological Innovation

Product Design and Technological Innovation

A READER

Edited by

Robin Roy and David Wield

at the Open University

Open University Press
Milton Keynes · Philadelphia

Open University Press
Open University Educational Enterprises Limited
12 Cofferidge Close, Stony Stratford, Milton Keynes MK11 1BY, England
and
242 Cherry Street, Philadelphia PA 19106, USA

First published 1986

British Library Cataloguing in Publication Data
  reader.
  1. Design, Industrial—Great Britain—
  Management
I. Roy, Robin   II. Wield, David
658.5′752′0941   TS57

ISBN 0–335–15110–8
ISBN 0–335–15109–4 Pbk

Library of Congress Cataloging in Publication Data
Product design and technological innovation.
  Bibliography: p.
  1. Technological innovations.   2. Design, Industrial.
I. Roy, Robin.   II. Wield, David.
T173.8.P725 1985   607′.2   85–13582

ISBN 0–335–15110–8
ISBN 0–335–15109–4 (pbk.)

Text design by Clarke Williams
Typeset by S & S Press, Abingdon, Oxfordshire
Printed in Great Britain by
M. & A. Thomson Litho Limited,
East Kilbride, Glasgow, Scotland

Contents

Local and Regional Initiatives

Section 6   Future Directions

The Pattern of Innovation

Controlling Technology

Preface

This book was produced as an integral part of a third-level Open University course, T362 *Design and Innovation*. It is also designed to be of relevance and interest to those who are not studying the T362 course.

The themes of the course and the contents of the book reflect the growing concern, in Britain, the USA and other industrialized countries, with the role played by product design and technological innovation in the performance of firms, industries and national economies. The articles and extracts included in this book were chosen to fill a gap in the literature, bringing together for the first time writings on *design* and *innovation*, and including the viewpoints of designers, engineers and technologists, as well as those of economists, management scientists, sociologists and policy-makers.

The *Design and Innovation* course itself consists of three related streams of study, namely, 'theory', 'case-studies' and 'project work'. Within this structure, the course is divided into blocks, each focusing on a key factor in the design and innovation process (the creative individual; the market; enterprise organization; government policy, etc.) and particular case-studies. The case-studies are drawn from the areas of renewable energy technology (water turbines, solar cells); information technology (Sinclair microcomputers, Prestel interactive videotex) and transport (British Rail's APT and HST high-speed trains).

This book forms the core of the course theory; it is divided into six main sections which parallel the course's block structure. Each section comprises a mixture of materials ranging from newspaper and magazine articles, which highlight the issues in a readily accessible form, to extracts from more theoretical and scholarly books and papers. Each section or subsection is preceded by an introductory essay written by the editors. These introductions are intended to do more than just review the articles; they also emphasize and comment on the themes discussed and fill in gaps in the coverage of the section.

It is worth noting that certain of the introductions refer to a publication entitled *Design and the Economy* by Rothwell, Schott and Gardiner. This booklet was originally produced to accompany an exhibition of the same title held at the London Design Centre in 1983, but has since been revised and is supplied to students as set reading for the T362 course. Further information on this and other course materials, including the case-studies, television and audio programmes, may be obtained from: Open University Educational Enterprises Ltd, 12 Cofferidge Close, Stony Stratford, Milton Keynes MK11 1BY.

Finally we should like to thank all those who helped in the preparation of this book. In particular we wish to mention David Elliott, David Walker and Ernst Braun for suggesting articles; Ernie Taylor for invaluable practical assistance; Sally Boyle for drawing diagrams, and several other course team members for their comments on drafts. Thanks are also due to John Taylor of the Open University Publishing Division for his advice, and to Olive Ainger and Nessie Tait for typing and retyping various drafts of the introductions.

Robin Roy

David Wield

Acknowledgements

1.1 Reprinted from T. Burns and G. M. Stalker, *The Management of Innovation*, second edition, Tavistock, pp. 22–32, 36, by permission of the authors and of Associated Book Publishers (UK) Ltd.

1.2 Reprinted from Georgia Technological Innovation Project, 'Introducing Innovation', in P. Kelly and M. Kranzberg (Eds), *Technological Innovation: A Critical Review of Current Knowledge*, San Francisco Press, 1978, pp. 1–7, 11–13, by permission of San Francisco Press.

1.3 Reprinted from C. Freeman, *The Economics of Industrial Innovation*, second edition, Frances Pinter, 1982, pp. 107–112, by permission of Frances Pinter (Publishers).

1.4 Reprinted from N. Cross, Unit 9, 'Design & Technology' in T262 *Man-made Futures*, Open University Press, 1975, pp. 9–13, 17–26, by permission of The Open University.

1.5 Reprinted from D. Pye, *The Nature & Aesthetics of Design*, Herbert Press, 1983, pp. 21–23, 70, 75–76, by permission of the Herbert Press Ltd.

1.6 Reprinted from Lord Caldecote, 'Investment in New Product Development', *Engineering*, June 1979, pp. 811–814, by permission of the Design Council.

2.1 Reprinted from M. Brown, 'Eureka! I've Found Some Money', *Sunday Times Business News*, 14 October 1984, by permission of Times Newspapers Ltd.

2.2 Reprinted from J. Wardroper, 'Dry Run in a Kitchen Sink' and J. Huxley, 'Hours Not to Reason Why', *Sunday Times Business News*, 11 December 1983 and 6 November 1983, by permission of Times Newspapers Ltd.

2.3 Reprinted from F. Amram, 'The Innovative Woman', *New Scientist*, 24 May 1984, pp. 10–12, by permission of *New Scientist*.

2.4 Reprinted from P. Kelly *et al*, 'The Individual Inventor-entrepreneur', in P. Kelly and M. Kranzberg (Eds), *Technological Innovation: A Critical Review of Current Knowledge*, San Francisco Press, 1978, by permission of San Francisco Press.

2.5 Reprinted from G. Glegg, *The Design of Design*, Cambridge University Press, 1969, pp. 18–24, by permision of Cambridge University Press.

3.1 Reprinted from C. Freeman, *The Economics of Industrial Innovation*, second edition, Frances Pinter, 1982, pp. 169–183, by permission of Frances Pinter (Publishers).

3.2 Reprinted from W. Evans, 'Japanese Management, Product Planning and Corporate Strategy', *Design Studies*, January 1985, by permission of *Design Studies* and Butterworth.

3.3 Reprinted A. Cottrell, 'Technological Thresholds', in *The Process of Technological Innovation*, National Academy of Sciences, 1968, pp. 48–57, by permission of the National Academy of Sciences.

3.4 M. Posner, 'Innovation in the Public Sector', Paper given to a British Association for the Advancement of Science Symposium *Investing in Innovation*, 2 May 1984.

3.5 Reprinted from E. B. Roberts, 'Generating Effective Corporate Innovation', *Technology Review* October/November 1977, pp. 27–29, 33, with permission from *Technology Review*, Massachusetts Institute of Technology, © 1977.

3.6 Reprinted trom M. Oakley, *Managing Product Design*, Weidenfield & Nicolson, 1984, pp. 49–55, 57–60, 73–76, 81, 89–92, by permission of Weidenfield (Publishers) Ltd and John Wiley and Sons Ltd.

3.7 Reprinted from R. Rothwell, 'The Role of Small Firms in the emergence of New Technologies', *Omega*, 1984, Volume 12, Number 1, pp. 19–22, 27, by permission of Pergamon Press Ltd.

3.8 Reprinted from D. Fishlock, *The Business of Science*, Associated Business Programmes, pp. 82–90, by permission of Associated Business Programmes.

4.1 Reprinted from J. Pilditch, 'Marketing: How Britain Can Compete', *Marketing*, December 1978, pp. 34–38, by permission of Haymarket Publishing.

4.2 Reprinted from C. Lorenz, 'Gut Feel is Market Research Too', *Design*, December 1983, pp. 31–32, by permission of the Design Council.

4.3 Reprinted from R. C. Bennett and R. G. Cooper, 'Beyond the Marketing Concept', *Business Horizons*, June 1979, pp. 76–79, 83, copyright © 1979, by the Foundation for the School of Business at Indiana University. Reprinted by permission.

4.4 Reprinted from A. E. Pannenborg, 'Technology Push Versus Market Pull — the Designer's Dilemma', *Electronics and Power*, Volume 21, part 9, May 1975, pp. 563–566, by permission of the Institution of Electrical Engineers.

4.5 Reprinted with permission of the Free Press, a Division of Macmillan Inc. from E. M. Rogers, *Diffusion of Innovations* (3rd edition), copyright © 1962, 1971, 1983 by the Free Press.

4.6 Reprinted from C. Freeman, 'The Diffusion of Innovations' in *Technology and Employment*, Industrial Development Research Series Number 20, Aalborg University Press, pp. 18–27, 29–32, by permission of Aalborg University Press.

5.1 Reprinted from E. Braun, *Wayward Technology*, Frances Pinter, 1984, pp. 123, 125–131, 134, 137–141, by permission of Frances Pinter (Publishers).

5.2 Reprinted from M. Walker, 'Britain's Self-inflicted Wound' *The Guardian*, 25 April 1982, by permission of Martin Walker.

5.3 Reprinted from R. Williams, 'British Technology Policy', *Government and Opposition*, Volume 19 Number 1, Winter 1984, pp. 30–40, by permission of *Government and Opposition*.

5.4 Reprinted from C. Lorenz, 'A Resurgence for UK Designers', *Financial Times*, 23 May 1984, by permission of Financial Times Business Information Ltd.

5.5 Reprinted from F. J. P. Clarke, 'Energy', in M. Goldsmith (ed), *UK Science Policy*, Longmans, 1984, pp. 82–84, 99–102, by permission of the Longman Group Ltd and of the Science Policy Foundation.

5.6 Reprinted from A. Moreton, 'Science Parks', *Financial Times*, 21 January 1983, by permission of Financial Times Business Information Ltd; R. Faux, 'High Technology in Silicon Glen', *The Times*, 27 April 1984, by permission of Times Newspapers Ltd; M. Cooley, 'Tapping London's Skill Resources', Times Higher Education Supplement 1 July 1984, by permission of the *Times Higher Education Supplement*.

6.1 Reprinted from W. J. Abernathy and J. M. Utterback, 'Patterns of Industrial Innovation', *Technology Review* June/July 1978, pp. 41–47, with permission from *Technology Review*, Massachusetts Institute of Technology, © 1978.

6.2 Reprinted from P. Hall, 'The Geography of the Fifth Kondratieff Cycle', *New Society* 26 March 1981, pp. 535–537, by permission of *New Society*.

6.3 Reprinted from G. Ray, 'Innovation in the Long Cycle', *Lloyds Bank Review*, Number 135, January 1980, by permission of *Lloyds Bank Review*.

6.4 Reprinted from D. Gabor, *Innovations: Scientific, Technological and Social*, Oxford University Press, 1970, pp. 1–3, 6–9, © Oxford University Press 1970, by permission of Oxford University Press.

6.5 Reprinted from D. Collingridge, *The Social Control Technology*, Open University Press, 1980, pp. 13–22, by permission of the author.

6.6 Reprinted from J. D. Davis, 'Appropriate Engineering', *Engineering*, December 1979, by permission of *Engineering*.

6.7 Reprinted from M. Cooley, 'Socially Useful Design', in R. Langdon and N. Cross (Eds), *Design Policy Volume 1: Design & Society,* Design Council, pp. 51–54, by permission of the Design Council.

6.8 Reprinted from Transport and General Workers Union, *A Better Future – Strategy For Arms Conversion,* Transport and General Workers Union, pp. 15–19, 24, by permission of the TGWU.

General Introduction

Robin Roy

This book is divided into six major sections, each of which is concerned with a key factor influencing product design and technological innovation. Section 1 shows the *practical* importance of gaining an understanding of the process and meanings of design and innovation, and how these activities relate to each other. The next four sections each focus on a different 'actor' influencing the design and innovation process. Section 2 looks at the creative individual inventor/designer/entrepreneur; section 3 considers innovative business organizations in the private and public sectors; section 4 examines market and user influences on the generation and diffusion of innovations; and section 5 deals with national and local government initiatives to promote design and innovation. The final part of the book (section 6) is concerned with past and future directions of design evolution and technological change, and considers whether there are alternatives to the present dominant patterns of design and innovation that might avoid some of the problems of modern technology.

Each of the sections or subsections is prefaced by a substantial introductory essay, which both reviews and covers areas missing from the reprinted articles. This General Introduction therefore attempts to set the rest of the book in context. It discusses the emerging awareness over the past century of the importance of both product design and technological innovation to economic success. The focus is on Britain, but as several of the articles in the book show, a similar awareness has developed in many other industrialized countries, in particular the United States and Japan.

The growing economic importance of design and innovation

Britain's share of world trade in manufactured products has been in decline since the late nineteenth century. Since 1945 this decline has accelerated, and in the past twenty years it has been accompanied by the increasing penetration of imported goods onto the British home market. By 1984, imports of manufactures exceeded exports for the first time. Not surprisingly, this situation has caused debate, analysis and concern, both inside and outside government for over a century.

Since the Second World War, most diagnoses and recommended cures have tended to focus on the *price* competitiveness of British goods, and the associated questions of how to increase productivity and improve industrial relations. In recent years, however, it has become apparent that price is not everything, and that the nature and quality of the *product* itself is a key determinant of competitiveness. During the past five years especially, a major interest has emerged in the links between price and 'quality' in competitiveness, and in what economists call 'non-price' factors. Non-price factors are discussed in more detail in the booklet *Design and the Economy* by Rothwell, Schott and Gardiner (1985, pp. 6–7), but broadly they encompass the factors affecting the value of the product as perceived by the customer plus marketing and after-sales efforts of the manufacturer.

One of the most important non-price factors, it has been recognized, is how well products are designed. This recognition was stimulated by two official reports, the Corfield Report on *Product Design* (NEDO, 1979) and the Finniston Report, *Engineering our Future* (Committee of Inquiry into the Engineering Profession, 1980). Both reports emphasized the importance of design in adding to the value of products, in particular to their technical performance and overall quality, as reflected in their appearance, finish, reliability, durability, safety, ease of use and maintenance. The Corfield Report concluded:

> The difference between the apparently more successful companies and countries and those less so is not in the *quantity* of work performed, but rather the *quality*. Not in the volume of final output, but in the value added to basic raw materials. This value is determined more by the quality of design and the way it is made to meet customers' requirements, than by other factors.

This conclusion was of course nothing new. There had been concern about the design of British products relative to European and American goods since the last century. In 1944, the Council of Industrial Design was established with the specific remit of promoting good design in British-made consumer goods. In the engineering field, too, the role of design had been recognized at least since the Fielden Report on *Engineering Design* (Department of Scientific and Industrial Research, 1963), one consequence of which was that by 1970 the Council of Industrial Design had become the Design Council, with responsibility for engineering products as well as consumer goods.

What *was* new was the interest in design at the highest levels of government. In January 1982, the Conservative Prime Minister, Margaret Thatcher, held a seminar on Product Design and Market Success at 10 Downing Street, to which a wide range of personalities from the design world were invited – from Zandra Rhodes, the fashion designer, and Terence Conran, founder of Habitat, to the Chairman of British Aerospace and the Vice-Chancellor of the Open University. In the May 1982 issues of the Design Council's sister journals *Design* and *Engineering*, Thatcher wrote:

> There are many ingredients for success in the market place. But I am convinced that British industry will never compete if it forgets the importance of good design. By 'design' I do not just mean 'appearance'. I mean all the engineering and industrial design that goes into a product from the idea stage to the production stage, and which is so important in ensuring that it works, that it is reliable, that it is good value and that it looks good. In short it is good design which makes people buy products and which gives products a good name. It is essential to the future of our industry.

The main outcome of the Prime Minister's concern was a modest Government-sponsored campaign called 'Design for Profit', launched in 1983. (Details are given in article 5.4 by Lorenz, in Section 5.)

So far I have discussed design, but not mentioned innovation. Clearly, designing reliable, safe, good-looking products is not enough to ensure competitive success, especially in rapidly changing technological industries. It is not surprising that parallel to the periodic concerns with design are similar concerns with the need to exploit the results of scientific research and technical development to produce industrial and technological innovations for social, economic and commercial benefit. Burns and Stalker in article 1.1 show that official concern with fostering technological innovation is as old as that with promoting good design: a Royal Commission investigated the problem in the 1870s, and in 1917 the Government established a Department of Scientific and Industrial Research. The Second World War gave another boost to those arguing for increased government support for science and technology, and in the 1950s the Board of Trade sponsored extensive studies of how science might be harnessed to industry (Carter and Williams, 1957). In 1963, Harold Wilson made his famous speech to the Labour Party conference referring to 'the Britain that is going to be forged in the white heat of this technological revolution'. This led to the establishment in 1964 of a Ministry of Technology, responsible initially for four key sectors – computers, electronics, telecommunications and machine tools – but later taking on the rest of manufacturing, including the less glamorous industries.

Although there was a period in the late 1960s and early 1970s during which the benefits of science and modern industrial technology were seriously questioned, by the late 1970s several studies, notably those conducted by the Science Policy Research Unit at Sussex University (Pavitt, 1980), had confirmed that international competitiveness depended on a nation's capacity for technical innovation in response to or in anticipation of changing market requirements and technological trends. Britain's weakness, as the articles in section 5 show, is that technology policy was not fully linked to industrial and trading strategy. There was another prime ministerial seminar, this time on 'Innovation', held in September 1983. Previously, the Finniston Report (1980) had noted that:

> the strengths of the advanced countries lie in inventing and exploiting new products and processes, incorporating high levels of human skill and knowledge, most of it at the leading edge of technology, and in continual incremental improvements to current products and processes through reducing production costs.

The above quotation neatly brings us to the point that the innovation, design and price aspects of a product are closely related. First, as is shown in section 1, what most designers do is in the area of evolutionary improvement or 'incremental' innovation of existing products. Secondly, designing products so that they can be manufactured economically, and introducing *process* innovations (such as robots and computer-controlled machine tools), are two ways in which technologically advanced, high-quality goods can be made and sold at a competitive price.

In recent years therefore there has been a convergence of opinion, at least in government and academic circles, that good product design combined with product and process innovation is the key to competitive success and economic recovery. None the less, many industrialists, managers and financiers remain to be sufficiently convinced to make best use of the skills of designers, engineers and innovators. This book is intended to provide some conceptual and practical guidelines for those attempting to improve the management and practice of design and innovation, both in British industry and elsewhere.

References

Carter, C. F. and Williams, B. R. (1957) *Industry and Technical Progress*, Oxford University Press.
Committee of Inquiry into the Engineering Profession (1980) *Engineering our Future* (the Finniston Report), HMSO.
Department of Scientific and Industrial Research (1963) *Engineering Design* (The Fielden Report), HMSO.
NEDO (1979) *Product Design* (The Corfield Report), National Economic Development Office.
Pavitt, K. (ed.) (1980) *Technical Innovation and British Economic Performance,* Macmillan.
Rothwell, R., Schott, K. and Gardiner, J. P. (1985) *Design and the Economy: The Role of Design and Innovation in the Prosperity of Industrial Companies,* 3rd edn, Design Council.

Section 1

Design and Innovation

Introduction: Meanings of Design and Innovation

Robin Roy

This section falls into two parts. The first is concerned with technological innovation and how it is influenced by social, technical, commercial and economic pressures. The second is concerned with an often unrecognised part of the innovation process, namely how new and improved products are designed and developed.

As the articles in this section demonstrate, there is a considerable diversity of meaning and interpretation of the terms 'design' and 'innovation'. Moody (1984) has pointed out that some of this diversity is because there are no unambiguous words in English for 'design' (which tends to be associated either with the activity of drawing, or with fashionable objects created by art-school-trained designers) or for 'innovation' (which tends to be associated either with the activities of inventors and scientific researchers, or with novel objects that emerge from research and development laboratories). This introduction therefore aims to clarify the meanings of design and innovation and to illustrate how they relate through other terms such as 'invention', 'research and development' (R & D), 'engineering' and 'technology'.

Technological Innovation

In the first article (1.1), Burns and Stalker tend to use the terms 'invention' and 'innovation' rather loosely, in the sense of any technologically new object or industrial process.

In the following article (1.2), Kelly and his colleagues are more precise, acknowledging that there are many different meanings of the term 'technological innovation'. They note that the most widely accepted definition is that of a 'milestone' in the process leading from *invention* (the first idea, sketch or model for a new or improved device, product, process or system) to *diffusion* (of the resultant product, process or system through a population of potential users). *Innovation,* in the words of Freeman in article 1.3, is the point of 'first *commercial* application or production of a new process or product'.

However, innovation is frequently also used to describe the *whole process,* from invention to the point of first commercial or social use. Technological innovation is defined in the 'Frascati Manual' (a document issued by the Organization for Economic Co-operation and Development as a proposed standard for measuring scientific and technical activities) as follows:

> Technological innovation is the transformation of an idea into a new or improved saleable product or operational process in industry or commerce . . . (OECD, 1981).

This definition is the one adopted by Kelly and his colleagues. It includes all the various activities – research, design, development, market research and testing, manufacturing engineering, etc. – involved in converting a new idea, invention or discovery into a novel product or industrial process in commercial or social use (see Figure 1).

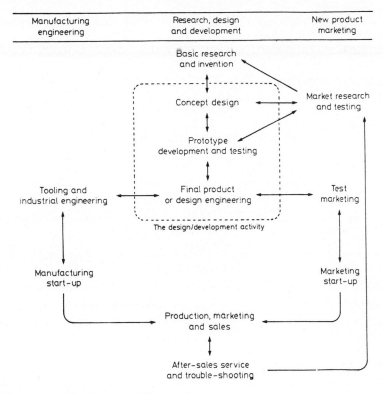

Figure 1 The process of technological innovation
Source: Roy and Bruce (1984).

Design and Development

Design and development may be seen from Figure 1 to be a part of the total innovation process. In article 1.6, Caldecote defines design as 'the process of converting an idea into information from which a new product can be made'. Design therefore is the activity in which ideas or market requirements are given specific physical form, starting from the initial sketches or conceptual designs, through prototype development, to the detailed drawings and specifications needed actually to make the product.

The essence of design, Cross notes in article 1.4, is that:

> it *separates designing from making* [and] . . . In its use of drawings, it contains *a formalized method for the abstract consideration of form*. This method enables new forms to be devised and to be tested in a modelled form before . . . the process of production.

In an attempt to distinguish design from research and development, in the context of the total innovation process, Christopher Freeman (1983) has identified four kinds of design activity:

1. *Experimental design:* the design of prototypes and pilot plants leading to the preparation of production drawings for the commercial introduction of a new product or process.

2. *Routine design engineering:* the adaptation of existing technology to specific applications (typical of the design work done by many engineering firms when installing new plant or equipment).
3. *Fashion design:* aesthetic and stylistic design of items ranging from textiles and shoes to chairs, car bodies and buildings. (This kind of design may result in novel forms, shapes or decorations, but often involves no technical change at all.)
4. *Design management:* the planning and coordinating activity necessary to create, make and launch a new product onto the market.

Engineering design and industrial design

The Corfield Report (NEDO, 1979) defines product design as including 'both engineering design and industrial (aesthetic) design'. However, in manufacturing industry a distinction is frequently made between the contribution of the engineer and the industrial designer.

The Fielden Report on *Engineering Design* (Department of Scientific and Industrial Research, 1963) defined mechanical engineering design as:

> the use of scientific principles, technical information and imagination in the definition of a mechanical structure, machine or system to perform pre-specified functions with the maximum economy and efficiency.

This definition suggests a conception of design in which products are regarded as an assembly of components and materials, the arrangement of which is determined by the imperatives of technical function. The selection of a particular design configuration emerges through the engineering designer trying out different arrangements on the drawing board (or CAD terminal), performing the necessary stress calculations, etc., and selecting components and materials according to their performance and cost. Although many engineering products are designed in this way, such an approach does not take into account the relationship of the product to its users. Moody (1984) provides a lucid explanation for the emergence of industrial design as an activity distinguishable from engineering design:

> Industrial design seeks to rectify the omissions of engineering, a conscious attempt to bring form and visual order to engineering hardware where the technology does not of itself provide these features. There are a few instances where technology has an intrinsic elegance: the steam-turbine rotor has complex symmetry which derives from the mechanics of fluids; the exterior of the modern aircraft fuselage has a continuous organic form which derives from its aero-dynamic purpose, the modern suspension bridge is the essence of structural simplicity . . .
>
> Industrial design seeks to relate the hardware to the dimensions, instinctive responses, and emotional needs of the user where these are relevant requirements. Through the conscious control of form, configuration, overall appearance, and detailing, industrial design is capable of conveying to the user the abstract characteristics of a product, e.g. robustness, precision. . . . It can arrange for controls to be comfortable, pleasant, and easy to operate. It is capable of imbuing a product with a distinctive ambience, style, and feeling of good quality which equates with the personal taste of the user.

In practice, the relative importance of industrial design and engineering varies considerably from product to product. There is a 'spectrum' of product design (as shown in Figures 3 and 4 of Caldecote's article) in which the relative contribution of industrial designers and the various branches of engineering depends on the particular blend of mechanical, electrical, electronic, ergonomic and aesthetic design required in the conception and development of the product.

Pye argues in article 1.5 that 'most design problems are essentially similar, no matter what the subject of design is'. This is true to the extent that all designing is iterative, using creativity and compromise to move from a field of possibilities to one unique solution (Figure 2). Nevertheless, the skills and knowledge required of different types of designer vary widely. Conway (1983), for instance, points out that, whereas a mechanical engineering designer needs the ability to visualize objects in three dimensions and understand the properties of materials and structures, an electronic designer works essentially in two dimensions and needs to know the logical consequences of arranging a particular set of components into a circuit design. Neither type of engineer tends to trust the industrial designer, since aesthetic skills and knowledge are not amenable to mathematical or logical analysis.

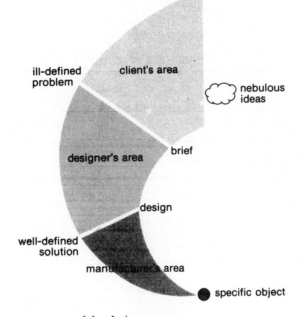

Figure 2   The convergent nature of the design process.
Source: Cross (1983).

Invention, design and innovation

Invention, design and innovation are frequently confused because all are *creative* activities. Invention and innovation, however, normally involve a technical advance in the known state-of-the-art of a particular field. Design normally involves making variations on that state-of-the-art. Thus, in article 1.5, David Pye distinguishes design from invention by noting that:

> Invention is the process of discovering a principle. Design is the process of applying that principle. The inventor discovers a class of system . . . and the designer prescribes a particular embodiment of it.

It follows that there usually exist a vast number of possible design configurations for every important invention or innovation. However, these different designs may be more than just variations in form or arrangement (Figure 3). Often they will incorporate evolutionary

basic product and component innovations	major product and component innovations	incremental product and component innovations	design variations and new models
Daimler internal combustion engine (1885)	Morris Minor (1948)		
	BMC Mini Minor (1959)	BL Mini Metro (1980)	BL Metro 'City' (1982)
Benz motor car (1885)	Vauxhall synchromesh gearbox (1932)		
	Moulton 'Hydrolastic' suspension (1962)	Moulton 'Hydragas' suspension	

(left margin: major change — right margin: minor change)

Figure 3   Spectrum of technological change.
Source: Roy and Cross (1983).

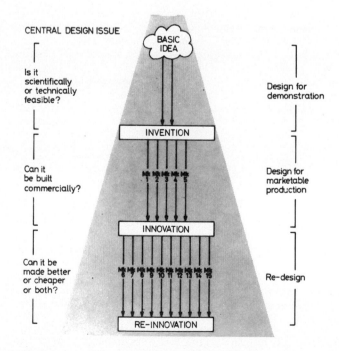

Figure 4   Design in the process of invention, innovation and evolution of a product.
Source: Adapted from Rothwell and Gardiner (1983).

improvements or 'incremental innovations' in components, materials and manufacturing methods (for example, a new model of washing machine with microelectronic controls).

Rothwell, Schott and Gardiner (1985) have pointed out in their booklet *Design and the Economy* (pp. 12–13) that the importance of incremental innovations and design improvements has been greatly underestimated. Attention tends to be focused on the research, design and development work involved in getting from an idea or invention to an innovation on the market for the first time. However, just as important are the processes of successive redesign, component improvement and evolution of the product to improve its performance and reduce its cost (see Figure 4).

The questions of how products evolve is taken up again in section 6 of this book. Here we may conclude that a better understanding of the various activities that contribute to the innovation process is an important step towards their better application in practice.

References

Conway, H. (1983) An engineer's view of design, *Journal of the Royal Society of Arts,* vol. 131, pt 5324 (July), pp. 454–69.

Cross, N. (1983) A review of design, T263 *Design: Processes and Products* (Unit 16), Open University Press.

Department of Scientific and Industrial Research (1963) *Engineering Design* (The Fielden Report), HMSO.

Freeman, C. (1983) Design and British economic performance, *mimeo*, lecture given at the Design Centre, London, 23 March. Science Policy Research Unit, Sussex University.

Moody, S. (1984) The role of industrial design in technological innovation, Ph.D. thesis, University of Aston in Birmingham. (Summarized in *Design Studies*, vol. 1, no. 6 (October 1980), pp. 329–39).

NEDO (1979) *Product Design* (The Corfield Report), National Economic Development Office.

OECD (1981) *The Measurement of Scientific and Technical Activities, Frascati Manual 1980*, Organization for Economic Co-operation and Development.

Rothwell, R. and Gardiner, P. (1983) The role of design in product and process change, *Design Studies*, vol. 4, no. 3 (July), pp. 16–9.

Rothwell, R., Schott, K. and Gardiner, J. P. (1985) *Design and the Economy: The Role of Design and Innovation in the Prosperity of Industrial Companies*, 3rd edn, Design Council.

Roy, R. and Bruce, M. (1984) Product design, innovation and competition in British manufacturing: background aims and methods, *Working Paper WP–02*, Design Innovation Group, Open University (September).

Roy, R. and Cross, N. (1983) Bicycles: invention and innovation, T263 *Design: Processes and Products* (Units 5–7), Open University Press.

TECHNOLOGICAL INNOVATION

1.1

The Social Context of Innovation

Tom Burns and G. M. Stalker

[. . .] Invention, even more than science, is a social phenomenon; in quite matter-of-fact ways, it is a human activity which can only be fulfilled when certain social conditions obtain, when the inventor inhabits a milieu which prompts him to devote himself to a specific line of work with the promise of rewards – in money, power, or even a secure livelihood, in fame, or even self-esteem – and which will thereafter support him economically and intellectually. The notion of the hermit genius, spinning inventions out of his intellectual and psychic innards, is a nineteenth-century myth, useful then, as myths may always be, but dangerous, as myths always are, once its period of usefulness is past.

If, as Whitehead said, the greatest invention of the nineteenth century was the invention of the method of invention, the task of the succeeding century has been to organize inventiveness. The difference is not in the nature of invention or of inventors, but in the manner in which the context of social institutions is organized for their support.

The social context of invention

The review of the past institutional context of industrial innovation which follows is designed, therefore, to underline the importance of that context and to point out the ways in which it has significantly changed. It is not offered as a history, even in a very abridged form, of the relationship between science and industry during the last two hundred years, but rather as a sketch of the phases of change in institutions of some importance to society. It forms the background to the succeeding account of the attempt by industrial concerns to digest the thing they have swallowed.

During the middle years of the last century the electrical industry was established on

T. Burns and G. M. Stalker, *The Management of Innovation,* 2nd edn, London: Tavistock Publications, 1961; extracts from Chapter 2.
Tom Burns was Professor of Sociology at Edinburgh University. G. M. Stalker was a research associate on the Management of Innovation Project at Edinburgh University; he previously worked at the Tavistock Institute for Human Relations.

the basis, largely, of supplying telegraph services. Within a few years the development of electric motors for tramways and stationary machinery led to very considerable expansion. As new applications multiplied, the need for heavier and more efficient generating plant and distribution equipment accelerated the process. By 1880 there was a flourishing, keenly competitive, electrical industry not only in Britain and in the United States, but also in Germany, France, and other European countries. It was an industry, moreover, in which the technological base was very recent – middle-aged men in the industry would be well aware of the first commercial applications – and in which new applications and design improvements followed each other extremely rapidly. Yet the two major innovations during the last twenty years of the century, incandescent electric lighting[1] and radio,[2] were the work of newcomers, of inventors and enterprises unconnected with the existing industry. No spectacular 'discovery' lighted upon by an individual genius was really responsible; electric lamps and wireless transmission were 'in the air' many years before the first commercial companies were floated.

Swan, a chemist, made experimental incandescent lamps in 1860 which employed the same high-resistance conductor, carbonized paper, as was used in the first commercial lamps marketed twenty years later. There were, by 1880, large industrial concerns manufacturing lighting and other electrical equipment; yet in the event it was Edison who, two years after becoming interested in the possibility, first developed the lamp and formed an independent concern to manufacture it.

Lodge, following upon Hertz's earlier experimental work, demonstrated wireless reception before the British Association in 1894, and two years earlier a physicist had written in the *Fortnightly Review* of the 'possibility of telegraphy without wires, posts, cables, or any of our present costly appliances', adding 'this is no mere dream of a visionary philosopher. All the requirements needed to bring it to within grasp of daily life are well within the possibilities of discovery, and are so reasonable and so clearly in the path of researches which are now being actively prosecuted in every capital of Europe that we may any day expect to hear that they have emerged from the realms of speculation to those of sober fact' (quoted[2]). Yet the development of this obviously profitable venture interested no commercial concerns for ten years.*

In the case of radio, it was the twenty-year-old Marconi who, on the basis of Hertz's work as described in an Italian journal, constructed home-made equipment which was sufficiently advanced after three years' work to communicate messages over eight miles and to bring the Marconi Company into being.

Anyone who has read accounts of technological advances, of inventions, during the nineteenth century will perceive this pattern of development as in many ways entirely typical. It is typical not only of the way in which invention then 'happened', but, even more, of the way people thought of invention as happening at any time. Invention was seen as the product of genius, wayward, uncontrollable, often amateurish; or if not of genius, then of accident and sudden inspiration. As such, it could not be planned for, organized as a part of the field of existing industry, the idea was intrinsically absurd. In nineteenth-century Britain the archetypal formula for the process of innovation was enshrined in the fantasy of Watt and the kettle.

The fitting of this latter myth to the key episode of the earlier technical revolution was

* In America the pattern was oddly repetitive, for in the 1870s Graham Bell had tried to interest telegraph companies in the new, and rival, method of communication by telephone which he had invented; after unavailing efforts he founded the American Telephone and Telegraph Company. By the 1890s this company was the 'most research-minded concern' in the industry. Yet it felt unconcerned about radio (apart from one brief and unsuccessful episode in 1906) until 1911, when the threat from wireless telegraphy was too strong to be any longer ignored.

itself characteristic. Of course, the myth of accident and inspiration did go some way towards accounting for the *nineteenth-century* facts. And the outstanding fact was the random distribution of scientific and technical information through the new journals, popular lectures, and societies. These diffusers, and the continued exploitation of major inventions by craftsmen, made it seem possible for any individual innovation to be produced by almost anybody, almost anywhere. Again, the disciplined attack on one difficulty after another, which is how the gap between the scientific idea and the ultimate product is bridged, was still intrinsic to the achievement, but the process was an individual, usually personal, enterprise. Often, as in the case of the electric lamp and radio, many individuals at great removes from each other were involved over a period of years in the development of a single invention.

Images and myths about the past had to fit these contemporary facts. So the boy Watt sat dreaming in front of a boiling kettle and later invented the steam engine. The essential condition of membership of a closely linked group of 'applied scientists', as they would now be called, in the Universities of Glasgow and Edinburgh, the especial circumstance of friendship with Joseph Black, whose discovery of latent heat lay at the bottom of Watt's improvements to the Newcomen engine, the inclusion of the industrialist Roebuck in the circle of personal acquaintanceship – these, the really significant factors, were simply left out of popular account. They were social circumstances which were no longer appropriate to the progress of technology.

Coteries and Clubs

In the latter half of the eighteenth century the Scottish universities were centrally involved not only in the primal discoveries of the industrial revolution in chemistry and engineering, but in the technical applications and commercial ventures which exploited them. The rapidity of technological development in so many fields which were being explored simultaneously in the laboratories of Edinburgh and Glasgow was the direct outcome of close personal association between persons with different expertises and different resources. But the association between people like Watt, Black, and Roebuck was founded not so much on their membership of a common profession or organization as on membership of a small, closely integrated society.[3] In the Scotland of the eighteenth century, for such men to be acquainted with each other was virtually inevitable.

Such circles of personal acquaintanceship served as a social medium for a further decade or so. By the beginning of the nineteenth century, fellow-students and friends sought to institutionalize their informal acquaintanceships. Clubs rather than learned societies, as the Lunar Society and the Royal Society of Edinburgh were, they and their offspring and kindred in Manchester and Newcastle, and the archetype in London, included the persons responsible for scientific advance, technical invention and, to a large extent, industrial innovation. In his life of Boulton and Watt, Smiles wrote:

> Towards the close of the last century there were many little clubs or coteries of scientific and literary men established in the provinces, the like of which do not now exist – probably because the communication with the metropolis is so much easier, and because London more than ever absorbs the active intelligence of England, especially in the higher departments of science, art, and literature. The provincial coteries of which we speak were usually centres of the best and most intelligent society of their neighbourhoods and were for the most part distinguished by an active and liberal spirit of inquiry. Leading minds attracted others of like tastes and pursuits and social circles were formed which proved, in many instances, the source of great intellectual activity, as well as

enjoyment. At Liverpool, Roscoe and Currie were the centres of one such group; at Warrington, Aiken, Enfield, and Priestley of another; at Bristol, Dr Beddoes and Humphrey Davy of a third; and at Norwich, the Taylors and the Martineaus of a fourth. But perhaps the most distinguished of these provincial societies was that at Birmingham, of which Boulton and Watt were among the most prominent members.

The object of the proposed Society was to be at the same time friendly and scientific. The members were to exchange views with each other on topics relating to literature, arts, and science; each contributing his quota of entertainment and instruction.[4] (Our italics.)

But the rate of expansion of science and technology was too rapid to be accommodated by adapting and multiplying the institutions of sociable intercourse, vigorous as they were in middle-class society at that time. The founding, in 1831, of the British Association, a self-conscious attempt to institute personal links between all scientists and technologists, may be regarded as marking the end of the period when a network of personal relationships on the necessary scale was feasible.

The diffusion of technical information

What took the place of the circle of people who were at once friends, fellow-scientists, and business partners, or the coterie whose common interests were at the same time scientific, technological, and financial? Instances of the kind quoted earlier, and the myth, still surviving today, of the lonely inventive genius, suggest that in the period roughly from 1825 to 1875 – succeeding the great days of the provincial Societies – information about scientific discoveries became available to a wide variety of people. Personal communication was replaced by mass communication. [. . .]

Not only did the books and journals appear at an accelerating rate, but clubs and institutes spread the new learning to the utmost limits of literacy in industrial Britain. Attendance at lectures by eminent scientists became as obligatory in the manufacturing towns as was attendance at church and chapel in the country; interest in science, even to the point of patronizing individual scientific workers or building a private laboratory, became gentlemanly. In short, the institutional process which we know as technology – the linking of (a) knowledge of the laboratory demonstrations which established scientific hypotheses with (b) knowledge of manufacturing operations, and of these two with (c) knowledge of existing or presumptive demand for goods and services – this process was spread at random among a very large proportion of the literate population. The fact that it was so spread meant that innovations might appear almost anywhere, might be lighted upon by almost anyone.

The incoherence of the social institutions of technology at the time was reflected in the rudimentary social forms by which innovations were socialized. Technological changes occurred largely through the birth and death of organizations, the simplest form of institutional change. Capital was plentiful, liquid, and diffused, and was readily available for the exploitation of a new device or product. But the institutional build-up around the invention was normally rigid and was identified closely with the line of application and development originally conceived. New concerns had a fairly restricted expectation of life, even if they survived the highly lethal period of early infancy; 'clogs to clogs in three generations' was a piece of proverbial wisdom current in the oldest factory area in the world. The new devices which arose to render the old ones obsolete were generally exploited by new concerns. As Elton Mayo has remarked, the small scale of business enterprises allowed this change to take place without too much dislocation of the social and economic order.[5]

Invention could, and did, make big fortunes in astonishingly short times. The supply of risk capital was relatively enormous. Technical limitations to large lot production were such that the build-up of production had to be slow, and the manufacture of a single device, progressively improved, could absorb all the energies and resources of its designer and backers over a period of years. In this milieu the economic, social and psychological pressures were all against the organization of research as an industrial resource, and against instituting invention in a professional salaried occupation. The anecdote of Ferranti's weaning from the post he entered at Siemens when he left school reflects the ethos of technical innovation at that time. Ferranti, at the age of seventeen, had invented his first alternator. A meeting with another engineer, Alfred Thompson, led to an introduction to a London barrister:

> 'And you mean to tell me you're content to be at Siemens', he said, 'earning £1 a week! Good God!'
> Lawyers see life on the seamy side; small wonder then that they become suspicious of all men.
> 'Ferranti', he said, 'if you continue at a job like that, I'll tell you what will happen. As soon as they discover you've got an inventive ability they'll offer you £5 a week and proceed to rob your brains. You'll do the inventing and they'll collect the cash.'
> This was rather bewildering, but it chimed in with certain thoughts that had arisen in Ferranti's own mind.
> 'Perhaps I'd better ask for a rise', he suggested.
> 'For God's sake, don't do anything of the sort', Francis Ince advised. 'Just clear out. That's no place for you. You might stay there till your teeth fall out and never get a dog's chance to doing anything. There's only one thing for you to do. You must start right away on your own.'
> Ferranti objected that he had no capital.
> 'Leave that to me', said his new friend.[6]

But even before Ferranti had his conversation with Ince, the situation was changing. The distribution of scientific information was rapidly becoming organized. The English provincial universities were founded in the second half of the nineteenth century; the major scientific and professional societies were created during the same period.

Professional scientists and technologists

By the end of the century, science was the province of groups of specialists working in and supported by universities or quasi-academic institutions. The unity of natural philosophy became separated into departments of chemistry, physics, geology, and later derivatives and hybrids. Information was organized in the form of textbooks and courses; traditions as to what was relevant and irrelevant were created under the authority of qualifying examinations. The intellectual segregation of scientific specialists was promoted by the way in which the new and reformed universities organized studies and teaching. Exchanges of the kind which had been characteristic of the earlier social milieux tended, outside the departmental enclave, to become attenuated and formalized in the meetings and journals of learned societies, where geologists produced papers for other geologists, physicists communed with physicists, and so on. By 1900 scientists were salaried professional men.

On the other hand, the situation of industrial technology was itself changing. When, before the middle of the century, the major scientific discoveries had been and could be the work of gifted amateurs and a few academic scientists, technically competent craftsmen like Maudslay, Nasmyth, and Whitworth had created the machine-tool industry. The engines and machines that were the showpieces of the 1851 Exhibition were largely the work of skilled mechanics and master men who had matched the opportunities presented all around them with the basic training of their apprenticeship, self-acquired mathematics, and a clear grasp of the principles of the new engineering. Yet even then, the development by improvement and new application was becoming a task beyond the capacity of men trained according to traditional craft methods. The outclassing of British products by European competitors at the 1867 Paris Exhibition made this quite explicit. The Royal Commission appointed thereafter to survey technical progress in a number of countries confirmed the impression that Britain had lost, or was losing, the technical lead established in the previous hundred years.

In Britain the answer to the problem was sought in improving and expanding the educational system. It is sought there now; it always is. One may remark at this point the different course followed in Germany. With social distinctions in many social, political, and economic fields more rigid and often more crippling (given the course then set for Western societies) than those prevailing in Britain, yet in one generation Germany overhauled and at many points outdistanced the technical advance of British industry. [. . .] Whatever the reason, there is little doubt that the rise of German industry was the consequence of the energy and enthusiasm with which academic scientists like Liebig and the members of the Berlin Physical Society preached their technical gospel and, in the case of the Siemens brothers, themselves created industrial empires.[7] Given this kind of liaison, the appropriate educational system followed. Without it, as in England, the educational system which was devised – in imitation, as it was thought – widened the breach between science and industry. [. . .]

The whole context of industrial innovation had changed. Before 1850 the worlds of science and industry, though separate, had not been distinct; the very existence, on such a large scale, of amateur scientific and technical enquiry demonstrates the ease of access to the world of science enjoyed by anyone with interests which might be satisfied by scientific information. By 1900 science and industry were distinct social systems, entered by different routes, and with very few institutional relationships by which people or information could pass between them. And by 1900, says Cardwell,

> The new applied science industries had left this country, or else had never been started here. In the natural sciences it could hardly be doubted that the lead was Germany's, while in technology the enormous possibilities of the internal combustion engine, for example, were being developed by the French, the Germans, and the Americans. . . . Lockyer, writing in 1901, compared our position at the beginning of the new century with what it had been in 1801, at the outset of the railway age – now, the chief London electric railway was American.[8]

The new technological system

Eventually, with a continuing need for the gap to be bridged, new social institutions have been developed. The gap became itself a new territory, explored, mapped, and eventually controlled by new specialists, the professional technologists, going by the name of applied scientists or industrial scientists.

Leaving out of account the prior development in Germany of a liaison initiated and purposefully maintained by the scientists themselves, the first successful institution set up to exploit this new territory was not only outside Great Britain, where the worst effects of the separation were experienced, but independent of both industry and established scientific institutions. Edison's Menlo Park Laboratory, employing a hundred workers, was established in 1870. But the entrepreneurial method followed contemporary practice: the concerns to manufacture the new devices were set up and financed as separate ventures.

This earliest model was not followed until the founding of the Department of Scientific and Industrial Research by the British Government in 1917. And until the years immediately before the First World War, very little had been done by industry, apart from chemicals, to provide the link itself.

Ever since 1918 the development of industrial research in Britain has depended on Government action much more than in the United States. This may be attributed, as it usually is, to the unenterprising character of British industry in the fields of technical development, to the unwillingness of entrepreneurs to divert resources to development work as being too risky. Yet there were exceptions between 1900 and 1938, notably in chemicals, the industry which had learned from German methods and technological organizations, and enjoyed most stability; and the case is now altered.

It is in the situation of industry as it was in the first decades of the century rather than in such hazy ineluctables as national character that the explanation lies; for such overcaution, such reluctance to take the profit-making opportunities latent in new scientific discoveries, can be explained only by the ignorance of the industrialists about the utility of contemporary scientific activity, by the lack of effective means of communication between the two worlds.

By the end of the First World War the need for such communication was publicly acknowledged. Since industry itself was not supplying the intermediary technologists, the Government set up, in 1917, the Department of Scientific and Industrial Research. Between the wars also, the Government supply of intermediary resources increased very considerably with the need to assure the translation of new inventions with military applications into manufactured weapons. It was from these sources that most of the industrial research and development effort in contemporary Britain has grown. [. . .] The work of producing innovations is now largely in the hands of salaried professionals employed in industrial firms, government establishments, or in institutions directly dependent on industry or government for funds. [. . .]

To sum up: two major changes have occurred in the social circumstances affecting the production of innovations. First, industrial concerns have increased in size: ever greater administrative complexity has brought a wide range of bureaucratic positions and careers into being; control has moved from owners to management. Their survival is therefore a matter of much more intense and widespread concern to themselves and to society; the chances of survival are improved if the technical innovations which might render its processes or products obsolete are developed within it and not by newcomers.

The other change has occurred in the form of institutional relationships and roles within which invention has been possible. The familiar and sociable relationships typical of the eighteenth century provided the ease of communication necessary for the major syntheses of ideas and requirements which introduced the early revolutionary inventions. The scale of scientific and industrial activity rapidly outgrew the social institutions within which the industrial revolution was generated; the syntheses which produced inventions and innovations tended to be random or opportunistic. Later in the nineteenth century, new institutional forms introduced barriers between science and industry, and between 'pure' and 'applied' science, as well as between departments of science. In the twentieth century the new and elaborate organization of professional scientists has been eventually matched by one of technical innovators into groups overlapping teaching and research institutions, Government departments and agencies, and industry.

Neither change is complete. Neither set of contrasts is clear. Few sectors of industry, outside chemicals, have fully accepted the changed situation. It was still possible, in the years between the wars, for a major innovation like the gas turbine to be developed in ways reminiscent of the classic days of nineteenth-century back-parlour invention. The jet engine's invention depended on an individual's persistence and enterprise, although the new massive organizations of government and industry were also involved.[9] On the other hand, the career of the most publicized inventor of the nineteenth century, Edison, reflects both the previous epoch, in the almost conscious exploitation of sociable contact with scientists and technologists, and the later, in the maintenance of development groups and the opening of professional careers in invention. Yet the process of change is now far enough advanced for the shape of the forms characteristic of the present system to be discernible.

References

1. A. A. Bright, *The Electric Lamp Industry: Technological Change and Economic Development from 1880 to 1947*, London: Macmillan, 1949.
2. W. R. Maclaurin, *Invention and Innovation in the Radio Industry*, New York: Macmillan, 1949.
3. A. Clow and N. Clow, *The Chemical Revolution*, London: Batchworth, 1952 (pp. 593–4).
4. S. Smiles, *Life of Boulton and Watt*, London: Murray, 1865 (p. 367).
5. E. Mayo, *The Social Problems of an Industrial Civilization*, London: Routledge, 1949 (p. 32).
6. G. Z. De Ferranti and R. Ince, *The Life and Letters of Sebastian Ziani de Ferranti*, London: Williams and Norgate, 1934 (pp. 51–2).
7. J. D. Bernal, *Science and Industry in the Nineteenth Century*, London: Routledge, 1954 (pp. 63–4).
8. D. S. L. Cardwell, *The Organization of Science in England*, London: Heinemann, 1957 (p. 147).
9. H. Whittle, *Jet*, London: Muller, 1954.

1.2

Introducing Innovation

Patrick Kelly, Melvin Kranzberg,
Frederick Rossini, Norman Baker,
Fred Tarpley and Morris Mitzner

Popular Views of the Innovation Process

Popular mythology presents only simplistic notions of the innovation process. Yet even this simplistic folklore manifests the complex nature of the process. For instance, we get a half-truth, albeit a very important one, from the comic strips. There the inventor – a somewhat eccentric fellow (after all, he walks around with a symbolic electric bulb suspended over his head) – receives a flash of inspiration (the electric bulb lights up), and lo and behold, an invention has been born. Such cartoons are true, because they depict the importance of individual imagination and ingenuity in innovation; but only in part, because inventive insight is only one in a series of developments necessary for a successful innovation.

'Necessity is the mother of invention' is another half-truth. As we shall see, demand is a strong motive force; but as a full explanation it fails, because many needs have not yet given rise to inventions, and many innovations arose from other causes. Besides, in many cases inventions require additional inventions in order to make the original one effective. Thus one could perhaps turn the adage around: 'Invention is the mother of necessity'. Yet there is also an important truth in the 'necessity' explanation, for it forces us to consider the social needs and human wants that help formulate the problems toward which inventors direct their attention.

Another old saw that stresses the same point is the saying attributed to Ralph Waldo Emerson, 'If a man . . . make a better mouse-trap than his neighbour, tho' he build his house in the woods, the world will make a beaten path to his door'. Although there are many 'better mousetraps' that never become successful innovations, Emerson's reputed statement is important because it focuses attention on the need to link together social needs with inventive activity.

P. Kelly and M. Kranzberg (eds.), *Technological Innovation: A Critical Review of Current Knowledge,* San Francisco Press, 1978; extracts from Chapter 1.
Professor Kranzberg and Drs Kelly, Rossini, Baker, Tarpley and Mitzner were all members of the Georgia Tech Innovation Project group, based in the Department of Social Science, Georgia Institute of Technology.

The 'captains of industry' are also often extolled in popular folklore as the moving force in the innovation process – linking social needs with technical responses in bold and imaginative syntheses. And, indeed, without the entrepreneurial contribution of organizational know-how, capital, production and marketing capabilities, etc., an invention is stillborn. The well-known 'captains of industry', however, have largely been replaced today by faceless organizations; the entrepreneur has become an entrepreneurial function. Further, an idea – an invention – must exist before the entrepreneur can develop it into an innovation. A half-truth 'writ large' in folklore is still a half-truth.

Another notion popular since the days of Sir Francis Bacon's *New Atlantis* is that technological innovations derive from scientific discoveries. If this generalization can be said to have any validity, it would hold only for relatively modern times – and even then, as we shall see, the situation is much more complicated than a simple causal relationship between science and technology. Nevertheless, this concept exerts great power – underlying, for instance, the argument for government support of basic research on the grounds that such research ultimately achieves utility.

Still another popular idea is that innovation can come about on command. This idea is based on the myth of technological virtuosity and the 'talent and money' syndrome. Nurtured on a diet of 'success' stories, the public retains great faith in the ability of technology to meet every challenge put to it. Although some recent scoffers might doubt that technology always triumphs, and although some humanists might decry that triumph (they plaintively, but properly, ask to what human ends and purposes), the general public seems, on balance, to regard technology as beneficent, or at least is not willing to forgo the advantages it provides. To that is added the belief that 'throwing money at problems' will induce technological innovation by the simple expedient of investing talent and money. Crash programmes, such as the atomic-bomb and synthetic-rubber projects during the Second World War and the space exploration programme, provide historical examples.

This idea of innovation as induced through the application of talent and money has come to the fore during the current 'energy crisis'. 'Project Independence', which, in its original form, postulated an investment in technological development to make the United States independent of outside energy sources by 1980, was based on a combination of several myths: necessity as the mother of invention, the invincibility of technology, and the efficacy of government financial support.

Although it can rightly be argued that such problems will not be solved without an investment in talent and money, and that technology has in the past responded on command with some spectacular success, the simplistic cause-effect relationship implicit in these notions wilts before the complexities they fail to address. Which technical approach (or combination of approaches) should the money be thrown at? What kinds of talents are needed and at which points in the innovation process? Will just any organizational structure do? What social, environmental, political, and human price will there be? Are all the second- and later-order consequences also desirable and, if not, are the undesirable ones acceptable? What, in fact, is the 'track record' of 'command innovation'; that is, have the failures as well as the successes been recorded? In brief, these apparently simple notions turn out to hide a maze of complex questions.

Yet such popular views of innovation turn out to be extraordinarily influential, because there seems to be something inherently attractive about simple explanations for complex phenomena. Even scholars engaged in innovation research are not entirely immune to this attraction. To serve as a corrective for the half-truths that might emerge from such one-sided approaches, we must endeavour to investigate the innovation process in all its ramifications. We are thus brought squarely to face the question: what do we know about innovation?

Definitions, Distinctions, and Usages

Researchers concerned with the process of technological innovation belong to a wide variety of disciplines and typically view the process from these diverse perspectives. As a result, there is little uniformity in their definitions and conceptual distinctions regarding the innovation process. For example, some economists reserve the word 'innovation' for the application of an 'invention', that is, its entrance on the market-place. Thus the economist Joseph Schumpeter (1939) regarded invention, the discovery of a new tool or technique, as the initial event; and innovation, the implementation of the new tool or technique, as the final event. Mansfield (1968, p. 83) also defines an innovation as the first application of an invention. To the US Patent Office, the interest is in the initial event – when rights are assigned to an individual – and there is no concern with the eventual application. As we shall see, cultural anthropologists, management scientists, and other specialists have different notions of what constitutes innovation and what are the various elements comprising it.

Scholars also differ in their use of the terms 'invention' and 'discovery'. According to Forbes (1958, p. 5), this difference is tied to a distinction between science and technology; scientific discovery usually recognizes or observes for the first time some natural object or phenomenon, whereas an invention is the creation of something technologically new that had not existed before. For example, man 'discovered' fire, but had to 'invent' means to start fire and use it for lighting and heating. But such a distinction can be misleading, for sometimes what we call inventions might also be classified as discoveries, such as Perkin's 'discovery' of aniline dye – which was at the same time the 'invention' of synthetic dyes.

In order to accommodate this wide range of specialized concerns, we have interpreted the term 'process of technological innovation' very broadly – as embracing the full range of activities from the initial problem definition and idea generation through research and development, engineering, production, and diffusion of new technical devices, processes, and products. It is tempting to identify these elements as phases in a linear sequential process. But, as we shall see later, this proves to be a speciously attractive, but unworkable view that simply does not fit the facts.

Implicit in both the popular myths and the scholarly literature we studied are three major points concerning the innovation process: (1) it is a response to either a need or an opportunity; that is, it is context dependent; (2) it depends on creative effort and, if successful, results in the introduction of novelty; and (3) it brings about or induces the need for further change.

Although most innovations are need induced, such needs do not necessarily express themselves specifically. For example, there might be a general need for quicker transportation, but that does not necessarily dictate that this need be met by supersonic planes. Furthermore, sometimes wants do not make themselves felt until after the innovation appears. For example, there is a general desire to make oneself attractive to members of the opposite sex, and the makers of underarm deodorants try to persuade consumers that their product fills this general want and does it better than alternative means, such as perfumes or soap and water. The 'social innovators' of Madison Avenue have proved especially skilful in turning generalized human desires into demands for specific products. In any case, an innovation promises to be in some way 'better' than existing means to the same end, perhaps by offering better quality at the same cost, the same quality at lower cost, faster production, etc.

In other cases, the stimulus event for an innovative effort may be a newly developed technical capability for which applications are then sought. The development of the transistor, for instance, with the potential for miniaturization it represented, led to a whole 'family' of innovations. Some data suggest that major advances are more likely to arise from a new technical capability, whereas smaller, incremental improvements are more likely to be need induced. In either case the innovation process is environmentally dependent from its very outset: that is, it is a response.

It should be noted that innovation is not the only response to felt needs or wants. Another response (a 'nonresponse') is simply to ignore the problem, to 'make do'; we tend to seek a way – often technical (the 'technological fix') – to solve a problem or fulfil a want. Some traditional Eastern religions and cultural systems, however, deny or dismiss such problems or needs rather than resolve them. A possible explanation is that those cultures view reality as primarily spiritual in nature and hence deny the importance or necessity of changing the material conditions of life.

Another type of response is not to innovate, but simply to apply more of the technology that is already employed. A typical example in twentieth-century America has been to build more freeways to resolve the problem of already clogged freeways.

Even when the response to a perceived need or want is to innovate, the innovation may not be a technological one. There are, for example, social innovations – 'new methods of inducing human beings to compete and cooperate in the social process' (Kuznets, 1962, p. 19) – such as the Peace Corps or advertising; these innovations are primarily social in nature and only incidentally technological. There are also economic innovations, which do not necessarily have technology as their major component, such as the introduction of double-entry bookkeeping in Renaissance Italy; or marketing innovations, such as the supermarket or installment buying. Political innovations would include new types of bureaucratic organization or new political systems, such as the corporate state and 'people's democracies'. Our investigation, however, is restricted to technological innovations, and will consider other forms only as this focus requires.

The elements of creativity and novelty also require comment. Whether the initial stimulus is a need or a technical opportunity, creativity is required – both in the initial idea for a response that might work, and at many steps in developing that response into a successful innovation. A host of contextual variables can facilitate or inhibit creative responses, or guide them along certain lines. Especially important in this regard is the fact that those from whom new ideas for technical innovations are typically expected to come – those in the R & D laboratory – are usually engaged in projects that are already under way. It is not at all clear that the environments most conducive to each are identical.

An innovative effort, if successful, results in the introduction of novelty. The patent law has something to say about novelty, but that is for determining legal priority of invention, not for defining the nature of innovation. To the Patent Office, an invention must be new and useful, and it must represent something more than a trivial improvement; it usually means 'a contribution over and above the exercise of mechanical skill', and in 1880 the US Supreme Court used the expression 'a flash of thought' to describe an essential attribute of invention, thereby stressing the role of creativity.

For our purposes, the notion of novelty requires further explication so as to avoid a semantic confusion that often appears in the literature. Something can be new only in relation to some frame of reference. In the innovation phase, the frame of reference is the current state of the art. In the diffusion phase, however, an innovation is new relative to the particular unit that is adopting it. It is this latter frame of reference to which Hodgen (1952, p. 45) refers when he speaks of a technical innovation 'as having taken place when a tool, a device, a skill or a technique, however unknown or well known elsewhere, is adopted by an individual in a particular community and is regarded as new by the members of that community'. Novelty, therefore, characterizes an innovation not only upon its first introduction into use anywhere, but also repeatedly as it is subsequently adopted for use in other contexts. The one refers to 'new under the sun', the other to 'new under this roof'.

Finally, whereas the social impact of technological innovation, that is, induced change, is only tangential to the scope of our investigation – our focus being on the process itself – some thinkers single out a purposive element in the process, that of effecting physical-environmental and social change through innovation. Tannenbaum (1970) has stated, 'Technological innovation is the novel application of physical knowledge and technique to make premeditated changes in the physical aspects of the environment', and Peter Drucker (1967) regards innovation as a conscious attempt to bring about, through technology, a change in the way man lives.

Although such statements implying the larger impact of innovation are correct, they neglect two major points in stressing its purposive nature. The first is that most innovations do not come about by such transcendental considerations of social change, but rather through more mundane calculations on the part of businessmen regarding profits, resource factors, costs, etc., or by technologists pursuing their own visions of technical efficiency, or by governments responding to various pressures, and the like. Second – and this is a point of major social concern today – innovations may bring about changes other than those envisaged by the innovators. Indeed, the unforeseen changes brought about by the large-scale application of such innovations as the automobile or DDT have given rise to the new art of Technology Assessment, which attempts to study the second- and higher-order effects of technological applications before they are applied (Hetman, 1973).

Nevertheless, whether motivated by narrow or broad considerations, whether consisting of a new mix of elements or the insertion of something old into a new context, innovation brings about changes (in varying measure) in what things people make and do, how they make and do them, how they work and live – and ultimately in how they think and act. Is it any wonder that we seek to understand the innovative process, both in theory and practice, in order to help us comprehend the present and, indeed, in order to make the future?

Classical Theories of Innovation

If we attempt to categorize the classical theories of innovation, we find that they polarize around two positions: deterministic and individualistic ('heroic', or 'great man') theories (Rae, 1967, p. 326). The determinist explanation holds that the innovation occurs when the conditions are 'right', and it stresses the role of social and other forces, principally military and economic, in bringing about technological change. The 'heroic' theory stresses the role of the individual and plays down the influence of external pressures.

These two major theses are not mutually exclusive; they are really matters of emphasis. No scholar has ever claimed that the individual innovator runs his own race entirely unbridled by any external pressures, and even the most ardent adherent of the deterministic school has recognized that the manifold and diverse forces operate through individuals. This accounts for the development of sophisticated, composite theories, embodying elements of both the deterministic and individualistic schools of thought.

A wide variety of exogenous elements – human wants, social needs, economic demands, military requirements, geographical and climatic constraints, and the like – can help determine the rate and nature of innovative activity, and these can express themselves in many different ways. For example, economic requirements might differ in relation to different resource factors, depending on labour costs, capital, fuel, materials, and the like. Such other external factors as geography and climate have been employed to explain why certain civilizations seem to be more innovative than others, and there are those who would ascribe technological creativity only to certain ethnic groups.

The importance of exogenous sociocultural factors is evident from the fact that different societies seem to be especially fruitful in innovations at different times in history. For example, only by reference to sociocultural elements can we explain why China in the Middle Ages displayed technological genius which far outshone that of the contemporaneous medieval West. Similarly, we must rely on a sociocultural explanation of why China failed to undergo an Industrial Revolution, and why the West took the leadership in both science and technology from the seventeenth century onward.

Sociocultural factors are also held to be responsible for multiple inventions. Given the widespread diffusion of technological knowledge in the modern world, the similarity of technical problems, and the apparently universal potentialities of the human mind, it is not surprising that many inventors find the same or similar solutions to technological problems at about the same time. Indeed, the great amount of patent litigation over priorities would be proof of that. Further, the number of patent applications turned down because they lack 'novelty' would indicate how often the human mind arrives at solutions for technical problems that others have already thought of – and already patented.

Yet, despite the importance of exogenous [external] social factors, we cannot do without the human element in innovation. Human beings define the problems, have the ideas, perform the creative acts of producing a device, do the research and development, and decide upon the application and diffusion of innovations.

Although the heroic theory of innovation does not deny the stimulus of economic needs and the influence of sociocultural conditions, it emphasizes the role of the individual hero in bringing about innovation. For example, scholars of the deterministic school would argue that eighteenth-century Britain was 'ready' for the steam engine, in terms of both economic need and level of technology, so that if James Watt had not invented the steam engine, someone else would have. Proponents of the heroic school, however, would claim that the characteristics and personality of Watt were primarily responsible for the steam engine as it actually came into being.

In his *Lives of the Engineers* (1861–2), a three-volume collection of biographies, Samuel Smiles attributed the landmark inventions and great engineering feats of the Industrial Revolution to heroic individuals. Not surprisingly, he found these men possessed of the standard Victorian virtues of his own time: self-discipline, self-help, devotion to duty, integrity, and perseverance. But such an enumeration of traits, like later and more scientific studies of technical creativity, merely transfer to another level the argument of individualism versus social determinism. After all, it can be argued, these virtues of the heroic Victorian inventors derived from the sociocultural forces of the time – and we are back to the old 'nature versus nurture' argument still waged by psychologists, educators, and sociologists on such new battlegrounds as the efficacy of school busing or the validity of IQ tests for ghetto children.

Both the individualistic and social deterministic theories of innovation may be introduced at yet another level, with reference to the entrepreneurial function, which was so greatly stressed by Schumpeter (1939, pp. 85–6). It is the energetic entrepreneur who is willing to take risks, who amasses capital to finance the invention, who sees the idea through to actual production and introduction to the market-place. He is the man who links together the social needs, frequently expressed in terms of profit potentialities, with the creative ideas of the inventor, thereby coupling the market-place with the invention.

Sometimes the inventor and the entrepreneur are combined in the same individual, as in the case of Edison or Elmer Sperry (Hughes, 1978), but more often they are different persons. For example, James Watt possessed remarkable technical ability, but lacked capital and business acumen; they were supplied by Matthew Boulton, who became the driving force for the successful introduction of Watt's steam engine. In contemporary business organizations, some of the entrepreneurial functions are carried on by the R & D manager, who brings together scientific knowledge, technical expertise, and knowledge of the market-place and of economic constraints in an effort to produce profitable innovations (Morton, 1971). Other entrepreneurial functions might be carried out by different people or by parts of a large modern corporation. Whether or not the inventor and the entrepreneur are the same person, different people, research teams, or corporations, it is the entrepreneurial *function* that marshals the resources – ideas, talent, technology, money – thus transforming the invention into a utilizable innovation.

In addition to stressing the importance of both individuals and the social environment in innovation, much is often made of the role of happenstance in history. In the case of technical innovations, the intervention of chance is called 'serendipity', a term coined by Horace Walpole from the Persian fairy tale, the Three Princes of Serendip (Ceylon), whose heroes often unexpectedly discovered something valuable. Two famous historical incidents illustrate serendipity at work: Charles Goodyear's invention of the vulcanization process, and W. H. Perkin's discovery of aniline dye.

These and other cases of 'accidental' discoveries or inventions are not quite so accidental as they might seem. On closer inspection they provide proof that innovation does not occur in haphazard fashion. In virtually every case of serendipitous invention, we find that the inventors were aware of the needs and problems, that they had already conducted persistent and careful searches for what they wanted, and that they were acute and perceptive enough to recognize when a happy accident gave them their answer. In other words, they could appreciate the significance of a chance occurrence and utilize it for practical purposes. Most innovative advances come as a cumulative result of answers to a series of closely directed questions; chance or accidental observations come as a bonus to the perceptive researcher who has already done his 'homework'. As Louis Pasteur observed, 'Chance favours only the prepared minds'.

It is apparent, then, that no monistic theory – social-deterministic, individualistic-heroic, or even blind chance – can serve by itself to explain the complexities of the innovative process. It is not begging the question, but simply stating a fact to say that innovation, like virtually every other creative activity, derives from the interaction between individuals and the sociocultural environment.

In his pioneer study of the organization of cultures, the anthropologist Dixon (1928) postulated a triad of factors in the background of every cultural innovation: opportunity, need, and genius. Almost four decades later, a study of the conditions fostering successful R & D, conducted by the Arthur D. Little Company (1965) for the Department of Defense, found a similar triad behind innovatory weapon systems: a clearly understood need; relevant ideas, information, insight, and experience; and the men and money to push through the job. Here were Dixon's anthropological factors translated into the context of the modern R & D laboratory.

One of the most influential theories of innovation, comprehending both the individualistic and the sociological points of view, was first presented by Abbott Payson Usher in the 1920s. Usher's theory, drawn from *gestalt* psychology,* regards innovation as a social process consisting of acts of insight of different degrees of importance and at many levels of perception and thought. This theory considers innovation as a four-step sequence: (1) the perception of a problem, meaning the recognition of a social need and of the

---

* See article 2.4 in Section 2 for details.

problems involved in its fulfilment; (2) the setting of the stage, involving the existence of a body of technical knowledge and of technological and financial capabilities; (3) the act of insight by which the essential solution of the problem is found; and (4) critical revision, in which the newly perceived relations are thoroughly mastered and effectively worked into the entire context of which they are a part. Today, we would comprehend Usher's 'critical revision' step in the 'development' part of modern 'research and development', and we also realize that Usher's four steps do not always occur in so neat a linear pattern.

Even so, Usher's theory is by no means complete. He was primarily concerned with the 'act of insight' in the inventive phase, probably because of the emphasis *gestalt* psychology places on the 'Eureka!' or 'Aha!' phenomenon. Usher helps tell us how some inventions occur, but not how inventions are translated into true innovations. He neglects both the risky economic decisions involved in the application and diffusion of inventions, and the feedback among the various steps. Above all, he fails to provide a detailed analysis of his crucial fourth stage of critical revision.

More recently, investigators have concentrated on this developmental stage neglected by Usher and other early theorists. At the same time, the development during the past half-century of a science-based technology has enabled them to augment the earlier phases of Usher's sequence by reference to the contributions of science to the innovation process. As a result, terms such as 'basic research' and 'applied science' enter increasingly into the literature. Indeed, in the popular mind the model of innovation emerges as a direct line from basic scientific discovery, through applied research and development, to the innovation – and even some scholars view the innovative process in those terms. Yet the relations between scientific and technological innovation are too varying and complex to justify such a simplistic model.

Of major significance to our understanding of the innovative process has been the detailed analysis of the crucial developmental stage to which early models had merely given passing attention. In analysing successful product innovations, Mansfield (1971, pp. 114–18) employed the following categories: applied research; preparation of project requirements and basic specifications for prototype or pilot-plant design, construction, and testing; production planning, tooling, construction, and installation of manufacturing facilities; manufacturing start-up; and marketing start-up. These categories were similar to those derived from their industrial experience by the members of the Charpie panel (1967): research – advanced development – basic invention; engineering and designing the project; tooling – manufacturing – engineering; manufacturing start-up; and marketing start-up. In these and similar studies, Usher's final stage has been extended and broken down into four or more separate categories of activities. [. . .]

Breakdown of the Linear-sequential Approach

When we first began our survey of the innovation literature, it seemed that the process could best be analysed in terms of phases in a linear and unidirectional sequence. This assumption was derived from a synthesis of the work of Usher, Schumpeter, Machlup, Gilfillan, etc., and our own initial propensities. Such a scheme, we thought, would allow us to focus on the key decision and/or leverage points in the process. Hence, we distinguished five separate functional phases: (1) problem definition and idea generation, (2) invention (i.e. the prototype device or process), (3) research and development, (4) application (meaning first use), and (5) diffusion (introduction into a context other than original application). A simple block diagram would suffice, showing a linear and unidirectional process, going successively from phase one through phase five. The totality constituted the process of innovation.

We were disabused of this linear-sequential notion rather quickly, however. That is, as we moved more deeply into our study, its inadequacy as an analytical framework became more and more obvious. It was a poor representation of the complexities we found.

Nevertheless, the concept of process phases is a valuable and valid structuring device that we shall use – not as identifying discrete elements in a fixed sequence, but as indicating loosely delineated clusters of activities that may overlap temporally and organizationally and that exhibit complex interrelationships.

The most basic distinction is between the phases of the process that consist of the development of a technological device, process, or product – from the initial conception of its possibility to the point of its first introduction – and the subsequent actions that result in the spread of that innovation to other contexts. The former, referred to as the *innovation* phase, the latter *diffusion*. It is somewhat awkward to use the same term, 'innovation', both generically to refer to the whole process and also to refer to one of its phases or elements, but it is a common practice in the literature and one that we have chosen to follow.

The relationship between even these broad phases of innovation and diffusion is complex. Logically, innovation precedes diffusion; i.e., that which does not yet exist cannot be diffused. Logical priority is not the whole story, however. It is not always possible simply to adopt an innovation; often it must also be *adapted* to its new context of use. Such adaptation, although a form of diffusion, also involves the process of innovation, since modifications are required. Thus diffusion may precede and bring into being new innovations, as well as the other way around.

The innovation and diffusion phases themselves may be divided into subphases. Innovation, for instance, includes problem definition and idea generation, research and development, engineering, and production. There is not widespread agreement on these distinctions. Some authors, as we have seen, identify a separate 'invention' phase. However, this term is fraught with difficulties; it requires a distinction between invention and discovery. For example, Charles Goodyear 'discovered' that rubber when sprinkled with sulphur and heated would not melt and would retain its resiliency; but this accidental discovery is also regarded as the 'invention' of the vulcanization process.

It is likely that such difficulties are semantic rather than substantive, but the more basic question is whether or not the identification of an invention phase, even if cleanly drawn and widely agreed on, is of much value in understanding the innovation process. We think not, for several reasons. First, for many people the term 'invention' carries the connotation of an 'event', when it is really part of a 'process'. Even if invention is disassociated from this 'event' connotation and is viewed as a phase of a larger process, it is still not a very convenient distinction except in the legal, patent-application sense. In terms of the way in which organized innovative efforts are structured (and as we shall see the innovation process has become increasingly institutionalized), the significant benchmarks lie elsewhere than at the point of patent application. A more important decision point, for example, lies at the point at which an idea has been submitted to management for consideration and funding; the process leading up to such submission is an important phase, but only a part of what might be called invention.

The process by which ideas thus submitted are reviewed by management and funded, rejected, shelved, or referred elsewhere in the organization is likewise a significant process phase, but one that would be lost or at least blurred by the choice of 'invention' as a phase concept. Finally, as we shall see, R & D is an extremely important and equally complex phase, both conceptually and organizationally, only a part of which could be treated by a phase labelled 'invention'.

Besides the problem of dividing innovation into process phases, there was also the difficulty, in our early attempts to apply the linear five-phase model, of picking out a starting point in the process. In the linear view, needs or opportunities in some portion of the exogenous world come to be recognized and trigger a problem-definition and idea-generation effort. But in many of the cases cited in the literature, the need, opportunity, or even the clearly defined technical problem arises from what the linear view treats as a 'later' phase; for example, R & D, engineering, production, or even diffusion. That is, while problem definition and idea generation are at times 'front-end' or logically initial activities, they may also be triggered by later phase developments. Problems are defined and ideas are generated in every phase, not just in the initial one. Examples of major innovations arising from an ongoing R & D effort rather than from 'beginning at the beginning' are nylon deriving from Carother's basic research in polymers at DuPont, and the development of the transistor by Shockley and his associates at Bell Laboratories (Jewkes *et al.*, 1969). [...]

Thus, in the relationship between invention and R & D, we encountered a problem that has frequently troubled scholars: when is something invented? When, for example, was nylon invented – when Carother's assistant first noticed the fibres, or when the properties of nylon were verified, or when a first sample in the form that it would be used was produced (Jewkes, *et al.*, 1969, pp. 25ff.)?

The question of when an invention actually takes place becomes more difficult to answer when we recognize the increasing complexity of technical devices and processes in our own times. For example, many people claim credit for the 'invention' of radar. How can honest men disagree on this question, particularly when radar is so recent and when we have ample evidence from those who actually participated in its development? The answer must be that radar, like a good many other modern innovations, is a highly complex aggregate of a series of innovations. It is foolish to argue over who added the last little bit from 'pre-radar' or 'almost-radar' to create radar, because all of the increments, or the subinventions, were essential to the creation of radar – and even when applied and diffused, radar was going through further refinement and development requiring whole new groups of innovations. It is clear that innovation is not a singular event, but a congeries of events, or, more correctly, a process.

As we reviewed the various theories and models, we began to realize that in almost every major innovation of recent times each functional phase is linked in some way to the others: every phase in our block diagram has lines connecting it to and from every other block in the diagram. Instead of a linear-sequential picture of a neat flow-chart, with single lines going from one block to another, we had a graphic portrayal of a plate of spaghetti and meatballs! [...]

The complexity of the process patterns we encountered led us finally to realize that the linear-sequential view described only a few cases. For the rest, the arrows led 'every which way', with the starting points being various and with each phase being linked to some, many, or all of the others. It was then that we began to consider innovation as a complex, highly interactive ecological system.

References

Charpie, R. L. (1967) *Technological Innovation: Its Environment and Management*, Washington DC: Department of Commerce, Report 0–242–376.

Dixon, R. B. (1928) *The Building of Cultures*, New York: Scribner.

Drucker, P. F. (1967) Technology and society in the twentieth century, in M. Kranzberg and C. W. Pursell (eds.), *Technology in Western Civilization*, vol. 2, New York: Oxford University Press, pp. 10–22.

Forbes, R. J. (1958) *Man the Maker,* New York: Abelard Schuman.

Hetman, F. (1973) *Society and the Assessment of Technology,* Paris: OECD.

Hodgen, M. T. (1952) *Change and History,* New York: Wenner Gren Foundation.

Hughes, T. P. (1978) Inventors: the problems they choose, the ideas they have and the inventions they make, in P. Kelly and M. Kranzberg (eds.), *Technological Innovation: A Critical Review of Current Knowledge,* San Francisco: San Francisco Press, pp. 166–82.

Jewkes, J., Sawers, D. and Stillerman, R. (1969) *The Sources of Invention,* 2nd edn, New York: Macmillan.

Kuznets, S. (1962) Inventive activity: problems of definition and measurement, in R. R. Nelson (ed.), *The Rate and Direction of Inventive Activity,* Princeton: Princeton University Press, pp. 19–43.

Mansfield, E. (1968) *Industrial Research and Technological Innovation,* New York: Norton.

Mansfield, E. (1971) *Technological Change,* New York: Norton.

Morton, J. A. (1964) From research to technology, *International Science and Technology,* vol. 29, pp. 82–104.

Rae, J. B. (1967) The invention of invention, in M. Kranzberg and C. W. Pursell (eds.), *Technology in Western Civilization,* vol. 1, New York: Oxford University Press, ch. 19.

Schumpeter, J. A. (1939) *Business Cycles,* New York: McGraw Hill.

Tannenbaum, M. (1970) A booming technology, a better environment: can we have both? *Bell Telephone Magazine,* May–June, pp. 25–6.

1.3

Successful Industrial Innovation

Christopher Freeman

[. . .] Jewkes and his colleagues (1958) have argued that the nineteenth-century links between science and invention were much greater than is commonly assumed. Certainly, the classical economists were well aware of the connection between scientific advances and technical progress in industry, in the eighteenth and early nineteenth centuries. Nevertheless, the evidence suggests that there were profound changes in the degree of intimacy and the nature of the relationship between science and industry.

The contact in the eighteenth and early nineteenth centuries was spasmodic and unsystematic. Few firms had scientists working for them, although it is true that scientists were consulted occasionally about industrial problems. There were in any case very few scientists. Many inventions and innovations in the textile, metal-working and railway industries owed little or nothing to scientific research. They were based far more on the practical experience of engineers and craftsmen. Scientist–entrepreneurs and inventor–entrepreneurs did play a very important part in the chemical industry and in some branches of mechanical engineering (Musson and Robinson, 1969). But with the growth of the dyestuffs industry and the electrical engineering industry in the second half of the nineteenth century there was a change in the pattern of relationships, which became more clearly established in the twentieth-century industries of electronics, synthetic materials and flow-process plant. Not only were the products and processes of these industries originally based almost entirely on scientific discoveries and theories, but even the day-to-day improvement and modification of the products and processes depended to an increasing extent on an understanding of scientific principles and on laboratory experiments. [. . .]

The new style of innovation in the industries which we are considering was characterized by professional R & D departments within the firm, employment of qualified scientists as well as engineers with scientific training, both in research and in other technical functions in the firm, regular contact with universities and other centres of fundamental research, and acceptance of science-based technical change as a way of life for the firm.

C. Freeman, *The Economics of Industrial Innovation*, 2nd edn, London: Frances Pinter, 1982; extracts from Chapter 5.
Professor Freeman is the former Director and currently Deputy Director of the Science Policy Research Unit, Sussex University.

Some of the firms had very strong scientific and technical resources, such as ICI, BASF, Du Pont, Hoechst, RCA, Marconi, Telefunken and Bell. An extreme case was the development of nuclear weapons and atomic energy.

Almost all of the major innovations we have considered were the result of professional R & D activity, often over long periods (PVC, nylon, polyethylene, terylene, synthesis of ammonia, hydrogenation, catalytic cracking, nuclear power, television, radar, semi-conductors . . .). Even where individual inventor—entrepreneurs played the key role in the innovative process (at least in the early stages) such individuals were usually scientists or engineers who had the facilities and resources to conduct sustained research and development work (Baekeland, Fessenden, Eckert, Houdry, Dubbs, Marconi, Armstrong, Zuse). Some of them used university or government laboratories to do their work, while others had private means.

Frequently university scientists or inventors worked closely as consultants with the corporate R & D departments of the innovating firms (Ziegler, Natta, Haber, Fleming, Michels, Staudinger, Von Neumann). In other cases special wartime programmes led to the recruitment of outstanding university scientists to work on government-sponsored innovations (the atomic bomb and radar). Intimate links with basic research through one means or another were normal for R & D in these industries and their technology is science-based in the sense that it could not have been developed at all without a foundation in theoretical principles. This corpus of knowledge (macromolecular chemistry, physical chemistry, nuclear physics and electronics) could never have emerged from casual observation, from craft skills or from trial and error in existing production systems, as was the case with many earlier technologies. [. . .]

So far we have discussed the new industries mainly in terms of the scientific basis of their new products and their manufacturing technologies, but it is impossible to disregard the pull of the market as an essential complementary force in their origins and growth. In many cases the demand from the market side was urgent and specific.

The strength of the German demand for 'ersatz' materials to substitute for natural materials in two world wars spurred on the intense R & D efforts of I.G. Farben and other chemical firms. The strength of the military/space demand in the American post-war economy stimulated the flow of innovations based on Bell's scientific breakthrough in semiconductors and the early generations of computers. The urgency of British wartime needs spurred the successful development of radar of all kinds while the German Government sponsored the development of FM networks, as well as radar.

Conversely, the absence of a strong market demand for some time retarded the development of synthetic rubber in the US, the growth of the European semiconductor industry, the development of radar in the US before 1940, or of colour television in Europe after the war.

This does not mean that only wartime needs and government markets can provide sufficient stimulus for innovations, although they were obviously important historically. A strong demand from firms for cost-reducing innovations in the chemical and other process industries is virtually assured, because of their strong interest in lower costs of producing standard products and their technical competence. The demand for process innovations is related to the size of the relevant industry and here again the American oil industry provided a key element of market pull for the innovative efforts of the process-design organizations. The market demand may come from private firms, from government or from domestic consumers, but in its absence, however good the flow of inventions, they cannot be converted into innovations.

Innovation is essentially a two-sided or coupling activity. It has been compared by Schmookler (1966) to the blades of a pair of scissors, although he himself concentrated almost entirely on one blade. On the one hand, it involves the recognition of a need or more precisely, in economic terms, a potential market for a new product or process. On the other hand, it involves technical knowledge, which may be generally available, but may also often include new scientific and technological information, the result of original research activity. Experimental development and design, trial production and marketing involve a process of 'matching' the technical possibilities and the market. The professionalization of industrial R & D represents an institutional response to the complex problem of organizing this 'matching', but it remains a groping, searching, uncertain process.

In the literature of innovation, there are attempts to build a theory predominantly on one or other of these two aspects. Some scientists have stressed very strongly the element of original research and invention and have tended to neglect or belittle the market. Economists have often stressed most strongly the demand side: 'necessity is the mother of invention'. These one-sided approaches may be designated briefly as 'science-push' theories of innovation and 'demand-pull' theories of innovation (Langrish *et al.,* 1972). Like the analogous theories of inflation, they may be complementary and not mutually exclusive. [. . .]

It is not difficult to cite instances which appear to give support to either theory. There are many examples of technical innovation, such as the atomic absorption spectrometer, where it was the scientists who envisaged the applications without any very clear-cut demand from customers in the early stages. Going even further, advocates of 'science-push' tend to cite examples such as the laser or nuclear power, where neither the potential customers nor even the scientists doing the original work ever envisaged the ultimate applications or even denied the possibility, as in the case of Rutherford. Advocates of 'demand-pull' on the other hand tend to cite examples such as synthetic rubber or cracking processes or photo-destruction of plastics, where a clearly recognized need supposedly led to the necessary inventions and innovations.

Whilst there are instances in which one or the other may appear to predominate, the evidence of the innovations considered here points to the conclusion that any satisfactory theory must simultaneously take into account *both* elements. Since technical innovation is defined by economists as the first *commercial* application or production of a new process or product, it follows that the crucial contribution of the entrepreneur is to *link* the novel ideas and the market. At one extreme there may be cases where the only novelty lies in the idea for a new market for an existing product,* at the other extreme, there may be cases where a new scientific discovery automatically commands a market without any further adaptation or development. The vast majority of innovations lie somewhere in between these two extremes, and involve some imaginative combination of new technical possibilities and market possibilities. Necessity may the mother of invention, but procreation still requires a partner.

Almost any of the innovations which have been discussed could be cited in support of this proposition. Marconi succeeded as an innovator in wireless communication because he combined the necessary technical knowledge with an appreciation of some of the potential commercial applications of radio. The Haber–Bosch process for synthetic ammonia involved both difficult and dangerous experimental work on a high pressure process and the development of a major artificial fertilizer market, stimulated by fears of war and shortage of natural materials. Despite their early complete underestimation of the market, IBM were the most successful firm in the world computer industry because they combined the

---

* Whilst this may be described as innovation, it cannot be legitimately described as *technical* innovation. Non-technical organizational innovations are extremely important and often associated with technical innovations, but they are not discussed here.

capacity to design and develop new models of computers, with a deep knowledge of the market and a strong selling organization. Firms such as General Electric and RCA with similar or greater scientific and technical strength, but much less market knowledge and market power in this field, in the end had to withdraw.

We may indeed advance the proposition that 'one-sided' innovations are much less likely to succeed. The enthusiastic scientist–inventor or engineer who neglects the specific requirements of the potential market or the costs of his product in relation to the market is likely to fail as an innovator. This occurred with EMI and AEI in computers and with several British firms in radar, despite their technical accomplishments and strong R & D organizations. Professionalization of industrial R & D means that there is now often an internal 'pressure group' which may push 'technologically sweet' ideas without sufficient regard to the potential market, sales organization or costs.

On the other hand, the entrepreneur or inventor–entrepreneur who lacks the necessary scientific competence to develop a satisfactory product or process will fail as an innovator, however good his appreciation of the potential market or his selling. This was the fate of Parkes with his plastic comb and of Baird with television. The failures may nevertheless contribute to the ultimate success of an innovation, even though the individual efforts fail. The social mechanism of innovation is one of survival of the fittest. The possibility of failure for the individual firms which attempt to innovate arises *both* from the technical uncertainty inherent in innovation *and* from the possibility of misjudging the future market and the competition. The notion of 'perfect' knowledge of the technology or of the market is utterly remote from the reality of innovation, as is the notion of 'equilibrium'.

The fascination of innovation lies in the fact that both the market and the technology are continually changing. Consequently, there is a kaleidoscopic succession of new combinations emerging. What is technically impossible today may be possible next year because of scientific advances in apparently unrelated fields. Although he developed the concept mainly in relation to invention rather than technical innovation, Usher's *Gestalt* theory probably comes close to representing the imaginative process of 'matching' ideas. What cannot be sold now may be urgently needed by future generations. An unexpected turn of events may give new life to long-forgotten speculations or make today's successful chemical process as dead as the dodo. Patents for a float glass process and for radar were taken out before 1914. The stone that the builders rejected is the corner-stone of the arch. The production of polyethylene was nearly suspended after the Second World War because the peacetime markets were thought to be too small. IG Farben offered to sell their synthetic-rubber patents to the natural-rubber cartel because they thought the synthetic product would not be able to compete in peacetime in price or quality. The early computer manufacturers expected that the market would be confined to government and scientific users. A century after early experiments stream-driven road vehicles were again seriously investigated by major automobile manufacturers. The apparently random, accidental and arbitrary character of the innovative process arises from the extreme complexity of the interfaces between advancing science, technology and a changing market. The firms which attempt to operate at these interfaces are as much the victims of the process as its conscious manipulators. Innovation works as a social process but often at the expense of the innovators.

These considerations lead to three conclusions of fundamental importance. First, since the advance of scientific research is constantly throwing up new discoveries and opening up new technical possibilities, a firm which is able to monitor this advancing frontier by one means or another may be one of the first to realize a new possibility. Strong in-house R & D may enable it to convert this knowledge into a competitive advantage. Secondly, a firm which is closely in touch with the requirements of its customers may recognize potential markets for such novel ideas or identify sources of consumer dissatisfaction, which lead to the design of new or improved products or processes. In either case, of course, they may be overtaken by faster-moving or more efficient competitors or by an unexpected twist of events, whether in the technology or in the market. Thirdly, the test of successful entrepreneurship and good management is the capacity to link together these technical and market possibilities, by combining the two flows of information.

Innovation is a coupling process and the coupling first takes place in the minds of imaginative people. An idea 'gels' or 'clicks' somewhere at the ever-changing interfaces between science, technology and the market. For the moment this begs the question of 'creativity' in generating the inventive idea, except to note that almost all theories of discovery and creativity stress the concept of imaginative association or combination of ideas previously regarded as separate. But once the idea has 'clicked' in the mind of an inventor or entrepreneur, there is still a long way to go before it becomes a successful innovation, in our sense of the term. Rayon was 'invented' 200 years before it was innovated, the computer at least a century before, and aeroplanes even earlier.

The 'coupling' process is not merely one of matching or associating ideas in the original first flash; it is far more a continuous creative dialogue during the whole of the experimental development work and introduction of the new product or process. The one-man inventor–entrepreneur like Marconi or Baekeland may very much simplify this process in the early stages of a new innovating firm, but in the later stages and in any established firm the 'coupling' process involves linking and co-ordinating different sections, departments and individuals. The communications within the firm and between the firm and its prospective customers are a critical element in its success or failure. As we have seen, in many cases the original idea may take years or even decades to develop, and during this time it continually takes on new forms as the technology develops and the market changes or competitors react. Consequently, the quality of entrepreneurship and good communications are fundamental to the success of technical innovations.

Summing up this discussion, we might conclude that among the characteristics of successful innovating firms in the industries considered were:

1. Strong in-house professional R & D.
2. Performance of basic research or close connections with those conducting such research.
3. The use of patents to gain protection and to bargain with competitors.
4. Large enough size to finance fairly heavy R & D expenditure over long periods.
5. Shorter lead-times than competitors.
6. Readiness to take high risks.
7. Early and imaginative identification of a potential market.
8. Careful attention to the potential market and substantial efforts to involve, educate and assist users.
9. Entrepreneurship strong enough effectively to coordinate R & D, production and marketing.
10. Good communications with the outside scientific world as well as with customers.

We might hypothesize that these are the essential conditions for successful technical innovation. [. . .]

References

Jewkes, J., Sawers, D. and Stillerman, R. (1958) *The Sources of Invention,* London: Macmillan.
Langrish, J. *et al.* (1972) *Wealth from Knowledge,* London: Macmillan.
Musson, A. E. and Robinson, E. (1969) *Science and Technology in the Industrial Revolution,* Manchester: Manchester University Press.
Schmookler, J. (1966) *Invention and Economic Growth,* Harvard: Harvard University Press.

PRODUCT DESIGN

1.4

The Changing Design Process

Nigel Cross

The unselfconscious process

Just as the presence of advanced technology is not necessarily synonymous with 'civiliza-tion', so the presence of any technological artefacts is not necessarily synonymous with the activity of 'design'. For almost all of human history there has not been a readily recognized class of persons regarded as 'designers'. Many complex, beautiful, functional, civilizing artefacts have been created without anyone consciously designing them. These artefacts are the many *craft-made* objects, tools and utensils.

By 'craft-made', I do not mean just any artefacts made by hand processes, nor the pro-ducts of the various 'craft' hobbies that you or I might indulge in with the help of our local 'Art and Craft' shop. I mean instead the products and processes of what must essentially be a *pre-industrial* culture. You or I could not make a craft object in this sense, because we have lost the unselfconscious approach towards 'design' which characterizes pre-indust-rial technology. Yet the pre-industrial craftsmen seem to have been capable of producing objects whose forms (or 'designs') tend to exhibit a remarkable subtlety within an appa-rent simplicity. This simplicity often masks the complexity of the relationship between form and function that the object embodies. To uncover this complexity, it will be neces-sary to analyse a craft product in some detail.

An example of this complex relationship between form and function in craft-made objects is provided in George Sturt's account (Sturt, 1923) of the practices in his wheel-wright's shop in the 1880s. In particular, Sturt tries to analyse the reasons for the peculiar 'dish', or concave shape, built into the traditional wooden cartwheel (see Figure 1). This feature of 'dish' puzzled Sturt, who was not himself a craftsman. He found one readily apparent reason, but this blinded him to a more important reason. (Indeed, there are a host of minor, but interrelated, reasons that Sturt himself hardly seems to understand.) The readily apparent reason was derived from the practice of shoeing the wheel rim with a con-tinuous iron tyre. The tyre was made to its 'finished' circumference, then heated so as to expand it and, whilst still red hot, slipped over the wheel. It was then cooled with water to shrink it on to and grip tight to the wheel. During this operation, the wheel, with some 'dish' already built in, was secured horizontally to the floor by a central screw. When the

N. Cross (1975) Design and technology, T262 *Man-made Futures: Design and Technology* (Unit 9, Section 2), Milton Keynes, Open University Press.
Nigel Cross is a senior lecturer in design at the Open University.

shoeing was completed and the screw was slackened, the wheel hub would rise up a little due to the forces exerted in the spokes by the shrinking-on of the tyre, thus increasing the 'dish'. This necessary tolerance was readily recognized by Sturt as a reason for dishing the wheel.

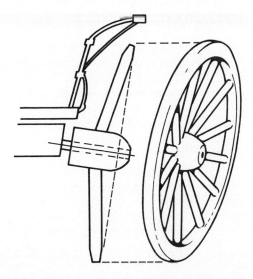

Figure 1   The 'dish' in cartwheels and a section through a waggon body, wheel and axle.

However, Sturt could not be entirely satisfied with this reason, since cartwheels had not always had a continuous tyre, and because he knew that a wheel with an *inadequate amount* of 'dish' would fail in use and collapse like a gale-inverted umbrella. There must, therefore, be some structural reason for 'dish'. Sturt eventually realized this more important structural reason when he observed the motion of a horse-drawn cart from behind. The gait of the horse produced a horizontal oscillation of the cart body on its axles; thus, at each stride of the horse, the cart body was being thrust, first to one side, then to the other, into the wheel hubs. Those thrusts must be countered by the inclined spokes of the dished wheel. If the spokes were not inclined sufficiently – if the 'dish' was inadequate – then the wheel would fail by the hub being pushed outwards by the force of the cart body.

However, dishing the wheel had other advantages, and also disadvantages, which were integrated into a total design of interdependencies. One immediate advantage to be gained was that the cart body could be wider across the top of its sides than across its floor, since the wheel trim was not in the same vertical plane as its hub. This increase in body width was an advantage in carting loads, and possibly may also have contributed to the comfort of passengers.

A serious problem was also induced by dishing the wheel, in that the spokes would be inclined at an angle from the vertical if the wheel was fitted to the cart on a simple horizontal axle. This would mean that the spokes would have to carry the weight of the cart by resistance to bending, rather than by resistance to direct compression. Although the spokes could have been thickened to cope with this, it was important that the total weight of the wheel should be kept as low as possible. (Sturt also describes how the spokes were shaved to their minimum possible section in order to save weight.)

This problem was solved, and additional advantages introduced, by fitting the cart axles inclined to the horizontal by a degree which would bring the lowest spoke of the wheel into the vertical. Thus as the various spokes gradually took up the load as the wheel turned, so they gradually came more towards the vertical until, at the point where each spoke in turn carried its maximum load, it became the 'upright spoke' and was in direct compression only. The additional advantages gained were: (a) inclining the axles raised the body of the cart, so helping to prevent the cart getting bogged down in deep mud; (b) the tops of the rims of each pair of wheels were thus spaced further apart, so that the cart body could be made yet wider across its top; (c) the inclination of the wheels allowed overhanging loads to be carried without their fouling the wheels; (d) the narrowed distance between wheel rims at ground level provided a smaller turning-circle for the cart.

But inclining the axle meant that the wheel had a tendency to slide down off the axle. This was countered by, firstly, a securing pin through the axle at its outer end, and secondly by pointing the axle's outer end slightly forwards as well as downwards. This 'foreway' tended to keep the wheel pressed back into the cart body as it moved forward. It must have also made the cart easier to draw along (although Sturt does not mention this, and was probably unaware of it), since 'foreway' is retained in modern vehicles – as 'toe-in' – so that they tend to steer in a straight line without wander. Hence wheels, axles, and cart body are an overall design solution in which the parts of the design are so integrated that to omit one part could mean failure of the whole design.

This rather lengthy analysis has been necessary in order to uncover some of the complexity that is embodied in a typical craft product which might otherwise, at a glance, seem a relatively simple object. In fact, any inspection of a craft object that is anything more than a glance will reveal subtle shapes, curves, alignments, etc., which begin to indicate the deeper complexities. Since the craftsman must invest a tremendous amount of effort in these subtleties, because of his limited resources of power, tools and materials, it is obvious that they must be incorporated for good reasons.

Of course, many industrially made objects also exhibit great complexity and subtlety in their forms. The wheel of a modern motor-car, for example, together with its sub-assemblies and linkages to suspension, steering and brakes, could probably be analysed to greater levels of complexity than the above analysis of a cartwheel. The relevance of the functional complexity of craft-made objects, to us in this context, is that it was achieved without conscious design effort; without separate 'designers' and without separate 'design' activity.

As J. Christopher Jones (who also relies on Sturt's account of craftwork) has commented:

> The surprising thing to us is that the beautifully organized complexity of the farm wagon, the rowing boat, the violin and the axe, should be achieved without the help of trained designers and also without managers, salesmen, production engineers and the many other specialists upon which modern industry depends. (Jones, 1970)

Jones goes so far as to suggest that the complexity of craft products gives them an 'organic' look, like that of 'plants, animals, and other naturally-evolved forms'. In this phrase lies the clue to the process by which this complexity is achieved: by a process, *over a very long time scale,* which is similar to that of the natural evolution of organisms, i.e. by a process of *very gradual* adjustment. This evolutionary process eventually brings forth a form for the object which is very well suited to its functional requirements, as the form of an organism is very well suited to its environment. [. . .]

The conclusion that I think it is helpful and important for us to draw from all this is that, as Jones points out:

> Neither the professional designer, nor the drawing board upon which the parts of a design can be adjusted relative to each other, are essential to the evolution of complex forms that are well fitted to the circumstances in which they are used. (Jones, 1970)

The selfconscious process

The professional designer and his drawing board are at the centre of the conventional modern design process. 'The drawing office' is also an essential feature of any manufacturing concern: it takes its place amongst all the other specialized offices which are complementary to the factory. Conventionally, then, design appears to be an integral function of the industrial process, and the designer is a specialist, just like any other office worker and just like any other production worker.

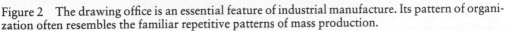

Figure 2   The drawing office is an essential feature of industrial manufacture. Its pattern of organization often resembles the familiar repetitive patterns of mass production.

Obviously, the process of design in its familiar drawing-board form of today replaced the 'unselfconscious' processes, just as all the other features of industrialization replaced craftwork. The development of the modern process of 'design' is one of the *organizational* changes which accompanied the other changes in technology during industrialization. [. . .] Let us consider how design, as a separate process, fits into the pattern of industrialization.

Factory production
The introduction of the factory system meant that the craftsman was no longer an independent agent; he would be unable to negotiate and discuss with his client the specific features of whatever object he was commissioned to produce. This 'design' function therefore had to be taken over by someone else, who took instructions from the client, reformulated them and passed them on to the production worker.

Division of labour
Each specialized task produces only one part of the final, complete product. There is, therefore, a need for a formalized method of, first, splitting the whole product into components and, second, ensuring that these components, when made, will indeed recombine into the

Figure 3   Computer-aided design systems are replacing the designer's traditional drawing board.
Part of Austin Rover's body engineering department transformed by the installation of VDUs.
Source: Austin Rover.

final product. This formalized method is embodied in the design drawings, in which each
component can be accurately specified, together with its immediate relationships to other
components.

Scientific management
The separation of designing from making means that each is considered in isolation. The
'design' of a craft product is contained, to a considerable extent, in the way that the
craftsman makes it; he does not know 'why' it has to have a particular shape, but only
'how' to make it. In the industrialized process of design, however, both the shape of the
product and the operations to be performed in shaping it can be considered from a 'scientific' viewpoint, and hence made more 'efficient'.

Mechanization
Formally splitting up the final product into small components not only facilitates the division of labour and scientific management, but also paves the way for mechanization. Each

component can now be considered in terms of how it can best be manufactured, and it can be designed to suit the manufacturing process.

Economies of scale

The new design process can readily take into consideration new design criteria, such as the manufacturing process (as in design for mechanization), the distribution process (design for ease of packing and transport), the sales process (design for display) and the demands of industrial and business economics (design for the multiple use of standard parts; design for obsolescence). The design process can itself, of course, be centralized in head office, just as other functions of the firm are.

It appears, then, that the industrialized process of design has two very strong features which make it an essential part of the overall pattern of industrialization. These are:

1. In itself, *it separates designing from making.* This separation undermines the craftsman's previous autonomy and authority in his work; it is a necessary aspect of the factory system and that system's subsequent development.
2. In its use of drawings, it contains *a formalised method for the abstract consideration of form.* This method enables new forms to be devised and to be tested in a modelled form before, and quite separately from, the process of production.

This conventional modern design process did not, of course, spring into being overnight as the willing partner-cum-servant of the Industrial Revolution. For instance, design drawings, of a kind, are known to have existed since at least 2800 BC (Figures 4–6). There have been individuals regarded as 'designers' since then, too. However, the design process as we know it today is clearly a significant feature of the wider process of industrialization.

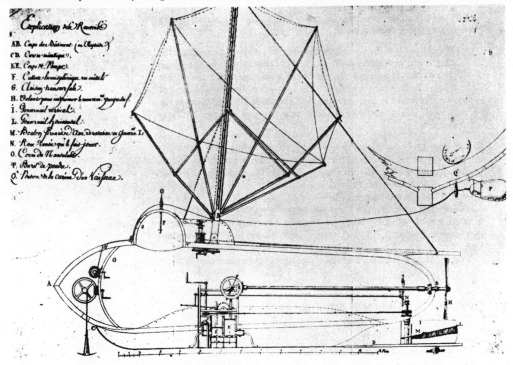

Figure 4   A drawing from the late eighteenth century: Robert Fulton's submarine, built and successfully tested in 1800.

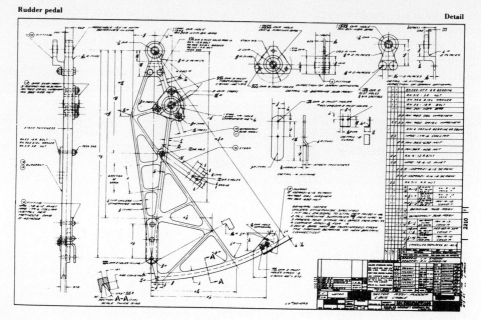

Figure 5

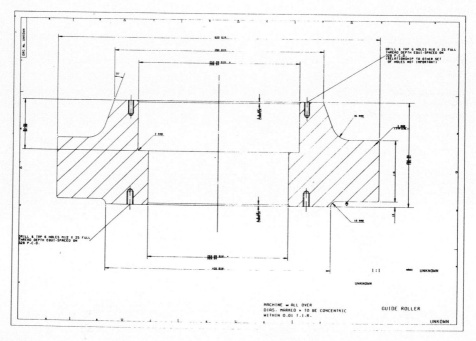

Figure 6

Figure 5–6   The development of design drawings. A typical twentieth-century engineering drawing (Figure 5) and its successor, a computer-produced drawing (Figure 6).

In his book *Design Methods,* J. Christopher Jones (1970) calls this conventional design process 'design-by-drawing', thereby acknowledging the importance and power of the scale drawing as the essential medium and method of the process. He suggests that drawings give the designer a much greater 'perceptual span' than the craftsman had: 'The designer can see and manipulate the design as a whole and is not prevented, either by partial knowledge or by the high cost of altering the product itself, from making fairly drastic changes in design.' In other words, the drawing is a simple means of modelling the proposed design in advance of its production and use.

This 'greater perceptual span' is part of what Alexander (1964) would regard as the gaining of selfconsciousness in design. That is, after the 'unselfconscious' processes of primitive and craft societies, industrial society has developed 'the selfconscious process'. [. . .]

Success, however, obviously is not guaranteed by the 'selfconscious' process. There are many 'design failures' experienced in everyday modern life – from the eye-level grill that spits hot fat in your eye, to the motorway fog disasters. It is this very lack of success in modern design, or the presence of so many 'misfit' situations, that concerns Alexander. By contrast with the 'unselfconscious' process of craftwork, he sees little hope for 'design-by-drawing':

> To achieve in a few hours at the drawing board what once took centuries of adaptation and development, to invent a form suddenly which clearly fits its context—the extent of the invention necessary is beyond the average designer. (Alexander, 1964)

This lack of success in modern design, the limitations of design-by-drawing, and the apparently still accelerating cultural 'context' – i.e. increasingly rapid rates of technological change – have led Jones, and many other designers, to open a search for new methods of design. They suggest that design-by-drawing is no longer wholly adequate, and that a new step in the selfconscious process is required in order to achieve, in Jones's words, 'collective control over the evolution of man-made things'.

The extended process

A number of factors have contributed to the feeling amongst some designers that there is a need to develop a new design process. In general these factors constitute a shift in the design context, claimed by some to be a shift that is comparable with the shift in context that the Industrial Revolution represented. As with the Industrial Revolution, rapid changes in both technology and social institutions are forcing, it is suggested, a reassessment of the nature and role of the design process. [. . .] Thus, the historical sequence of society, technology and design could be:

Pre-industrial society:	Craftwork; the unselfconscious process
Industrial society:	Design-by-drawing; the selfconscious process
Post-industrial society:	A post-industrial design process?

[. . .] Donald Schon (1969) reviews the pattern of industrial and technological changes that are generally regarded as indicative of emerging post-industrialism. He insists that 'this transition has deep implications which we can already feel. It forces a virtual revolution in our concepts of the design process and the design profession. . . .' Especially, Schon sees a series of elemental shifts:

– from component to system and to network;
– from product to process;

– from static organizations (and technologies) to flexible ones;
– from stable institutions to temporary systems.

The key concepts in this list appear to be those of *system* and *process*, replacing component and product. In industrial terms, this means:

> The corporation does not commit itself to a single product line or even to a single technology, its commitment is to a major human function, and to the changing technologies and organizational relations required to carry it out. (Schon, 1969)

As examples, this would mean a business corporation recognizing that is should diversify from producing, say, motor-cars to transportation systems; or from houses to shelter systems. The implications for the designer are clear; he will be called upon to generate systems innovations, rather than just product modifications.

> The obsolescence of products (not the obsolescence of particular products but of products as the unit of design) creates the requirement for a new kind of design – namely, the design of the systems and sub-systems . . . to which business systems are coming to respond. And this, in turn, transforms the concept of the designer's role and his place in the firm. Design becomes indistinguishable from systems design and development. Systems design becomes a central corporate function rather than a peripheral stage in product development. (Schon, 1969)

To some critics, the idea of design becoming 'a central corporate function' will sound alarming. It is the fulfilment of the technocracy, with the experts and the businessmen in joint control of technological change, and their aims being the aims of the modern corporation: economy, efficiency, production and consumption. Yet, as I tried to show above in my analysis of the role of a separate design process in industrialization, design is already, in many senses, a 'corporate function'. In post-industrial society will it merely become more 'central'? [. . .]

A slightly different analysis of the 'need for new methods' in the design process is presented by J. Christopher Jones. His starting-point is not so much that a changing design context is presenting new opportunities for the evolution of the design process, but that the conventional design process (design-by-drawing) seems to be creating as many problems as it does solutions. He suggests that a notable limitation of the scale drawing is that it is a very weak model of the product-in-use situation (Figures 7–8). That is, the scale drawing is a means of designing a product in isolation from its manufacture and use, and although this was once an advantage (to the manufacturer), it now begins to present disadvantages (to the user). For example, incompatibilities may arise between different products which a user wants to juxtapose, or when a product is used on a mass scale (e.g. the mass use of the motor-car). These are what Jones calls the *external* compatibilities of a product. The scale drawing only resolves *internal* compatibilities, such as the relative locations, dimensions, etc., of separate components in the product.

Jones points out that:

> When considering the external, as opposed to the internal, compatibility of a new product the designer gets no help from the drawing and has to rely, in the main, upon his experience and imagination. . . . (Jones, 1970)

Figures 7–8   The design drawing is an unreliable model of reality. Figure 7 shows the type of drawing used to present the designer's vision of how his product will be used. Figure 8 is the reality of the same product – instead of a sunny, happy, communal space in which mothers can safely leave their babies, it is cold, bleak, padlocked and the haunt of vandals.

Figure 7

Figure 8

In novel situations, such as designing in new materials or for new environments, experience is often irrelevant and imagination may be inadequate. So the conventional design-by-drawing process seems to have a major, and increasingly important, shortcoming. This shortcoming (of failing to deal adequately with external compatibilities) manifests itself in the 'design failures' that abound:

> Perhaps the most obvious sign that we need better methods of designing and planning is the existence, in industrial countries, of massive unsolved problems that have been created by the use of man-made things, e.g. traffic congestion, parking problems, road accidents, airport congestion, airport noise, urban decay and chronic shortages of such services as medical treatment, mass education and crime detection. These need not be regarded as accidents of nature, or as acts of God, to be passively accepted: they can instead be thought of as human failures to design for conditions brought about by the products of designing. Many will resist this view because it places too much responsibility upon designers and too little upon everyone else. If such is the case then it is high time that everyone who is affected by the oversights and limitations of designers got in on the design act. (Jones, 1970)

This sounds like another call for wider design participation, and, indeed, Jones elsewhere suggests that a common feature of the new design methods that he advocates is that they *externalize* the design process, or 'make public the hitherto private thinking of designers'. This has the advantage, he argues, that 'other people, such as users, can see what is going on and contribute to it information and insights that are outside the designer's knowledge and experience'.

However, Jones's principal recommendation is that the design process should be extended from its concern with products, to include the design of systems [Figure 9]. This is very much the same as Schon's conclusion of what is already happening in industry anyway. This new level of design activity is a kind of 'thinking-before-drawing', or 'deciding what should be designed', which can be seen as another step back, or another level of abstraction, like the step back from making to designing. 'Such an extension of the design process', Jones suggests, 'is at least as great as that from craftwork to design-by-drawing'.

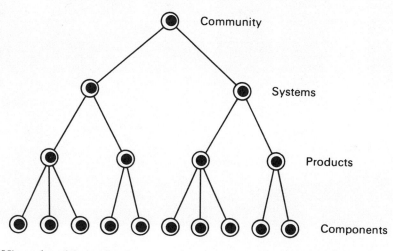

Figure 9   Hierarchy of design levels.
Source: Jones (1970)

References

Alexander, C. (1969) *Notes on the Synthesis of Form*, Harvard University Press.
Jones, J. C. (1970) *Design Methods: Seeds of Human Futures*, Wiley (2nd edn, 1981).
Schon, D. A. (1969) Design in the light of the year 2000, *Student Technologist*, Autumn. (Reprinted in N. Cross, D. Elliott and R. Roy (eds.) (1974) *Man-made Futures*, Hutchinson, pp. 255–63).
Sturt, G. (1923) *The Wheelwright's Shop*, Cambridge University Press.

1.5

The Nature of Design

David Pye

Invention and design distinguished

Although in many fields designers quite frequently make inventions, designing and inventing are different in kind. Invention is the process of discovering a principle. Design is the process of applying that principle. The inventor discovers a class of system – a generalisation – and the designer prescribes a particular embodiment of it to suit the particular result, objects, and source of energy he is concerned with.

The facts which inventors discover are facts about the nature of the world just as much as the fact that gold amalgamates with mercury. Every useful invention is a discovery about the way that things and energy can behave. The inventor does not make them behave as they do.

'A system of this kind' means, 'this way of arranging things'. There is a principle of arrangement underlying each class of device. In some cases, for instance Supports or Enclosures, it is extremely simple. In others it is not. An inventor's description of the essential principle of a more intricate device, might be worded as follows: 'If you have a wheel, and if at any place except the centre you fix to it a pin standing at right angles to the plane of the wheel, then (provided always that the system is properly designed) the wheel can be turned by the piston rod of a reciprocating engine. It must be linked to the piston rod by a connecting rod which is longer than the distance from the centre of the wheel to the crank pin. One end of the connecting rod must be hinged to the piston rod so that it is free to swing, and the other end must be pivoted on the pin so that the pin can rotate freely in it.'

The piston rod, connecting rod, and crank pin are 'the device'. The wheel is 'the object'.

For the sake of brevity this description of an invention for converting reciprocating motion into rotary motion has not been worded to cover all conceivable instances; and would no doubt have been worded better by a patent agent – whose job it is to write such things – but it will serve to illustrate the essential point that while the description is virtually complete and comprehensive, Figure 1 is very incomplete although easier to understand. The description describes the essential principle of the device, which is purely a matter of its arrangement. Almost every conceivable instance is covered by the description. Figure 1, on the other hand, merely describes one particular embodiment of the invention. Someone failing to guess the essence of the innovation from the diagram might suppose

D. Pye, *The Nature and Aesthetics of Design*, Herbert Press, 1983; extracts from Chapters 2, 3 and 8.
David Pye was for many years Professor of Furniture Design at the Royal College of Art.

that the swelling in the connecting rod was essential, whereas the connecting rod can be of almost any shape provided it is properly designed. Shape is among the least important properties of a connecting rod, not much more important than colour. But for that matter shape is not all important even to a ball bearing. No two are the same shape and certainly none of them is spherical.

Figure 1

The description says nothing about the shapes of the components of the system. 'A wheel' it says. The wheel may be a Geneva wheel, of a complicated star shape. The pin could be triangular in section. The connecting rod could well be in the shape of a dragon. The hinge could be made of eel-skin like a flail's. And still the thing could work. The description says distinctly how the parts are related to each other in arrangement without saying what they are like. The relations between them which it describes are those which determine in what direction energy can and cannot be transmitted from one to another. Quite a complicated relation is often implied by one word. The word pivot, for instance, implies a pair of things one of which embraces the other but cannot, except by friction, transmit turning forces to it.

It is really rather remarkable that, while anyone can tell whether a thing is a pocket-knife because, presumably, anyone can recognise the principle of arrangement which constitutes the similarity between all pocket knives, no one can *visually* abstract that arrangement. We recognise it when we 'see' it embodied, we can describe it disembodied, but we cannot visualise it disembodied.

The six requirements for design

When a device embodying some known essential principle of arrangement such as we have discussed is to be adapted and embodied so as to achieve a particular result, there are six requirements to be satisfied:

1. It must correctly embody the essential principle of arrangement.
2. The components of the device must be geometrically related – in extent and position – to each other and to the objects, in whatever particular ways suit these particular objects and this particular result.
3. The components must be strong enough to transmit and resist forces as the intended result requires.
4. Access must be provided (this is a special case of 2 above).
    These four together will be referred to as *the requirement of use*.
5. The cost of the result must be acceptable.
    This is *the requirement for ease and economy*.

6.   The appearance of the device must be acceptable.
     This is *the requirement of appearance*.

Design, in all its fields, is the profession of satisfying these requirements. [...]

The requirements conflict. Compromise

The requirements for design conflict and cannot be reconciled. All designs for devices are in some degree failures, either because they flout one or another of the requirements or because they are compromises, and compromise implies a degree of failure.

Failure is inherent in all useful design not only because all requirements of economy derive from insatiable wishes, but more immediately because certain quite specific conflicts are inevitable once requirements for economy are admitted; and conflicts even among the requirements of use are not unknown.

It follows that all designs for use are arbitrary. The designer or his client has to choose in what degree and where there shall be failure. Thus the shape of all designed things is the product of arbitrary choice. If you vary the terms of your compromise – say, more speed, more heat, less safety, more discomfort, lower first cost – then you vary the shape of the thing designed. It is quite impossible for any design to be 'the logical outcome of the requirements' simply because, the requirements being in conflict, their logical outcome is an impossibility. It must however be remembered that by the use of magic, that is to say by unknown forms of energy (of which electrical energy was recently one) impossibilities can be designed; but now that we have nuclear fission and fusion we may have come to the end of magic.

Of the many inevitable conflicts between the requirements of economy the crudest is that between durability and low first cost. The design of consumer goods according to 'built-in obsolescence' is arbitrary indeed. At other levels there are inevitable conflicts between high speed and low maintenance, high speed and low first cost, high speed and low running cost; light weight (for, say, low fuel consumption) and high strength (for durability and safety); more daylight (through large windows) and more quiet, with even temperature in sunny weather; more cargo capacity and more speed; a keener edge and a lasting edge; and as many more as you like. [...]

It is said that by the aid of computers we can arrive at the correct solution in such cases with certainty. They are clever, these computers! They are going to show us the cheapest answer. But if they think their clients are going to be satisfied with that, they are not so clever as they think.

The fact that compromise is inevitable in so many kinds of design has led theorists to classify design as a 'Problem-solving activity', as though it were nothing more than that. It is a partial and inadequate view.

Most design problems are essentially similar no matter what the subject of design is, but while the discerning layman understands that in the design of large constructions, a new town or an airport, the problems are overwhelming, he probably does not realise so clearly that there are problems just as pressing and difficult for the designer in the design of almost any trivial product. A bad town will do more harm than a bad toothbrush but the designer of either will experience his job as the necessity to make a series of decisions between alternative courses of action, each affecting the decisions which come after it; and if no life hangs on the outcome of the series of decisions about the toothbrush, the livelihood of several people does. The designer can do harm enough if he does not take care. He cannot shrug off the decisions. He has to take the problem-solving aspect of his job seriously.

There are times when the problems are so intractable that they absorb all the designer's attention and seem to leave him little or no choice about the appearance of what he is to design. He feels that he has been lucky to arrive at even one solution which goes most of the way to meet the requirements and as though no alternative to it were conceivable.

There is, I believe, only one circumstance in which that can ever be nearly true. It is the case where the designer has committed himself beforehand always to choose the cheaper alternative, wherever alternative solutions are feasible. Whether it can ever be right for a designer to commit himself absolutely to such a course is very doubtful because a point will always be reached at which a cheaper alternative can only be found by settling for something worse: something less inherently durable, or less useful, or less safe, or having a less durable finish so that its appearance and perhaps its strength deteriorate more rapidly. The washing machine that embodies parts which are not rust-proofed is a familiar example. There is always a worse way of making a job and it is nearly always cheaper.

Even in this extreme case alternative details or finishes will in reality be available which are equally cheap, so that the designer will still be in a position to make the job look better or worse.

The designer always has more freedom of action than appears at first, and particularly in the matter of detail and finish. That matter is of great importance and the quality of the appearance of anything designed depends very largely on it, as also it does on workmanship, which is the extension of detail and finish down to a scale at which the designer has no power to specify appearance directly. In design as in all art the difference between good and bad may be very slight, yet absolute.

Design is not all a matter of problem-solving, neither is it all a matter of art. It is both. The two parts of it are inseparable.

1.6

Investment in New Product Development

The Viscount Caldecote

[. . .] Britain more than any other industrial country depends on international trade to create prosperity and employment. For instance, about 24% of our Gross Domestic Product is devoted to international trade compared to about 17% in France, 12% in Japan and 6% in the USA. And the harsh fact is that our share of international trade expressed as a percentage of total exports from all countries has fallen from about 22% ten years ago to under 10% now.

There is much that can and must be done to correct this unhealthy situation and I want to examine one very important factor, investment and new and improved product design and development, with the objective of driving home both its vital importance and some of the problems associated with it. But first let me set the scene and clear away some misconceptions.

The role of research

The objective of the manufacturing industry is to make things which will sell profitably in the market for which they are intended. This requires a continuing programme of investment in the design and development of new and improved products, which will only be successful if a clear market need has been identified and a specification drawn up to meet it.

In discussing investment of this type it has become the custom to talk about 'research and development' (R & D), as if this was a single activity with one objective. Perhaps this may approximate to the truth in, for instance, the chemical and pharmaceutical industries; but in the wide field of manufacturing hardware for the markets of the world, which

Extracts from a lecture delivered to the Royal Society of Arts on 11 April 1979; published in *Engineering*, June 1979, pp. 810–14.

Lord Caldecote is a former Chairman of the Design Council and of the Delta Metal Co. He is currently Chairman of Investors in Industry.

Figure 1   The Flying Bedstead [research vehicle designed to test the concept of a hovering structure]

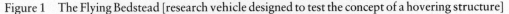

includes the increasingly complex engineering industry, contributing some 40% of our visible exports, research is one thing, design and development quite another.

The prime objective of research is to generate knowledge: while the manufacture of some hardware will often be involved, it is not a marketable product. Thus applied industrial research will contribute towards, but will not itself achieve, the successful launching of a new product. For example, the Flying Bedstead (Figure 1) was built principally to study the problems of stability and control of a hovering structure supported by jet thrust. It was never conceived as a saleable product but the principles and knowledge obtained were very relevant to the success of the Harrier aircraft (Figure 2).

Figure 2   The Harrier [a practical vertical take-off and landing aircraft]

The market specification

The first stage in creating a new product is to discover with as much certainty as is practicable what the market needs, or will need by the time sales commence. From this a specification can be drawn up; this is a description of the product in terms understandable by the designer, to which he works. It will include some or all of the following: performance, maintainability, standards of reliability, intended working life, ergonomic standards, appearance, cost, and proposed rate of production. The latter is important in many products because it can have a profound influence on methods of manufacture and so on detailed design.

Some preliminary design work will start before the specification is finalised to help the iterative process of harmonizing the needs of the market with what is possible both technically and financially.

The design process

Let us then look first at what is involved in the design process, and then at the development phase leading up to the proving of the new product in a form suitable for manufacture to meet the specification.

Design is the process of converting an idea into information from which a product can be made. The output of the designer has traditionally been drawings and specifications, but today these may be to a large extent replaced by magnetic tapes or discs, as we shall see later, from which hardware can be made directly.

Development is closely related to design and it is, therefore, sensible and highly desirable to talk of 'design and development' (D & D). For development is the process of proving a design, by for instance, making prototypes and subjecting them to rigorous testing, or trying out different manufacturing methods to minimise cost, to confirm that the final product meets the original specification, is reliable and, in short, meets all the requirements to ensure that it will sell profitably in the market for which it is designed. In complex products, such as an aircraft, it is a long and very expensive process, in which the design team is intimately involved. As we move along the spectrum of design through less complex products, such as machine tools, diesel engines, refrigerators, to toys and textiles, the paramount importance of the engineering designer in ensuring technical excellence gives place to the skills of the industrial designer, who is more concerned with ergonomics and aesthetics.

There is still unfortunately much confusion about the scope of design for it means different things to different people. A textile is created by an individual designer specialising in textile design. He or she would have been educated at a college of art and design or in the art and design department of a polytechnic and would have particular skills in aesthetics and the man/product interface. But some knowledge of technology is required to ensure that the textile can be manufactured on the machinery available and to assess the effect of the texture on the design.

And today the computer can also add enormous strength to the industrial designer by, for instance, enabling different patterns to be rapidly compared and in setting up the production process.

The designer responsible for creating the display unit of a marine echo sounder would be an electronic or systems engineer who is familiar with the problems of acoustical echo ranging and electronic circuitry. However, the skill of an industrial designer will be

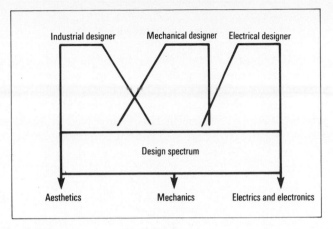

Figure 3   The design spectrum

required for the display, the layout of the controls and lighting, and for the design of the case to allow the unit to be fitted in any odd corner on a ship's bridge.

Figure 3 illustrates the design spectrum in another way. It shows that some products (e.g. textiles) are almost entirely the prerogative of an industrial designer while other products (e.g. a submarine cable) are the sole responsibility of an engineering designer. Some products (e.g. a telephone) have approximately equal industrial and engineering design content. Figure 4 shows the order of magnitude of the relative costs of the engineering design and industrial design for a range of products. You will see from this that if the engineering design costs for a typical product are, say, £1m then it is probable that £10 000 will be required for the industrial design aspects.

Another aspect of the design spectrum, relates to design skills, from aesthetics through ergonomics, structures, mechanics, electromechanics and electronics to systems. The design spectrum is now so broad and complex that it is beyond the capability of any one person to achieve professional competence throughout its entire extent, but some breadth of understanding is essential in a designer. A 'T'-shaped ability profile – a combination of depth and breadth – is needed by a designer if he is to make his best contribution to the work of the design team, while retaining an appreciation of, and respect for the contributions of his colleagues.

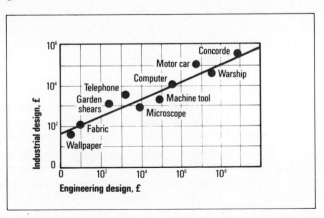

Figure 4   Comparison of industrial design and engineering design costs for a variety of products

The development phase

As soon as the design has progressed sufficiently, one or more prototypes or pre-production models must be made. In more complex products the prototypes then start an intensive testing programme to prove the design which will involve modifications to it. In simple products little more is required than to confirm the suitability of the design for the intended manufacturing process. Thus the designer is closely involved in this development phase which varies widely in time and cost, depending on the complexity and on the extent of the technical advance.

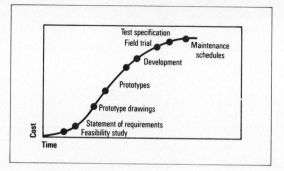

Figure 5   Cumulative expenditure against time for the design and development of a new engineering product

A graph of typical, cumulative expenditure against time for the design and development of a new engineering product up to the completion of the fully proved design is given at Figure 5. The beginning of the curve may well include an element of research leading on to design and development. However, this graph takes no account of other expenditure on production tooling, purchase of piece parts, marketing or sales launch, which all require further investment. The graph represents an optimum expenditure – if things go well. If problems arise the straight centre part of the graph can be extended almost indefinitely – expenditure escalates and serious, perhaps catastrophic, delays occur.

It is significant that typically the first prototype appears before the halfway point, implying that at this milestone the designer's task is barely half completed in terms of both time and cost – with the more difficult half yet to come.

Financial considerations

The overall effect on cash flow (both negative and positive) covered by investing in the launch of a new product is shown at Figure 6. This graph shows the cumulative effect on cash flow through the design and development phases, to the build up of stock and work-in-progress in the early stages of production, when there is no balancing in-flow of cash from sales, to the phase of profitable sales which bring the cash in-flow. It should be noted that the moment when the curve starts to rise above the point of maximum investment, and therefore the extent of this investment, is critically dependent on completing development and being able to start profitable sales. If development problems delay the latter, the curve will plunge downwards both because of increased development expenditure and because there is no cash in-flow from sales. When these investments are large in relation to the com-

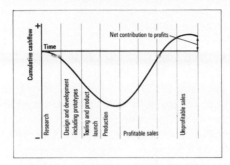

Figure 6   The product's overall effect on cash flow throughout its life

pany's resources or where profits from products in current production are already declining, such delay may threaten the financial stability of the company, as occurred in the case of Rolls-Royce, in the development of aircraft engines.

An aircraft engine is an example of a product at the extreme top end of the complexity and risk spectrum and it is obvious that the shape of this curve varies widely with the type of product, as shown below.

| | | *Time from start to:* | |
| | *Maximum* | *Maximum* | *Breakeven* |
Product	*cash outflow*	*cash outflow*	*on cash*
New executive aircraft	£80 million	5 years	10 years
New family car	£500 million	3–4 years	5–8 years
Fork-lift truck	£500 000	3 years	5–6 years
Cooker	£750 000	15 months	3 years
Toy	£30 000	2 years	3 1/2 years
Electric kettle	£10 000	10 months	15 months

Despite the growth of design aids based on computers, design and development costs are steadily increasing – even though discounted for inflation. There are many reasons for this including the greater and therefore more costly skills required, increasingly complex constraints and specifications imposed on the designer, the implications of legal liability and pollution control and the problems caused by multiple assessment often required by different major customers. It follows that, in general, D & D timescales are becoming longer. In other words the stake is becoming higher and the outcome less certain.

It must also be appreciated that in a product of any appreciable complexity there is no certainty that the design and development process will be successful in producing a product which will sell at a profit. It is extremely difficult to estimate accurately the development cost and timescale, simply because the need for proving the design through development implies a degree of uncertainty and ignorance about the outcome. Thus the design and development of any but the simplest new products is a risky and expensive business. Investment in it is very different to investment in new plant and buildings, since fixed assets have a residual value even if they cannot be used for their originally intended purpose. But the resources invested in an unsuccessful development programme are virtually valueless. For this reason it is now accepted accounting practice for the expenditure on D & D to be written off as it is incurred. [. . .]

Encouraging profitable investment

Now I would like to consider ways of encouraging investment in the development of new products as a contribution to restoring Britain's position in world trade.

Above all, of course, must come the commitment of the Board and Chief Executive of a company. This leads naturally, as recommended in the Corfield Report (NEDO, 1979), to making one director responsible for stimulating such investment and ensuring that advantage is taken of every new relevant technique. From this will follow the allocation of proper priority to investment in new products from the resources available.

But remembering the point made earlier about the high risk involved, three further steps are important: first to study the market in as much detail as possible, so that when all the money has been spent, the development successfully completed and the specification met in every detail, the product can be sold at a good profit.

In most cases development of a new product will be initiated in response to an observed market requirement, which will largely determine the specification. But sometimes a significant advance in technology may itself create a new market, as did the advent of semiconductors to the small-transistor-radio market, or the application of lasers in medicine and in production processes. But whatever the origin of the idea for a new product, the importance of drawing up a specification which will meet a market need is paramount.

Secondly, select and appoint the best designers available, not only because they will have the most creative ideas but also because their designs are most likely to be right first time and so to need the least modification during developments: and they will also create products which are easy and quick to make, requiring the minimum investment in plant and work in progress. [. . .]

Thirdly, provide the design-and-development team with the best possible facilities you can afford. One example of this today is computer aids to design, combined with numerically controlled machine tools, or other production equipment. These will enable the design to be completed quickly and efficiently and will also help to speed up the manufacture of prototypes or pre-production runs.

Often several years have been spent, in the past, checking designs and laboriously making and modifying models of the product by hand. Now the design can be completed and the models made in as little as two or three weeks, during which time modifications can be quickly and easily incorporated. This very substantially reduces the time required for design and development and so reduces the overall investment, and the interest accruing on it.

Reference

NEDO (1979) *Product Design* (the Corfield Report), London: National Economic Development Office.

Section 2

Creative Individuals

Introduction: The Creative Process

Robin Roy

This section is concerned with that crucial part of the technical innovation process, the creation of new ideas, concepts and solutions to problems. As shown in section 1, both invention and design are creative activities, even though their outcomes may often differ. A creative designer, like the one who produced the novel 'Chronalog' clock described in article 2.2 (b), may conceive a new *form* for an existing artefact. A creative inventor, like the lorry driver described in article 2.2 (a) who devised a way of suppressing wheel-spray, may conceive a novel *principle* as well as designing a device embodying it.

How then do such ideas come about? Is there a common pattern to creative thinking? Brian Lawson (1980) in his book *How Designers Think* has summarized various descriptions and theories of the creative process and has identified a common set of five stages. These stages are outlined in Figure 1.

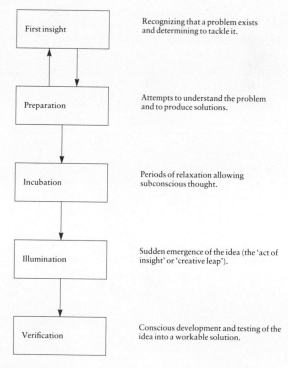

First insight	Recognizing that a problem exists and determining to tackle it.
Preparation	Attempts to understand the problem and to produce solutions.
Incubation	Periods of relaxation allowing subconscious thought.
Illumination	Sudden emergence of the idea (the 'act of insight' or 'creative leap').
Verification	Conscious development and testing of the idea into a workable solution.

Figure 1   Five-stage model of the creative process
Source: Adapted from Lawson (1980).

Lawson's model has much in common with the *gestalt* theory of the creative process proposed by the historian Abbott Usher and summarized in article 1.2 in section 1 and article 2.4 in this section. Usher's four-stage creative process comprises:

1. perception of the problem ('first insight');
2. setting the stage ('preparation' and 'incubation');
3. the act of insight ('illumination');
4. critical revision ('verification').

Some writers on creativity (e.g. Schon, 1967) have emphasized the importance of *chance* in creative thought. Indeed there are many examples (such as Pearson's discovery of the effect of light on semi-conducting materials, which led to the silicon solar cell) where important inventions have arisen as a result of an unexpected discovery, a chance observation, or simply a sudden crazy idea. 'Chance', however, 'favours only the prepared mind' and article 2.5 by Glegg emphasizes the point that 'illumination' or the 'act of insight' does not usually occur without preparation, possibly over many years. As was noted in article 1.2, 'chance or accidental observation come as a bonus to the perceptive researcher who has done his "homework" '. Normally an inventor or designer has first to recognize and commit him or herself to solving a problem. Then he or she has to 'saturate the mind' with the problem and with thinking about (and sketching) possible solutions. With luck the 'creative leap' will occur after the subconscious mind has had time to work on the problem.

Glegg notes:

> The secret of inventiveness is to fill the mind and the imagination with the context of the problem and then relax and think of something else for a change. . . . If you are lucky, [the] subconscious will hand up into your conscious mind . . . a picture of what the solution might be. It will probably come in a flash, almost certainly when you are not expecting it.

However, Glegg is careful to point out that having creative ideas is only the first step in producing innovative products and processes. There is much hard (and often creative) work to be done in developing the idea into prototype designs, testing and modifying these and getting the product into production. Different people, Glegg points out, may be suited to different stages of the process, and those with highly developed creative abilities may not always be best at the logical and detailed tasks required as the product reaches its final form.

Although there is substantial agreement about the basic phases of the creative process, there are many points of disagreement regarding other aspects of creativity. Two in particular are examined in the articles in this section. First, how important is the personality of the individual relative to the features of the environment in creative thinking? And second, how important is the *individual* inventor or designer today, when so much technical development is performed by teams working in large organizations?

Kelly and colleagues in article 2.4 argue that, in fostering creative thinking, *both* the individual and his or her environment are important. No defined set of characteristics of the 'creative personality' has yet been isolated, but according to Garrett (1963) and Lawson (1980) creative individuals tend to share a certain mix of traits. For example, such individuals tend to be problem-oriented, self-centred, unorthodox, intelligent and mentally restless, and have the ability to concentrate on a task, but to shift their attention when necessary.

The influence of the environment on the creative individual is sharply illustrated in article 2.3 by Amram entitled 'The innovative woman'. First, Amram notes that women's inventions, like men's, are strongly influenced by their immediate social and working environment. Thus historically women's inventions tended to be concerned with improvements to household appliances, clothing and agricultural implements. Secondly, the fact that less than 2 per cent of American patents have been issued to women is not because women are inherently less creative than men, but because of the social expectations, constraints and lack of opportunities for women to engage in technically inventive activities. Amram notes: 'one can hardly be surprised that women inventors are so few, what is surprising is that they exist at all and that their inventions are not only patented, but commercially developed'.

Kelly and colleagues also address the question of whether there is still a role for the individual inventor and designer in the modern world of organized research, design and development teams. They note that, although the *independent* 'backyard' inventor/ designer has become much less important, creative *individuals* are still the most important source of original ideas. Such individuals may work in large organizations, but the modern equivalents of the nineteenth-century inventor/entrepreneur are research and design consultancy firms and the small technology-based firms founded by entrepreneurial scientists, inventors and engineers.

Where does this leave the old-style 'backyard' inventor, whose difficulties in obtaining finance to exploit his or her 'good ideas' is a source of much media comment (e.g. Hope, 1978)? The experiences of one such individual are recounted in article 2.1 by Malcolm Brown. Brown shows that independent inventors tend to view the lack of finance for developing inventions as due to the weaknesses of the venture-capital system in Britain. He also shows that financiers tend to view 'one-man-band' inventors with suspicion, mainly because they often lack the business skills needed for innovation. Kelly and colleagues note that this situation is not new. Inventive individuals like James Watt only succeeded in getting their ideas developed because of their association with someone with the necessary business and entrepreneurial skills, while the most successful innovators were those, like Edison, who were inventor/entrepreneurs, rather than merely producers of creative ideas.

None of the articles properly discuss the question of how much prior knowledge and experience is needed to produce original and workable ideas. Jewkes, Sawers and Stillerman (1969), in their classic book *The Sources of Invention*, have emphasized the importance of the independent inventor and the 'untutored mind', unconstrained by too much conventional wisdom, in the creation of major inventions. They cite several examples to support this view, such as the inventor of the ball-point pen, who was an artist and journalist; the inventor of the pneumatic tyre, who was a vet; and the inventor of the safety razor, who was a commercial traveller. However, in more highly specialized and technical fields, one could cite many more counter-examples, such as the transistor, in which possession of the necessary theoretical and scientific knowledge was essential to the invention (Braun and McDonald, 1978).

Lawson (1980) notes that in many fields the creative individual needs a 'reservoir' of experience and technical knowledge on which to draw, but at the same time needs to retain the flexibility of mind to allow that knowledge to be patterned and combined in new ways. Koestler's classic work *The Act of Creation* (1964) argues that what is vital to creative thinking is the ability to *synthesize* ideas and skills from normally unassociated areas of knowledge. As an example Koestler cites Gutenberg's invention of the printing press, which involved first perceiving the analogy between printing with wax seals and with cast type, and then combining the technology of the wine press with that involved in casting seals and coins.

A more modern example is the invention and development of the silicon solar cell, mentioned earlier. As Flood (1985) describes, it arose as a by-product of a major research programme into semi-conductors that followed the invention of the transistor at Bell Telephone Laboratories. In this case, the invention arose from a combination of systematic research, chance and synthesis of ideas from different fields.

References

Braun, E. and McDonald, S. (1978) *Revolution in Miniature*, Cambridge University Press.

Flood, M. (1986) Invention: water turbines; solar cells, T362, *Design and Innovation* (Block 1), Open University Press.

Garrett, A. B. (1963) *The Flash of Genius*, Van Nostrand.

Hope, A. (1978) It's a wonderful idea, but . . ., *New Scientist*, vol. 78, no. 1105 (1 June), pp. 576–81.

Jewkes, J., Sawers, D. and Stillerman, R. (1969) *The Sources of Invention*, 2nd edn, Macmillan.

Koestler, A. (1964) *The Act of Creation*, Hutchinson.

Lawson, B. (1980) *How Designers Think*, Architectural Press.

Schon, D. (1967) *Technology and Change*, Pergamon.

2.1

The Independent Inventor: Eureka! I've found some money

Malcolm Brown

Leighton Evan's latest invention – an ingenious device to stop thieves forcing door locks – could change his life. Hoteliers from Yorkshire to Dorset, including some of the biggest names in the business, are queueing up to fit it and the money is rolling in. Evans ought to be a happy man. Instead he is bitter and resentful, almost to the point of obsession, about the way he has been crushed by 'the system'.

Evans, a 55-year-old former art teacher and freelance designer, has 40 inventions to his credit, but, with the sole exception of his security device, Lockit, not one of them has yet got beyond the workshop.

He blames the money men.

'My advice for anyone who has a brilliant idea with massive potential', he says, 'is to forget it.'

'There is no financial support from any source – high street banks, government departments, EEC funds, venture capital companies. If you ask any of these establishments how many new ideas they have given financial assistance to, to enable the idea to be developed and patented, the answer will be none.'

It's like hitting a brick wall, says Evans, who is only managing to buy patent protection for Lockit – £6000 already paid and another £4000 due shortly – by raiding the kitty for money already put down by eager customers.

There are thousands of Evanses around. The men with the money mark their files 'One Man Band' and put them into the 'pending' or the 'out' tray, almost never the one marked 'action'.

Which is all very well if they slip away quietly into obscurity, but when, against all odds, they succeed as Evans may well do, they point up an enormous paradox in the system. The paradox is this: historically, the individual inventor/entrepreneur has provided society with some of its best inventions, so much so that the most progressive big companies today are desperately searching for ways to make their research and development act more like these rugged individualists; yet almost nobody controlling risk capital is prepared to put

M. Brown, Eureka! I've found some money, *Sunday Times*, Business News, 14 October 1984.
Malcolm Brown is a freelance correspondent for the *Sunday Times*.

Figure 1   Inventor Leighton Evans blames the British venture capital system for stifling new ideas, like his 'Lockit' security device.

a penny on the very people whose characteristics those big companies are trying so hard to emulate.

The banks, says Evans, talk about risk capital 'but it's only risk provided there's no risk involved'. Public sector funds can be painfully difficult to extract from the bureaucratic purse (though, to be fair, some public bodies, like Evan's particular *bête noir,* the Welsh Development Agency, do now offer seed capital to fledgeling innovators). And the venture capital houses shun the small man almost completely. 'You have to have a track record of three or four years.' says Evans.

The caution of the public agencies and banks is no great surprise. But why are the venture capitalists, whose whole reason for existence is risk taking, not taking this kind of risk?

Critics believe there are now serious structural faults in the venture capital system. There are, say the critics, three main constraints:

1.   *Money.*   Not lack of it but how it is used to make it cost effective. There is a lower limit below which venture capital business isn't worth writing – a limit which knocks for six people who, initially at least, may only be looking for tens of thousands of pounds to test feasibility or build a prototype.

They are caught in a 'Catch 22' situation. They can't put a 'package' up to the risk capital houses because they won't take the initial risk. Most venture capital houses won't do business worth less than £50,000 and often double, or even quadruple, that.

The reason? The venture capitalist has to do almost as much investigation for a £100,000 investment as he would do for one worth £50,000. Pound for pound the bigger deal is a far better investment for the venture capitalist.

2. *Time.* This is the one thing that all venture capitalists are short of. Most venture capital organisations are made up of just a handful of people (a dozen would put it in the big league) but they have to scrutinise hundreds of projects. It is not worth their while looking at £75,000-worth of business when, for the same investment of their scarcest resource, *time,* they could be looking at a £750,000 project.

3. *Conservatism.* Most venture capitalists come from a financial rather than a technological background. Instead of asking 'How does your perpetual motion machine work and what can we do with it?' they are, by and large, much more interested in the last three years' accounts, market projections for the next three years, and an assessment of cash flow over the next 12 months. They are not, on the whole, gamblers.

The government's Business Expansion Scheme should have taken up some of the slack at the lower end of the venture capital scale, but in practice it hasn't. One of its primary aims was to bridge the equity gap, give businesses capital amounts up to £50,000 or even £100,000 to help with their early development. The reality is that because of the high returns available under the scheme, much BES money is competing with conventional investment sources available to established businesses, rather than providing risk capital.

Derek Allam, chief executive of the venture capital company Prutec, has little sympathy with the critics. He says that it simply isn't true that individuals with bright ideas are being scorned by the venture capitalists. Prutec sees 1000 submissions a year and gives them a fair hearing. But there are limits.

'We frankly haven't the time to take somebody and educate him. In most cases it's probably too late anyway. If you are talking to a 40-year-old entrepreneur who's been spending the last five years trying to sell a left-handed widget, its going to be rather difficult to pick him up and train him.'

And the payoff may simply not be worth the effort: 'We have made an investment in one small company,' says Allam. 'I made this in the early days in the belief that small companies – one man bands – should be supported, and the time that we've spent with that person far exceeds, *far exceeds,* the time that we've spent with multi-million pound investments.'

In fact Allam takes a diametrically opposed view to people like Evans, about the British venture capital system. It's not the venture capitalists who are wrong but the customers – they are not putting up good enough cases and more often than not it is the human element which is the weakest link.

A first class management team may make a success out of a less than perfect market opportunity, but no matter how good the opportunity a poor management team is unlikely to succeed.

'We've seen about 4000 opportunities in the last four years,' says Allam . 'If you compare these 4000 with the sort of opportunities that come out of North America, for example, there, for a start, individuals are much more rounded and they tend to collect around them a team of people. They have a much more powerful team approach, whereas we tend to get people with technical ideas, no idea of the market and a one man band – in some cases not willing to share his ideas and usually unaware of the need for help in all these other areas.' In America, he says, the quality of both ideas and management are far superior to those in Britain.

'The ideas are generated because they have a number of critical masses. If you go to places like Silicon Valley and Route 128, they've developed a critical mass which is very like an atomic bomb – it's going to go; there's no stopping it . . . and the whole thing is mushrooming, which we don't see in Britain. I can't think of a single critical mass in Britain.'

2.2 (a) Creative Individuals

Dry Run in a Kitchen Sink

John Wardroper

Alan Buckley is a lorry-driver who cares. He used to hate throwing a daunting and danger-ous shower of spray over other road users. An anti-spray idea flashed into his mind when he was in his kitchen using one of those pan-scourers made of a plastic network. Now the lorry he drives is amazingly inoffensive in the rain.

His idea was so impressive that he was brought on to Technical Committee AUE/13 of the British Standards Institution to help draft a specification for 'spray-reducing devices for heavy goods vehicles'.

Buckley's idea was that if he attached layers of plastic mesh to the undersides of his mud-guards, spray flying off his tyres would be trapped and thrown back down to the road in big drops instead of swirling up in his slipstream in the opaque clouds that motorists know so well [Figure 1]. He did some experiments with a hose in his garden at Rochdale, then got a lorry fitted up and went to the makers of the pan-scourer, Netlon at Blackburn. Now Netlon is going into production with a new product, spray-suppression kits, for lorry operators who care about their image – such as Buckley's employer, Mark's & Spencer's Fashionflow delivery service.

Most hauliers, however, won't fit anything unless they are compelled to. Indeed, the trailers of most juggernauts don't even wear minimal mudguards above eight or twelve big tyres. An official survey this year showed that the thing about lorries that bothers motorists most is spray. After 20 years of looking at the problem, the Department of Transport is about to act. The question to be decided is: how stiff should the standard be?

Tests show that at 50 mph the Netlon system stops nearly 70% of the water flying free as spray. However, some people on the committee, which comprises a couple of dozen men from all parts of the lorry industry, have been pressing to let fittings qualify at as low as 50%. Netlon's marketing director, Roger Gibson, says: 'If the pass-mark is as low as 50%, after all the work we've been involved in, our product won't be worth it as a commercial venture.' Road hauliers would go for something cheaper – 'a dimpled sheet of rubber'.

The committee is thought likely to settle for 60%. Netlon has several competitors in this coming market. The best-established is a Monsanto product, Clear Pass, using plastic material inspired by Astroturf. It is already fitted to 10,000 vehicles, including many police Range Rovers and the lorries of image-conscious firms such as Cadbury Schweppes,

J. Wardroper, Dry run in a kitchen sink, *Sunday Times*, Business News, 11 December 1983.
John Wardroper is a freelance correspondent for the *Sunday Times*.

Figure 1   Creative invention. Alan Buckley and spray-reducer. As the diagram shows, fine spray flies up (1), hits the Netlon-lined mudguard and flap (2), and falls to the road in heavy drops (3).

Reckitt & Colman, Bass-Charrington, Scottish & Newcastle, Sketchley, Shell and Texaco.

Clear Pass's spray-suppression rating is a little below Netlon's. But another product claimed to be superior to both is Cat's Whiskers, marketed since May by Schlegel UK, of Leeds, makers of vehicle door seals and other devices to control liquid flows. Cat's Whiskers are rows of soft polypropylene bristles hanging in a fringe along each edge of the mudguard. They too throw the water to the road in large drops. Schlegel UK says Cat's Whiskers, though dearer, are lighter and will never clog with slush. [. . .]

2.2(b)

Hours not to reason why

John Huxley

If you want to know the time ask a . . . draughtsman. James Goodchild, a lecturer in product design at Glasgow School of Art, has discarded the conventional clock face, dumped the more modern digital, and come up with what is claimed to be the first truly rectangular time piece.

The Chronalog comprises two lines of continuous bands. One is horizontal and travels down the face in 12 hours. The other, vertical, line moves from left to right in one hour. Where they cross indicates the time, read off from the two axes [Figure 2].

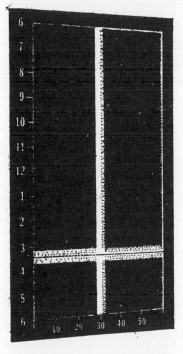

Figure 2   Creative design. James Goodchild's 'Chronalog' clock

J. Huxley, Hours not to reason why, *Sunday Times,* Business News, 6 November 1983.
John Huxley is the industrial editor of the *Sunday Times*.

Goodchild says the Chronalog is easier to read (for example, the 15-minute mark does not appear as '3') and indicates the passage of time visually. But he admits that its prime attraction is aesthetic. 'It gives architects, town planners and developers far more design options.' Chronalogs can come in any shape and size, colour and lighting arrangement.

The idea for the Chronalog came when Goodchild asked his students to design a new clock. All retained the round face, changing only the cosmetics. Goodchild then demonstrated that time didn't have to go round in circles.

Now he's working on a three-dimensional clock: 'That is the big one.' Too big, he fears. Although it is possible to incorporate a third (that is, seconds) hand, he feels it could make a clock too difficult to understand.

2.3

The Innovative Woman

Fred Amram

About 1200 American patents are issued each Tuesday at noon. Of these, about 20 (1.7 per cent) bear the name of a woman. While over four million patents have been issued since the US Patent Office opened in 1790, women have received only about 60,000 (1.5 per cent). Why have women made so little progress as inventors? What sort of women constitute that small percentage who do invent? And what sort of inventions have they patented? Patent office records offer a unique opportunity to explore these questions and thereby to shed light on a little known area of women's activities.

Before the creation of the US Patent Office in February 1790, American colonists went to Europe to record their inventions. The very first patent issued to a resident of the colonies was British Patent No. 401, to Thomas Masters of Pennsylvania in 1715. It was entitled 'A new invention found out by Sybilla, his wife, for cleaning and curing the Indian corn growing in the several colonies in America' (Figure 1). While Mrs Masters was in England acquiring her patent, her husband built a mill in Philadelphia incorporating her invention; he subsequently became the mayor. Sybilla Masters' invention of 'A new method for working and staining straw for hats, bonnets etc.' won the second patent granted to a colonist – again it was her husband who was granted 'the whole use and benefit of the said invention during the space of 14 years'.

Not until 1809 was the first patent issued by the US Patent Office to a woman in her own right. She was Mary Kies of Windham County, Connecticut, and her invention was for 'straw weaving with silk or thread'. By the end of the century, the number of women inventors was increasing, as women themselves recognised. The centenary of the US Patent Office was marked by the publication of a newspaper called *The Woman Inventor*, which published the achievements of innovative women up to 1890. But by 1910 the number of women's patents was still only 8596, just 0.8 per cent of the total.

Why were they so few? Some of the answers are obvious. Invention usually requires money, materials and the opportunity to share ideas. Historically, few women have been financially independent, and most have been excluded from sources of education and intellectual stimulation. More importantly, however, society's expectations of women have simply not included technical innovation. Women themselves have absorbed this view and adopted the belief that to be an inventor is neither possible nor appropriate. An old edition of the *Encyclopedia Americana* volunteers the information that 'Minors and women and

F. Amram, The innovative woman, *New Scientist*, no. 1411, 24 May 1984, pp. 10–12.
Professor Amram is coordinator of speech communication at Minnesota University.

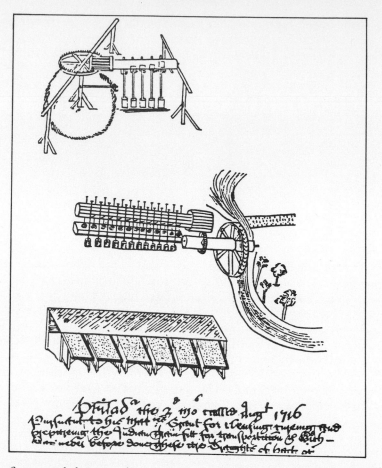

Figure 1   The first recorded patent of a woman's invention: Sybilla Masters' machine for cleaning corn. But the prestige went to her husband – he eventually became mayor of Philadelphia.

even convicts may apply for patents under our law' – hardly encouraging company. [. . .]

Our cultural history leaves little encouragement for the woman thinking of becoming a creative scientist. Women need a new perspective. In *The Underside of History: A View of Women Through Time*, Elise Boulding writes: 'Women think of things in small scale, and if they have to turn attention to large-scale systems, they have to train for it . . . women have to think of themselves as potential innovators.' Unfortunately, even those who manage to make this conceptual leap still face problems. Louise Kiel invented a rocking wheelchair and applied for a bank loan to produce it. Her six patents were in order, the market potential was clear, and the bank was impressed. 'But I won't give you the loan,' said the manager, 'unless you have a man with you.'

One can hardly be surprised that women inventors are so few. What is surprising is that they exist at all, and that their inventions are not only patented, but commercially developed. In 1954, a survey found that although women received only 1.5 per cent of patents, the profits they received from them were on average higher than those of men. The range of women's inventions is now impressive; models are certainly available for young women if only we bring them to light. For example, Patsy Sherman, commercial products development manager for the 3M company, contributed to 15 important patents including several for 'Scotchgard', the widely used stain repellent for fabrics.

The nature of women's inventions, like that of men's, is a function of time and place. If a creative woman's horizons are bounded by the household or the farmstead, then she will invent domestic or agricultural implements. The mid-nineteenth century saw a preoccupation among early feminists with dress reform; they longed to abandon their corsets and crinolines. A radical journal called *The Revolution* proclaimed in 1868 that '(man) not only prescribes woman's sphere but how she shall dress in that sphere. Now one of the rights we claim for women is to wear a bifurcated garment and be sailors and soldiers and whatever they choose . . .' [. . .]

Many women designed ingenious new products to ease their domestic burdens, apparently without training in physics or materials science. Mary Florence Potts exhibited a sad-iron at the 1876 Centennial Exhibition that was remarkable in its sophistication and proved popular in various versions for the rest of the century (Figure 2). Her iron was double-pointed so it could be used longer without having to be reheated. It also had a removable wooden handle and a non-conducting core, both of which made the business of ironing a great deal cooler.

Authors usually write about what they know; inventors derive problems as well as solutions from their experiences. Another example comes from the role women play in time of war. Women have usually supported war efforts through the manufacture of arms and ammunition. Is it any wonder then, that women have been inventors of weaponry? Table 1 is a selection of the patents granted to women during the First World War.

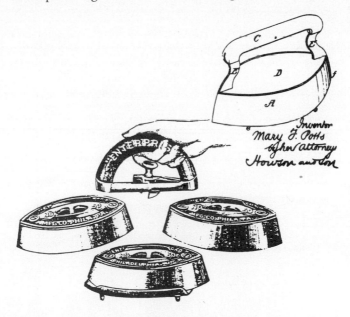

Figure 2   Mary Pott's sad-iron was a sensation at the 1876 Centennial Exhibition; developments of it were popular for years. One version had a removable handle so that one iron could be used while the other was heating.

Success in a man's world

Since women have had access to education, laboratories and finance, they have begun to demonstrate their creativity in a far wider range of activities. Kate Gleason made history as the first female mechanical engineer. Her design of a worm-and-gear won her election

Table 1   Patents granted to women during the First World War – a selection

Automatic pistol	Percussion and ignition fuse
Bomb-launching apparatus	Primer
Cane-gun	Railway torpedo
Cartridge tube filter	Rear sight for guns
Flashlight attachment for firearms	Resilient missile
Front sight for firearms	Single trigger mechanism
Incendiary ball	Submarine mine
Loading device	Top for powder cans
Woven carriage carrier	Torpedo guard

as the first woman member of the American Society of Mechanical Engineers in 1914. More recently Elsie F. Harmon patented 'a hot die-stamp method of infusing silver conductors on polymerised thermoplastic and thermosetting materials'. Marguerite She-wen Chang invented a triggering device for an underground nuclear test, and Mary Olliden Weaver was co-inventor of a starch-graft polymer known as the 'Super Slurper'. For this she and her colleagues were awarded the 1977 National Inventor of the Year award – the first time a woman had received it. Two years later a woman won the title in her own right: Barbara S. Askins received the award for her method of obtaining clearer pictures from old negatives. Her technique also makes it possible to reduce X-ray exposure times.

The history of women inventors will always remain incomplete. Many of their inventions were patented by men because women had no right to property or because they were embarrassed to take on an unusual or unacceptable role. *The Woman Inventor* raised this question as early as 1890. For example, Mrs A. H. Manning of Plainfield, New Jersey, invented a clover cleaner and a mower and reaper, both of which were patented by her husband in the early 1830s. Eli Whitney's cotton gin, a key invention in America's economic history, was the brainchild of his landlady, Catherine L. Greene. Whitney simply helped to build the prototype.

Women may have had difficulty obtaining recognition and commercial success for their inventions, but for black women the problem was multiplied. Ellen Eglui of Washington DC invented a clothes wringer, but she sold her invention to an agent for $18 in 1888. He

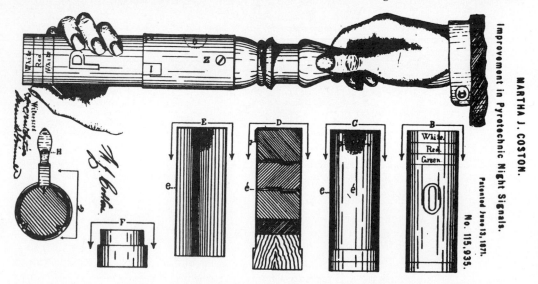

Figure 3   Martha Coston invented a flare that earned her $20 000 during the American civil war.

subsequently made a great deal of money from it. A writer on *The Woman Inventor* asked her why she sold it for so little. She replied: 'You know I am black and if it was known that a Negro woman patented the invention, white ladies would not buy the wringer; I was afraid to be known because of my colour, in having it introduced into the market, that is the only reason.' [. . .]

We know relatively little about the type of personality that leads to inventions. There is little or no difference between men and women on tests of 'creativity'. Inventors do have exceptional drive and a willingness to take risks. Most have sporting inclinations; they seem to be both athletically able and competitive. [. . .]

Almost all the women inventors I have spoken to identify with their fathers, most without doubt or hesitation. Apparently these men were themselves physically or intellectually adventurous. Several women hoped fervently that the next generation would have adventurous mothers to identify with. In invention, as in mountain climbing, a new woman is emerging.

2.4

The Individual Inventor-Entrepreneur

Patrick Kelly, Melvin Kranzberg,
Frederick Rossini, Norman Baker,
Fred Tarpley and Morris Mitzner

Importance and Types

The process of technological innovation in this century has been characterized by an increasingly strong trend towards the 'institutionalization' of all its phases, including that cluster of activities that often bear the generic label 'invention'. The emergence and rapid growth of the corporate R & D laboratory, especially in the years after the Second World War, is the major manifestation of this trend, which may be referred to as a move from 'random' to 'deliberate' creativity. However, this phenomenon should not cause us to lose sight of the role of the individual inventor.

Despite the growth of institutionalized innovation in R & D laboratories, some solo work is still being done. Many investigators discount the role of the 'lone-wolf' inventor in contemporary innovations, yet from time to time claims are made that the individual inventor still maintains a major role in the innovation process. In the 1960s this view gained support from the Jewkes, Sawers, and Stillerman (1969) study of fifty important twentieth-century inventions, showing that over half stemmed from independent inventors or small companies.

From there the argument moved to a quantitative basis with the counting of patents – or a pseudo-quantitative basis which allowed differing interpretations. The proponents of the individual inventor argued against those who claimed that group inventive effort produces more innovations by pointing to the fact that the total annual issue of US patents in

P. Kelly and M. Kranzberg (eds.), *Technological Innovation: A Critical Review of Current Knowledge*, San Francisco Press, 1978; extracts from Chapter 3.

Professor Kranzberg and Drs Kelly, Rossini, Baker, Tarpley and Mitzner were all members of the Georgia Tech Innovation Project group, based in the Department of Social Science, Georgia Institute of Technology.

the 1960s was no greater than 30 or 50 years previously, and that in terms of patents per unit of population, the number was less in 1960 than in 1870, despite an annual rise of 10–20 per cent in R & D expenditures in the half century from 1910 to 1960. They also pointed out that the number of patents had not grown in proportion to the increased number of scientists; in other words, the research force was growing far faster than the number of patents produced by that force.

Their opponents – the believers in the efficacy of R & D laboratories in producing innovation – first discounted the number of patents as a true index of the nature, amount, and quality of innovative efforts. Furthermore they claimed that current patents, though fewer in proportion to the population, are individually longer and more technical, and a larger percentage of them were being worked than formerly. In addition, an increasing proportion of inventions were being made by government employees or were in weaponry, and so would be less likely to be patented. They also stated that it was becoming more difficult to make patentable inventions as time went on, because of a tendency of the proportion of basic inventions to shrink while that of minor, unpatentable improvements grew.

The high point of the argument on behalf of the independent inventor was reached in the Charpie Report of 1967. This report, the product of a panel of private citizens convened by the US Secretary of Commerce, was officially titled *Technological Innovation: Its Environment and Management*, but is usually referred to by the name of the panel's chairman, Robert A. Charpie, then president of Union Carbide Electronics. The thrust of the report was that the government, primarily through tax concessions, must ease the way for the backyard or garret inventor and for the small company. Its major recommendation – a White House conference on Understanding and Improving the Environment for Technological Innovation – has not been taken up, and few of its other recommendations ever took hold; perhaps just as well, because the Charpie Report was based upon a static and hence an unpredictable database.

History is a dynamic process. Individual inventors in 1959 still accounted for 40 per cent of the new mechanical patents, 35 per cent in electricity and electronics, and 30 per cent of new chemical patents; that was still far less than in previous years. The Jewkes, Sawers, and Stillerman study (1969), constantly referred to by the Charpie panel, dealt with inventions back to the turn of the century, when the structure and nature of technological innovation were far different from what they have become. Furthermore, although the original ideas for many of the inventions cited in that study might have come from individual inventors, their actual development had gone nowhere until they were put into the hands of large corporations possessing industrial research laboratories that could develop them into into commercially feasible and saleable innovations.

By focusing on a limited number of innovations, by looking upon the innovatory environment as static rather than dynamic, and by failing to distinguish among the different elements entering into the innovation process, the Charpie Panel had diagnosed – and prescribed for – a situation which was at least a quarter of a century out of date. At the very time the Charpie Panel was carrying on its deliberations, the percentage of significant inventions made by independent inventors, even measured by their beloved patent count, was dropping markedly. The lone inventor was giving way to the group worker in the organized research laboratory.

Nevertheless, the individual inventor cannot be ignored. With such examples drawn from recent history as Edwin Land (Polaroid camera) and Chester Carlson (xerography), it is obvious that the solo inventor is by no means obsolete and that he can be responsible for major innovations – although Carlson's original invention had to undergo much development, carried on in a structured R & D situation, before it achieved successful application.

In this section we are concerned with individual inventors whose activities have not been carried on in the context of R & D organizational structures. We can distinguish three such types. First, there is the inventor who is fertile in imagination and technical ingenuity and who can produce an inventive idea or even a prototype device, but who lacks the entrepreneurial capacities to carry it through to an innovation. James Watt at an earlier date and Chester Carlson of today's xerography are representative of this type. Second are those whom Thomas Hughes had characterized as inventor-entrepreneurs, who embody the characteristics of both the independent inventor and the entrepreneurial capitalist. In many cases, such as Thomas Edison and Elmer Sperry, they work alone at the beginning of their careers, but later establish their own firms, sometimes working within the context of an organized laboratory. And finally, there are research scientists and engineers who begin their careers in a corporate context and later establish new technologically based 'spinoff' firms.

Individual Creativity

Investigation of the role of the individual inventor brings up complex problems of the wellsprings of technical creativity. Creativity has been studied from two methodological viewpoints, which correspond, interestingly enough, to the social-deterministic and individualistic approaches employed for interpreting the innovation process: sociological and psychological. The psychological approach to creativity tends to focus on forces within the individual, concentrating on such factors as intelligence, personality, and attitudes. The sociological approach, though not denying the importance of those elements, claims that they derive from various types of social background and conditioning. In other words, this is the old 'nature versus nurture' argument applied to innovative creativity.

One of the earliest scholars who sought to explain creative genius was Francis Galton (1870), who found heredity a primary determinant of eminence. Other pioneer psychologists also found the explanation for creativity in 'native genius'.

More recent studies have relegated heredity to a minor role, although not discounting it completely. Ann Roe (1953) showed other factors to be of major importance, such as the intellectual atmosphere of the home, childhood interests, and position in the birth order. Not until 1955 was a conference on the identification of creative scientific talent held, and then scholars placed differing emphasis on various demographic, cultural, religious, and personality attributes (Taylor, 1956). Within a few years, however, it was evident that 'profiles' of eminent scientists did not necessarily shed much light on the creative process itself (Anderson, 1959).

Samuel Smiles had made out his great inventors to be the most reasonable and virtuous of men (Hughes, 1966), but some iconoclastic thinkers of the late Victorian and the Edwardian eras were endeavouring to show that creativity resided primarily in certain choleric and splenetic individuals who refused to adjust to the world about them and who did not adopt its values. Not surprisingly, the psychoanalysts, with their emphasis on the neurotic and irrational elements in the human mind and behaviour, came forth with theories relating creativity to emotional disturbances. More recent investigators have abandoned the popular cliché of linking creative genius with a light touch of madness; they now tend to view creativity and psychological health not only as compatible, but as mutually supportive (Rogers, 1964).

Although one body of opinion holds that the creative act is basically the same in every field of endeavour (Coler, 1963), other studies distinguish among various types of creativity and link them to various kinds of activities and goals. Some creative individuals (especially composers, expressionist painters, sculptors, and writers) are simply expressing their inner states; others direct their creativity to meet externally defined needs and goals; and a third type cuts across the first two.

Nevertheless, such studies of creativity would seem to distinguish between an independent inventor and one operating in the context of an organized innovative effort. This failure to distinguish between the two is justified by our analysis, in which the inventor emerges as a function, not as a person. Before the institutionalization of innovation, the individual and function were merged; in the organized R & D effort, the individual is submerged in the function.

Because invention is a function, discussion of the individual inventor's characteristics becomes virtually irrelevant. For our purpose it is much more important to look at the environmental context in which innovation takes place. Yet we can perhaps learn something about the innovation process by briefly focusing on the independent inventors in their different roles.

The Inventor *qua* Inventor

Some light may be cast on the role of the inventor *qua* inventor by viewing this function in the light of Usher's *gestalt* theory of invention (Usher, 1959). As outlined in article 1.2 of this book, that theory predicated a four-step sequence: perception of a problem, setting of the stage, act of insight, and critical revision. Typically, the independent inventor is strong on the first three of these steps, but his critical revision is frequently lacking in the elements necessary to make the invention into a successful innovation.

James Watt's invention of the steam engine might serve as an exemplar of the Usher theory. When given a model of a Newcomen engine to repair in 1763, he soon perceived the problem of inefficiency caused by heat loss in the cylinder's wall. For two years he set the stage by tinkering with the cylinder and trying wooden rather than brass cylinders. Then Watt tells us the act of insight which occurred to him 'on a fine Sabbath afternoon' in 1765 while strolling on the Glasgow green, for it was then that he hit upon the idea of condensing the steam not in the operating cylinder as Newcomen had done, but in a separate condensing chamber.

Although Watt conceived his brilliant idea of the separate condenser in 1765, it was not until 1769 that he obtained his first patent, and it was more than a decade later, in 1776, that the first Watt engine was brought into commercial use. What happened during the eleven years between Watt's act of insight and the first successful commercial installation of his engine proves the importance – and difficulty – of the developmental stage in transforming an idea (and a model, made within three days after the Sabbath afternoon walk) into a practicable innovation. Here the question of defining the technical problems and viewing them in their economic context was to prove crucial; that was to be largely the work of Matthew Boulton, not of Watt who had the original idea.

For the fact is that Watt lacked sufficient capital to devote full-time effort to scaling up his model to an efficient and reliable machine, which involved the solution of many additional technical problems. Watt also lacked the requisite managerial and entrepreneurial expertise. Boulton became the driving force making for the successful introduction of Watt's steam engine; he provided the capital and also brought together the market demand with the creative ability of Watt.

Chester Carlson in more recent times exhibits the same inventive imagination as did Watt. But he too lacked both the capital and entrepreneurial skills to transform his basic concept of xerography into commercial application, and was forced to rely on organized

R & D establishments for the critical revision and marketing phases of his invention.

It would seem that the contribution of the entrepreneur is sometimes equal to that of the inventor in arriving at an innovation. However, the entrepreneur need not be an individual, for we are really talking about a function, not an individual. That function – which includes risk taking, the provision of capital, the development from idea or prototype to operational status, and the coupling of the market place with the inventor's concept – can be and is increasingly performed by a corporate entity. But it is also possible for the inventor himself to have entrepreneurial qualities and to do the entire job from perception of need through development and marketing. That unique individual is the inventor-entrepreneur.

The Inventor-Entrepreneur

In his prize-winning biography of Elmer Sperry, Hughes (1971) utilized the concept of the inventor-entrepreneur to explain the process of innovation. He was offered the following definition and characterizing generalizations of the inventor-entrepreneur in the history of American technological development.

> Inventor/entrepreneurs are inventors who preside over the innovative process from its origins as a problem to, at least, the introduction of the invention into use. The usual reason that inventor/entrepreneurs were not simply inventors was that they were determined to have their invention used, and to achieve that they realized they would have to take the initiative not only in the early phases of innovation but in research and development and marketing. In essence they were inventors; in effect, they had to be entrepreneurs. The evidence, however, is that they found their work more satisfying when identifying problems and inventing solutions, not when presiding over and promoting the other phases. The evidence also tends to support the generalization that in America before 1930 most successful inventors were in fact inventor/entrepreneurs.

The pre-eminent example of the inventor-entrepreneur – and indeed America's most spectacular and prolific inventor (some 1093 patents to his name) – was Edison. Perhaps the best illustration of Edison's ability to bring an innovation to completion – from definition of the problem to profitable application and diffusion – is electric lighting. Hughes has pointed out that 'only the naive inventor assumes that the challenge is to invent an arc lamp, an electrical generator, a streetcar, or an automobile'. Edison saw things in their entirety, and one of the major reasons for his success was that he realized that the problem was to develop an electrical lighting *system*, not just to devise an incandescent bulb (Figure 1).

The scientific principles and technical requirements of a viable electric light bulb were known as far back as 1860 and had been tried by many inventors. Practical electric generators were already at work, and there were arc-lighting systems employing generators, transmission lines, and lamps. The technological level was thus at a stage where further steps could be taken. At the same time Edison possessed the capital to undertake the creation of a complex system because he was the owner of a considerable fortune derived from his previous successful inventions. What is equally important was Edison's well-equipped and well-staffed Menlo Park laboratory, which provided him with resource requirements – shop facilities, instruments, a library, specialized personnel, etc. – for a high level of inventive activity. The mere existence of such facilities illustrates the essential entrepreneurial underpinning for this type of innovative project. Above all, Edison defined the problem in large terms and was fully cognizant of the economic constraints involved: he was attempting to develop an entire lighting system that would compete with gas illumi-

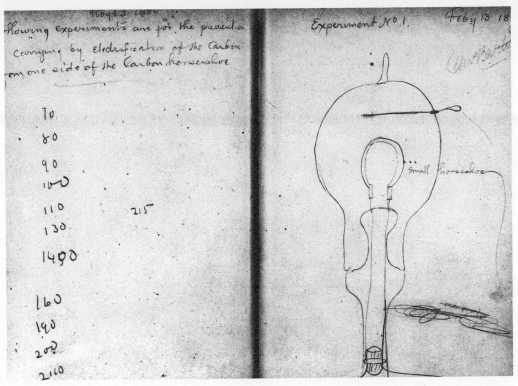

Figure 1   A page from one of Thomas Edison's notebooks showing the results of experiments on a carbon-filament electric lamp. He realized, however, that the problem was to develop the whole electric lighting *system*, not just the electric lamp. [Figure inserted by the editors.]

nation. Edison's achievement was as much a triumph of entrepreneurial ability, managerial expertise, and economic reasoning as of technical ingenuity.

Unlike Edison's work on a whole complex system, Elmer Sperry focused on bottlenecks, or 'reverse salients', in rapidly expanding areas. Such areas were chosen because capital was available and there was a probable market for his invention. Hughes (1978) writes:

> Identification of critical problems was a professional capability of Sperry's and one crucially important for his survival as an independent inventor. When Sperry's numerous patents are examined and their claims considered, it becomes clear that he did not invent dynamos, arc lights, streetcars, or automobiles, though the title of his patents might lead to such superficial conclusions. In the case of his dynamo patents, he claimed automatic controls; in connection with the arc light, he invented a regulator for the feed of carbons; and when inventing for streetcars he contributed an operation control. His patent claims in these instances and in numerous others show that he solved very specific problems, which can be aptly labeled 'critical problems'.

Sperry's pattern, at least until 1910, was to identify such an area in which his special competencies would apply, move in and concentrate on its weakest point, make his contribution, and then move to something else rather quickly.

Sperry seems to have lost interest in a field after about five years, which suggests to Hughes 'that an inrush of inventors, engineers, managers, and corporations brought by capitalization, growing market, and size convinced him that his special characteristics and circumstances could best be employed elsewhere' (Hughes, 1978). He may well have realized that after a field has experienced a period of rapid change it then enters a 'mopping-up' phase in which incremental improvements are worked out. His own talents were less well suited for that phase.

The pattern of Sperry's activities prior to 1910 allows the anticipation of a point to be discussed later in this book (Section 4). One of the crucial elements in the problem definition/idea generation phase of the innovation process, as conducted in the corporate R & D lab, involves the identification of *market needs*. In fact, most innovations are stimulated in this way. That seems to have also been the case in Sperry's career as an independent inventor. How did Sperry go about acquiring need information?

> Sperry's letters, memoranda, notebooks, and other records reveal thtat he identified critical problems by close study of technical journals, patterns of patent applications, the patents of others, attendance at professional engineering society meetings, conversations, and his intimate knowledge of expanding technological systems. Articles in the technical journals often told Sperry of the interests of other inventors and therefore of problems on which they were working; in weekly reports of patents granted, like those published in the *Scientific American*, he could discern a pattern of concentration on certain problems (for instance a bevy of patents on arc-light regulation). By a close reading of the claims of these patents he could delineate the problems of focused attention more precisely, and the Official *Patent Office Gazette* provided regular summaries of all patents. Sperry regularly attended the sessions of the engineering society meetings, for there he might gather from fresh reports and papers more intelligence pertaining to critical problems on which other inventors were working. (Hughes, 1978)

The typical modern corporation, with all the information-gathering potential at its command, scarcely acquires need information as carefully, systematically, and with such result as did Sperry working alone.

Once a critical problem had been identified, Sperry 'tried to discern the weakest point and strengthen it'. After 30 years, this 'hit and run' pattern of inventing in response to the weakest point in an expanding field and then moving on to another was finally altered with the establishment of the Sperry Gyroscope Co. in 1910. During the early period we can see most clearly another characteristic of Sperry's inventing behaviour, and at the same time note two of the characteristics of the innovation process itself that require an entrepreneurial response.

With the idea for an invention in mind (often the problem definition/idea generation phase in our terminology), Sperry would embark on a process in which 'invention merged unperceptibly into development', as he successively 'scaled up' towards the environment of intended use. The successive test environments 'involved more variables, altered parameters, and new factors, and these revealed the need for successive modifications of the first idea by revision and new invention'. Such scaling up of course requires resources and the organization of various specialized testing activities, which in turn imply entrepreneurial commitments that may be substantial indeed, far beyond the resources of an individual. Such was the case with Sperry's work on a marine gyrostabilizer, the development of which required access to the US Navy's experimental model basin, and ultimately sea trials aboard the destroyer *Worden*. In such cases arrangements have to be developed in which an independent inventor shares the entrepreneurial commitment, or consigns the development of his invention to an organization with more adequate resources.

However, the independent inventor intersects with and in fact depends on the entrepreneurial activities of others in an even more basic sense. We recalled that Sperry sought out the fields that were undergoing rapid change and looked for the reverse salients, or bottlenecks, to further progress within them. Implicit in this approach is a dependence on the vigorous activities of the entrepreneurs within the field who had brought about the very progress that revealed such weak points and would reward an independent for inventions that strengthened them. Thus the independent inventor, if not an entrepreneur himself, is doubly dependent on those who are entrepreneurs: he requires the progress they impel and organize, and ultimately the developmental capabilities they possess.

One further point should be noted in this connection. As Gilfillan (1935) has pointed out, inventions are themselves changes in a system which necessitate further invention. They introduce perturbations into the ecological system to which new adjustments must be made. Although this consideration poses a constraint on the independent inventor, and may ultimately influence the acceptability of his invention, it is not his primary concern. But the concern of the entrepreneur is with the market fate of a larger system, of which the critical problem addressed by the independent inventor is only a part.

[A summary of different views of the invention/innovation process of Usher, Gilfillan and Hughes is shown in Table 1, taken from Hughes (1978).]

Table 1   Concepts of the innovation process

General	*Usher* Emergence of novelty	*Gilfillan* Principles of invention	*Hughes* Inventor–entrepreneur
Problem identification	Incomplete pattern Perception of problem	Growth and chance evokes system need or imbalance	Reverse salient in expanding technological system Congenial critical problem identified by inventor
	Setting of state	Evolutionary accretion of response components	Consideration of prior unsuccessful solutions Experimentation
		Nontechnological factors stimulate response	
Idea response Invention	Act of insight	Institution or person organizes congenial synthesis of responses	Discern weakest point Invention
Research and development			Development toward complexity of the use environment
Introduction into use			Introduction into use
	Critical revision		Postinnovational development

[Inserted by the editors from Hughes (1978)).

The Business of Independent Inventing and Developing

Mention of Sperry's need to employ resources in his innovations which were greater than even a prosperous inventor could command helps to account for the development of another recent phenomenon: firms specializing in inventions which develop them up to the point of application. Despite their corporate nature, functionally these firms perform the tasks which were previously the province of the independent inventor or of the inventor–entrepreneur (Machlup, 1962).

Some examples of this kind of specialized inventive research and development are well known: Arthur D. Little Co., Battelle Memorial Institute, Stanford Research Institute, Denver Research Institute, Midwest Research Institute. Depending on the contracts offered them by government or private corporations, these firms can perform all the functions of invention from perception and definition of the problem through development, up to the point of application. Sometimes, as in the case of Chester Carlson and xerography, the inventor comes to them with basic idea, and their task becomes the critical revision and development of the concept to the commercially applicable phase. In the case of the Research Corporation of America, the task becomes the weeding out and patenting of the ideas of inventors – typically university professors – and sales and licensing of the patent to companies which will exploit the invention commercially.

Still another type of such independent inventing firm is a number of the high-technology research firms which sprang up about Route 128 on the outskirts of Boston in the 1960s. These firms were sometimes spinoffs of larger corporations, originated by highly talented scientific and technical people who felt constrained within the organizational structures of larger corporations [see article 3.7 in this book]. Their specialty was the 'critical problem' which had attracted the attention of men like Sperry at an earlier date. Regardless of their antecedents, organization, and field of concentration – the Route 128 firms specialized in electronics and computer technology – these firms were selling 'know-how'. As we have pointed out, technology is a form of knowledge, and these firms specialized in particular forms of scientific knowledge and technical expertise requisite for today's complex process of innovation.

Such firms constitute today's 'independent inventor'. Indeed the term 'independent' can lead to misconceptions. Unless invention is regarded as more than an idle and engaging pastime, the 'independent inventor' means little more than an 'unsalaried professional'. But once invention is regarded as a function, as an activity to provide a useful solution to an existing problem, then it must be coupled with the entrepreneurial function, as in the case of Watt-Boulton, or Edison and Sperry.

Yet, there comes a time in today's highly complex and scientifically connected technology when the requisite knowledge and entrepreneurial base go far beyond what a single individual or a small group can muster. Innovation then requires the efforts of numerous individuals and some type of organized endeavour. At that point the individual inventor 'becomes' a team, a research institute, or a specialized R & D firm, which is increasingly the case in modern innovations.

References

Anderson, H. H. (ed.) (1959) *Creativity and its Cultivation,* New York: Harper.
Charpie, R. L. (1967) *Technological Innovation: Its Environment and Management,* Washington, DC: Department of Commerce, Report 0–242–376.
Coler, M. A. (ed.) (1963) *Essays on Creativity in the Sciences,* New York: New York University Press.
Galton, F. (1870) *Hereditary Genius,* New York: Macmillan.
Gilfillan, S. C. (1935) *The Sociology of Invention,* Chicago: Follett Publishing.
Hughes, T. P. (1966) *Lives of the Engineers: Selections from Samuel Smiles,* Cambridge: MIT Press.
Hughes, T. P. (1971) *Elmer Sperry: Inventor and Engineer,* Baltimore, Md: Johns Hopkins Press.

Hughes, T. P. (1978) Inventors: the problems they choose, the ideas they have and the inventions they make, in P. Kelly and M. Kranzberg (eds.), *Technological Innovation: a Critical Review of Current Knowledge*, San Francisco: San Francisco Press, pp. 166–82.

Machlup, F. (1962) *The Production and Distribution of Knowledge in the United States*, Princeton: Princeton University Press.

Jewkes, J., Sawers, D. and Stillerman, R. (1969) *The Sources of Invention*, 2nd edn, New York: Macmillan.

Roe, A. (1953) *The Making of a Scientist*, New York: Dodd, Mead.

Rogers, C. R. (1964) Toward a theory of creativity, *ETC Review General Semantics*, Vol. 2, pp. 249–56.

Taylor, C. W. (1956) *First Conference on the Identification of Creative Scientific Talent*, Salt Lake City, Utah, New York: McGraw-Hill.

Usher, A. (1959) *A History of Mechanical Inventions*, Boston: Beacon Press.

2.5

The Design of the Designer

Gordon L. Glegg

Creative design is an essentially personal achievement. There is nothing automatic about it. [. . .] Whether he is conscious of it or not, the mind of the designer has three realms of activity. [. . .] For this purpose we will consider that the creative mind can be subdivided into the inventive, the artistic, and the logical or rational. [. . .]

The inventive

Everyone wants to be an inventor. We dream of creating some wonderful new machine and living in luxury ever afterwards. There is no reason why this dream should not come true but the process is likely to involve much more hard work than was mentioned in the dream. Rarely do inventions fall like a bolt from the blue; they have to be conjured up from the conscious and subconscious mind. You cannot command your mind to invent something, but you can encourage it. The best way to do this is by saturating your mind with all the elements of the problem. Study everything you can; try to find the feel of the job.

When I was a small boy, a neighbour of ours called Mrs Roe said that her husband always wanted to spend his holidays in the same way. He sat on the top of a cliff at the seaside and watched the seagulls. He studied them for days on end. We all thought this to be highly eccentric, but it was really highly sensible. It enabled him to be a pioneer in aircraft design. He was A. V. Roe of the Avro aeroplane.

The secret of inventiveness is to fill the mind and the imagination with the context of the problem and then relax and think of something else for a change. Perhaps you could read a book, play a game, or climb a mountain and thus release mental energy which your subconscious can use to work on the problem. If you are lucky, this subconscious will hand up into your conscious mind, your imagination, a picture of what the solution might be. It will probably come in a flash, almost certainly when you are not expecting it. This is true of all creative thinking whether in engineering or not.

History tells us that in a selection of fifteen creative artists in various fields from music to mathematics, their key inspiration came suddenly and unexpectedly and never when

G. L. Glegg, *The Design of Design,* Cambridge: Cambridge University Press, 1969; extracts from Chapter 2. Gordon Glegg is a consulting engineer and formerly a lecturer in engineering at Cambridge University.

they were working at it. This is what they were doing at the time:

Half asleep in bed	4
Out walking or riding	3
Travelling	3
In church	2
At a state dinner	1
Sitting in front of fire	2

Concentration and then relaxation is the common pattern behind most creative thinking.

It is also important to realize that our subconscious minds will hand up their suggestions in the form of symbols or pictures. The subconscious has no vocabulary. To encourage communication between the conscious and subconscious, we should practise their only common language, which is in three-dimensional pictures. That is why all engineers should learn to do three-dimensional sketches.

The artistic

The next subdivision of the designer's mind is the artistic and it is much the most difficult of the three to define. The sense of the artistry of engineering is invaluable but cannot be formally stated. A machine may look artistic in the normal meaning of that word without being good engineering. A bridge may look nice but fall down. An efficient high-voltage insulator is often ornamental to look at, but if you designed it merely for perfection of form it would not necessarily be a good insulator. The artistry of engineering is essentially a matter of style and this is always a problem to put into words. We find the same difficulty in talking about style in music, literature or art. The definition is often only appreciated after the style has been recognized. Writing about the world of science, Sir Arthur Eddington observed, 'We sometimes have convictions which we cherish but cannot justify; we are influenced by some innate sense of the fitness of things.' Perhaps this 'sense of the fitness of things' is the nearest we can get to a positive definition of engineering artistry too.

Surely a self-aligning roller bearing commends itself as good in principle! And is there not a simplicity of style in the design of a squirrel cage motor? I sometimes wonder if such a humble thing as an umbrella is not a remarkable example of structural engineering. Very few structures can be erected or pulled down so quickly.

We must be careful to distinguish between a good principle of working and good workmanship in applying a principle. Probably most of us will remember the first time we prised off the back of a watch with a pocket knife. We were immediately struck by the fascinating delicacy and precision of the diminutive components, but this of itself said nothing about the essential style of the working principle. Did we go on past this perfection of workmanship and appreciate the economy of style in the principle? The precise metering of energy by a small spring from a bigger one is an engineering joy. The total energy that could only sustain a child's top for less than a minute, despite its low friction bearing, will run a watch for a day and, in total contrast to the top, at an exceedingly uniform angular velocity.

For transporting its own weight over rough ground, a child's hoop is artistically good while a motor bicycle and sidecar, even though designed to a high state of engineering sophistication, is clumsy in principle.

This distinction is important, for it is nearly always true that some new breakthrough in engineering design may initially appear less mechanically sophisticated than the highly developed traditional one it will soon replace. Style will always win in the end.

The rational

The third realm of activity is the rational, which represents disciplined thinking applied over the entire field of design from theoretical analysis to economic realities. The inventive and the artistic, the inspirational and the intuitive must all be impartially scrutinized. The rational must hold the power of veto over them all.

The reason for this is that all machines and structures are inherently rational; they always work exactly to theory. They obey with 100 per cent accuracy the circumstances of their construction and environment. A machine never makes a mistake here, it could pass any exam in its own subject. Theory is practice. Thus, paradoxically, the more expert we become in theoretical analysis, the more we approach the actual working of a machine and so the more practical we become. A machine does not guess. We must reduce our guesswork to a minimum, and so safeguard ourselves against emotional or illogical decisions.

For instance, the artistic style of a roller bearing may tempt us to use it indiscriminately, but in some circumstances it is useless. Logic reminds us that the higher the rotational speed, the greater is the centrifugal self-loading of the rollers themselves. So that the bearing will not overload itself by internal forces it must be made smaller and smaller as its speed increases. At 80,000 rev/min a quarter-inch-diameter bearing uses up nearly all its strength in avoiding flying to bits. To avoid self-destruction at much over 100,000 rev/min the bearing has to be so small that it almost ceases to exist and therefore is not a very useful component. Neither our sense of style nor inventiveness would, of itself, erect the necessary frontiers. The logical is the watchdog of design.

Vital as this is, we must beware of going to the other extreme and regarding all design as a strictly logical exercise. It is no substitute for the inventive or the artistic. Logic may decide between alternatives but cannot be relied upon to initiate them. [. . .]

In the realm of the logical, we must extend its scope beyond that of being merely a mathematical referee to include the economic. It would be nice if engineers didn't have to worry about money; very nice indeed. Unfortunately the real position is exactly the reverse. In general, the purpose of engineering design is to spend money in order to make it – to make a great deal of it if possible – and, most important of all, to raise the standard of living for the community. Often we would like to construct a machine just for the fun of seeing an ingenious principle at work, and sometimes we are tempted to think that the more engineeringly subtle a machine is the more money it is likely to make. The logical part of our thinking is the great enemy of all such wishful thinking and that is sometimes why we are reluctant to use it. I knew of a firm whose managing director devised what he thought to be a marvellous labour-saving machine. His firm employed a large number of men who, working in pairs, assembled large sheets of material in mesh form. The managing director ordered the machines to be built and these did the assembling nearly as quickly as the men. Unfortunately the finished article needed careful handling and no facilities for mechanizing this had been thought out. The two operators therefore had to be retained to lift the product off the assembly table. The design had used capital, slowed output and left labour costs slightly higher than they were before. [. . .]

Something further needs to be said about the design of a designer to avoid giving the impression that all competent engineers must be equally at home in all branches of design. It is probably rather rare to have a design mind equally at home everywhere; and from a career point of view facility in one realm only is quite enough. You can have unbounded opportunities provided you know what you are good at and go where you have an opportunity to exploit it.

Most larger engineering firms have three internal departments dealing with new designs. Although not always called by the same names, these are 'the design or project department', 'the development or prototype department' and 'the production department'.

The design department is responsible for new ideas; development clothes them in mechanisms and then, if the prototype works well enough, production takes over to refine and streamline. It follows that these three categories of industrial organization broadly correspond to the three categories of engineering thinking. So, if you are an inventor by nature, head for the research and design departments; if artistry of style fascinates you, go into development; and if you have a logical or mathematical mind, they will need you in the production of the final machine.

Section 3

Innovative Organizations

Introduction: Organizational Strategies and Practices for Innovation

David Wield

The division of sections 2 and 3 reflects a 'classic' division of the literature between study of how creative *individuals* are born or made (section 2), and what it is that makes *organizations* innovative (section 3). Section 3 focuses on those institutions (that is, industrial firms) where arguably most design and innovation takes place, and that are responsible in the main for making innovations work and introducing them to the market place. Out of the vast range of literature emanating particularly from organization and management theory, but also from technology, design, economics and other disciplines, a selection of material illustrating several different issues has been made. The section is divided into two parts: the first concentrates on strategies at corporate level, and the second on organization and project management within innovative companies.

Corporate Strategies

First, there is the issue of what motivates organizations to innovate. Freeman puts it bluntly in article 3.1 when he writes that for a firm 'not to innovate is to die'. Its death, however, can be slow and painful. In 1982, the chemical company ICI reported on an investigation about the results of stopping the introduction of new products. The answer that came from its model was that profits would decline very slowly for around 15 years, before falling very sharply. ICI also asked itself another question, which has important implications for companies that delay innovation until they reach a profits crisis point. It checked what would happen to its model if, at the 15-year crisis point, it suddenly reacted by magically increasing the rate of product innovation by three times what it was before it ceased. The model predicted that it would take another 25 years for profit to recover to the level achieved before the introduction of new products was stopped. In the meantime, profit would have fallen to about 60 per cent of its original value (Suckling, 1984).

As might be expected, few (if any) students of innovation disagree with the slogan 'innovate or die'. There is little agreement, however, on the second issue: 'What innovation strategy should a firm adopt?'

Freeman makes a six-fold classification of innovation strategies, ranging from 'offensive' strategies designed for technical and market leadership, through 'defensive', 'imitative', 'dependent' and 'opportunistic' strategies, to 'traditional' strategies of producing unchanging products for unchanging markets. Freeman believes that most spending on innovation is within what he calls a 'defensive' strategy – highly research-intensive, not wishing to take the risk of being the first in the world, but wanting to be in the first division at least.

Corporate strategy consultants have recently made something of a fad of Japanese business success. A good deal of time has been spent on looking for the attributes allowing Japanese manufacturers to 'outsmart' and 'overtake' the rest of the world in producing new consumer (and industrial) products. Some point to Japanese persistence after the failure of the early 'imitative' products for Western markets that gained the reputation of being 'cheap and nasty'; others to the emphasis on developing new markets and educating consumers about new products like the 'Walkman' portable stereo and 'Watchman' pocket TV. Most observers have pointed to high investment, linked to long-term global strategies for high-technology products, up to 15 years in some cases. This strategic approach has been contrasted with the slavish adherence to conventional market research, combined with cost cutting, low investment, and short-term approaches in the US and Britain. Article 3.2 by Bill Evans discusses a number of these issues, linking a questioning attitude to some of the myths of Japanese economic development with astute observations on corporate design strategy. He sees the success as no miracle, but based on hard work and thoroughness, linked to strategic planning. Design is central to Japanese corporate policy and careful product planning is done for new products. Evans argues that design acts as the 'glue' linking basic technological ability with production techniques, and as *agent provocateur* taking product planning and marketing departments into new areas. He contrasts this against the typical concern in British firms with existing technologies and current market demands, rather than with future possibilities for opening up new demands with new design concepts and technologies.

By far the most successful published study of the attributes of innovative and commercially successful companies is that of Peters and Waterman (1982). Their book has sold 3 million copies, making it the best-selling business textbook ever. The authors use popular language to describe 'lessons from America's best-run companies'. These attributes are listed in Box 1.

Box 1 Peters and Waterman's attributes for commercially successful innovative companies.

---

1. A bias for action (for example, by developing prototypes and trying them out on customers within weeks).
2. Sticking close to the customer (for example, by listening intently and regularly to customers).
3. Encouraging autonomy and entrepreneurship (for example, encouraging internal innovation by turning a blind eye to unauthorized product development 'on the side').
4. Respect for individuals and open-door management policies.
5. 'Hands on, value driven' management with a common 'culture' of common attitudes and shared beliefs throughout the organization.
6. 'Stick to the knitting' (stay in a familiar business you know how to run).
7. Simple form, lean staff (for example, with a small headquarters office).
8. 'Simultaneous loose-tight controls' (giving autonomy down to shopfloor-level, but insisting on certain 'core values').

Peters and Waterman's results echo some of the conclusions from the studies of Japan.

> Our findings were a pleasant surprise. The project showed, more clearly than could have been hoped for, that the excellent companies were, above all, brilliant on the basics . . . These companies worked to keep things simple in a complex world. They persisted.

They insisted on top quality. They fawned on their customers. They listened to their employees and treated them like adults. They allowed their innovative product and service 'champions' long tethers. They allowed some chaos in return for quick action and regular experimentation. (Peters and Waterman, 1982, p. 13)

The third issue, discussed by both Cottrell and Posner, is that of selectivity. In article 3.3, Cottrell uses the concept of a scientific and technological 'Olympic Games' to criticize those in small countries who wish for national excellence in all fields. He suggests that there is an upward limit of 2–3 per cent of gross national product (GNP) that a country can spend on research and development. Cottrell also looks at the question of project choice at firm level, and postulates that large organizations need to spend around one-half of trading profits on innovation. From these calculations, he concludes that a country could only spend about 10 per cent of GNP on both R & D *and* plant and machinery. He points to the danger of a country like Britain being a 'jack of all trades and a master of none'. In article 3.4, Posner calls for a brutal selection of areas where government support for innovation is really worth while, and gives a few examples of instances where public-sector enterprises in Britain have made 'technologically sweet', but financially disastrous, decisions on what innovations to fund. Unfortunately his rules for what innovations to select are all negative ones – what not to support, rather that what should be supported! Nevertheless, it is not clear that his rules are being used in recent government support in Britain for projects in fifth-generation computing, cable television or biotechnology.

Company organization and project management

The remaining extracts are about organizational requirements for design and innovation within companies. Roberts's article (3.5) summarizes work on corporate management of industrial research and development undertaken at MIT. He believes that different types of staff are required for successful innovation, and that the role of the 'creative scientist or engineer' has been overemphasized in comparison with other roles like those of the 'project manager' and the 'sponsor' who protects the ideas in their early stages. Roberts's work shows the need for team work and planning for successful corporate innovation.

The importance of integration and multidisciplinarity is also emphasized in the design and development of new engineering products. The Corfield Report (NEDO , 1979), mentioned in the General Introduction to this book, points to the need for the marketing, production and finance functions in a company to be linked to all stages of product development (just as Evans described in Japanese companies). Once again, Corfield points to the need for a corporate-level design strategy with careful monitoring of the design process at all phases. He details eleven phases (see Box 2).

Oakley's two extracts (articles 3.6(a) and (b)) take up the difficulties of organizing design activities from concept to production – though what constitutes a successful design team seems even more slippery than most corporate decision-making concepts. Oakley introduces the classic analysis of Burns and Stalker (1961) into his argument on design management. Burns and Stalker's study analysed the attributes of innovative companies. They argued that 'mechanistic' (formal and hierarchical) organizations are satisfactory in stable conditions, but that 'organic' (informal, less hierarchical) teams were more appropriate where flexibility and innovation were required. Oakley takes these concepts and, applying them to design, suggests that product design departments should model themselves on 'organic' lines.

A myth, opened to critical appraisal in article 3.7 by Rothwell, is that small firms are crucial for technological innovation to occur, and should therefore be encouraged to assist

Box 2   Managing Design: the eleven phases of innovation and launch

Phase	Activity	Responsibility
1.	Identify need or want	Marketing with engineering assistance
2.	Specification	Marketing and engineering
3.	Relevance of product	Marketing, production, financial and legal
4.	Conceptual design	Engineering
5.	Preliminary cost estimate	Production
6.	Evaluation	Finance and marketing
7.	Detail design	Engineering
8.	Prototype	Engineering
9.	Manufacture	Production
10.	Product launch	Marketing
11.	Product review	Marketing, engineering and financial

Source: NEDO (1979, p. 32–4)

national industrial regeneration. He finds advantages and disadvantages in both large and small companies. He shows in two areas of technological development – semi-conductors and computer-aided design – that large companies made the initial discoveries, and small companies, founded by former employees of those large companies, did the development work enabling successful marketing as novel products. He suggests that different-sized companies often relate to each other in these and many other ways.

This section concludes with article 3.8 on the management of large technologically innovative projects. It is surprising how seldom lessons are learned from 'failed' projects. It is always easier to look to future ideas, than to monitor the past. Fishlock points out some painful lessons of not controlling the natural inclinations of creative engineers to innovate without consideration of finance. He uses examples of projects like the hover-train and the Advanced Passenger Train, as well as several aerospace and nuclear power schemes, to demonstrate that engineers tend to grossly underestimate development costs.

To summarize, the articles in section 3 cover a range of issues. In the first part (Corporate strategies), the themes include what motivates organizations to innovate; what makes successful innovation strategies; and, the need for selectivity in research and development. The second part (Company organization and project management) is more focused on project management in corporations. Its themes include successful management of innovative projects; the relative success of small and large firms; integration of design and other corporate functions in successful projects; and the dangers of not controlling costs in large new projects.

References

Burns, T. and Stalker, G. M. (1961) *The Management of Innovation,* Tavistock Publications.

NEDO (1979) *Product Design* (the Corfield Report), National Economic Development Office.

Peters, T. and Waterman, R. (1982) *In Search of Excellence: lessons from America's best-run companies.* Harper and Row.

Suckling, C. W. (1984) Long range strategy in product planning in high technology, in R. Langdon (ed.), *Design Policy Vol. 2; Design and Industry*, Design Council, pp. 7–10.

CORPORATE STRATEGIES

3.1

Innovation and the Strategy of the Firm

Christopher Freeman

[. . .] Any firm operates within a spectrum of technological and market possibilities arising from the growth of world science and the world market. These developments are largely independent of the individual firm and would mostly continue even if it ceased to exist. To survive and develop it must take into account these limitations and historical circumstances. To this extent its innovative activity is not free or arbitrary, but historically circumscribed. Its survival and growth depend upon its capacity to adapt to this rapidly changing external environment and to change it. Whereas traditional economic theory largely ignores the complication of world science and technology and looks to the market as *the* environment, changing technology is a critically important aspect of the environment for firms in most industries in most countries.

Within these limits, the firm has a range of options and alternative strategies. It can use its resources and scientific and technical skills in a variety of different combinations. It can give greater or lesser weight to short-term or long-term considerations. It can form alliances of various kinds. It can license innovations made elsewhere. It can attempt market and technological forecasting. It can attempt to develop a variety of new products and processes on its own. It can modify world science and technology to a small extent, but it cannot predict accurately the outcome of its own innovative efforts or those of its competitors, so that the hazards and risks which it faces if it attempts any major change in world technology are very great.

Yet not to innovate is to die. Some firms actually do elect to die. A firm which fails to introduce new products or processes in the chemical, instruments or electronics industries cannot survive, because its competitors will pre-empt the market with product innovations, or manufacture standard products more cheaply with new processes. Consequently, if they wish to survive despite all their uncertainties about innovation, most firms are on an innovative treadmill. They may not wish to be 'offensive' innovators, but they can often scarcely avoid being 'defensive' or 'imitative' innovators. Changes in technology and in the

C. Freeman, *The Economics of Industrial Innovation*, 2nd edn, London: Frances Pinter, 1982; extracts from Chapter 8.
Professor Freeman is the former Director and currently Deputy Director of the Science Policy Research Unit, Sussex University.

market and the advances of their competitors compel them to try and keep pace in one way or another. There are various alternative strategies which they may follow, depending upon their resources, their history, their management attitudes, and their luck (Table 1). [. . .]

Table 1   Strategies of the firm

Strategy	In-house scientific and technical functions within the firm									
	Fundamental research	Applied research	Experimental development	Design engineering	Production engineering -quality control	Technical services	Patents	Scientific and technical information	Education and training	Long-range forecasting and product planning
Offensive	+	++	++	++	+	++	++	+	++	++
Defensive	−	○	++	++	+	○	+	++	+	+
Imitative	−−	−	○	+	++	−	−	++	○	○
Dependent	−−	−−	−	○	++	−−	−−	○	○	−
Traditional	−−	−−	−−	−−	++	−−	−−	−−	−−	−−
Opportunist	−−	−−	−−	−−	−−	−−	−−	++	−−	++

Range ++to−−indicates weak (or non-existent) to very strong. Key ++=(very strong) +=(strong) ○=(average) −=(weak) −−=(very weak/non existent).

[. . .] We consider six alternative strategies, but they should be considered as a spectrum of possibilities, not as clearly definable pure forms. Although some firms recognizably follow one or other of these strategies, they may change from one strategy to another, and they may follow different strategies in different sectors of their business.

'Offensive' strategy

An 'offensive' innovation strategy is one designed to achieve technical and market leadership by being ahead of competitors in the introduction of new products. Since a great deal of world science and technology is accessible to other firms, such a strategy must either be based on a 'special relationship' with part of the world science–technology system, or on strong independent R & D, or on very much quicker exploitation of new possibilities, or on some combination of these advantages. The 'special relationship' may involve recruitment of key individuals, consultancy arrangements, contract research, good information systems, personal links, or a mixture of these. But in any case the technical and scientific information for an innovation will rarely come from a single source or be available in a finished form. Consequently the firm's R & D department has a key role in an offensive strategy. It must itself generate that scientific and technical information which is not available from outside and it must take the proposed innovation to the point at which normal production can be launched. A partial exception to this generalization is the new firm which is formed to exploit an innovation already wholly or largely developed elsewhere, as was the case with many scientific instrument innovations. The new small firm is a special category of 'offensive' innovator. [. . .]

The firm pursuing an 'offensive' strategy will normally be highly 'research-intensive', since it will usually depend to a considerable extent on in-house R & D. In the extreme case it may do nothing but R & D for some years. It will attach considerable importance to patent protection since it is aiming to be first or nearly first in the world, and hoping for substantial monopoly profits to cover the heavy R & D costs which it incurs and the failures which are inevitable. It must be prepared to take a very long-term view and high risks. Examples of such an offensive strategy are RCA's development of television and colour television, Du Pont's development of nylon and Corfam, IG Farben's development of PVC, ICI's development of Terylene, Bell's development of semiconductors, Houdry's development of catalytic cracking, and the UK Atomic Energy Authority's development of various nuclear reactors. It took more than ten years from the commencement of research before most of these innovations showed any profit, and some never did so.

The extent to which an offensive strategy requires the pursuit of in-house fundamental research is a matter partly of debate and partly of definition. From a narrow economic point of view it is fashionable to deride in-house fundamental research, and to regard it as an expensive toy or a white elephant. Certainly it can be this, and the advice of many economists and management consultants to leave fundamental research to universities has a kernel of good sense. But it may be too narrow. Certainly some of the most successful 'offensive' innovations were partly based on in-house fundamental research. Or at least the firms who were doing it described it as such, and it could legitimately be defined as research without a *specific* practical end in view (the definition of applied research). However, it was certainly not completely pure research in the academic sense of knowledge pursued without *any* regard to the possible applications. Perhaps the best description of it is 'oriented fundamental research' or 'background fundamental research'. A strong case can be made for doing this type of research as part of an offensive strategy (or even in some cases as part of a defensive strategy).

The straightforward economic argument against in-house fundamental research holds that no firm can possibly do more than a small fraction of the fundamental research which is relevant, and that in any case the firm can get access to the results of fundamental research performed elsewhere. This over-simplified 'economy' argument breaks down because of its failure to understand the nature of information processing in research, and the peculiar nature of the interface between science and technology. There is no direct correspondence between changes in science and changes in technology. Their interaction is extremely complex and resembles more a process of mutual 'scanning' of old and new knowledge. The argument that 'anyone can read the published results of fundamental scientific research' is only a half-truth. A number of empirical studies which have been made in the United States indicate that access to the results of fundamental research is partly related to the degree of participation (Price and Bass, 1969). [. . .]

We may conclude both from the results of Price and Bass and from our own survey (SPRU, 1972) that the performance of fundamental research, whilst not essential to an offensive innovation strategy, is often a valuable means of access to new and old knowledge generated outside the firm, as well as a source of new ideas within the firm. Whilst ultimately all firms may be able to use new scientific knowledge, the firm with an offensive strategy aims to get there many years sooner. Even if it does not conduct oriented fundamental research itself it will need to be able to communicate with those who do, whether by the performance of applied research, through consultants or through recruitment of young postgraduates or by other means. This has very important implications for manpower policy as well as for communications with the outside scientific and technological community.

But although access to basic scientific knowledge may often be important, the most critical technological functions for the firm pursuing an offensive innovation strategy will be those centred on experimental development work. These will include design-engineering on the one hand, and applied research on the other. A firm wishing to be ahead of the world in the introduction of a new product or process must have a very strong problem-solving capacity in designing, building and testing prototypes and pilot plants. Its heaviest expenditures are likely to be in these areas, and it will probably seek patent protection not only for its original breakthrough inventions but also for a variety of secondary and follow-up inventions. Since many new products are essentially engineering 'systems', a wide range of skills may be needed. Pilkington's were successful with the 'float glass' process and IG Farben with PVC, largely because they had the scientific capacity to resolve the problems which cropped up in pilot plant work and could not be resolved by 'rule of thumb'. The same is even more true of nuclear-reactor development work.

There has been a great deal of confusion and misunderstanding over expenditure on R & D in relation to the total costs of innovation. It became fashionable to talk of R & D costs as a relatively insignificant part of the total costs of innovation – at most 10 per cent. This view is not supported by any empirical research and is based on a misreading of a United States Department of Commerce report frequently quoted and re-quoted. The small amount of empirical research which has been done on this question indicates that R & D costs typically account for about 50 per cent of the total costs of launching a new product in the electronics and chemical industries. As in so many aspects of industrial innovation it is Mansfield and his colleagues (1971 and 1977) who got down to the hard task of systematic empirical observation and measurement, rather than plucking generalizations from the air. Their results were confirmed on a larger scale by the Canadian surveys of industrial R & D and more recently by German work. [. . .]

'Defensive' innovation strategy

Only a small minority of firms in any country are willing to follow an 'offensive' innovation strategy, and even these are seldom able to do so consistently over a long period. Their very success with original innovations may lead them into a position where they are essentially resting on their laurels and consolidating an established position. They will in any case often have products at various stages of the product cycle – some completely new, others just established and still others nearing obsolescence. The vast majority of firms, including some of those who have once been 'offensive' innovators, will follow a different strategy: 'defensive', 'imitative', 'dependent', 'traditional', or 'opportunist'. It must be emphasized again that these categories are not pure forms but shade into one another. [. . .]

A 'defensive' strategy does not imply absence of R & D. On the contrary a 'defensive' policy may be just as research-intensive as an 'offensive' policy. The difference lies in the nature and timing of innovations. The 'defensive' innovators do not wish to be the first in the world, but neither do they wish to be left behind by the tide of technical change. They may not wish to incur the heavy risks of being the first to innovate and may imagine that they can profit from the mistakes of early innovators and from their opening up of the market. Alternatively, the 'defensive' innovator may lack the capacity for the more original types of innovation, and in particular the links with fundamental research. Or they may have particular strength and skills in production engineering and in marketing. Most probably the reasons for a 'defensive' strategy will be a mixture of these and similar factors. A 'defensive' strategy may sometimes be involuntary in the sense that a would-be 'offensive' innovator may be out-paced by a more successful offensive competitor.

Several surveys (Nelson, Peck and Kalachek, 1967; Schott, 1976) have shown that in all the leading countries, most industrial R & D is 'defensive' or 'imitative' in character and concerned mainly with minor 'improvements', modifications of existing products and processes, technical services and other work with short time horizons. Defensive R & D is probably typical of most oligopolistic markets and is closely linked to product differentiation. For the oligopolist, defensive R & D is a form of insurance enabling the firm to react and adapt to the technical changes introduced by competitors. Since 'defensive' innovators do not wish to be left too far behind, they must be capable of moving rapidly once they decide that the time is ripe. If they wish to obtain or retain a significant share of the market they must design models at least as good as the early innovators and preferably incorporating some technical advances which differentiate their products, but at a lower cost. Consequently, experimental development and design are just as important for the 'defensive' innovator as for the 'offensive' innovator. Computer firms which continued to market valve designs long after the introduction of semiconductors could not survive. Chemical contractors which attempted to market a process which was technically obsolescent could not survive either. The 'defensive' innovator must be capable at least of catching up with the game, if not of 'leap-frogging'.

In an interesting study of the computer market, Hoffman (1976) maintains that IBM has mainly followed a 'defensive' innovation strategy, although with some 'offensive' elements, while Sperry Rand (Univac) has pursued a more consistently 'offensive' strategy and Honeywell an 'imitative' strategy. Since IBM spends far more on R & D than Sperry Rand in absolute terms, this illustrates the point that the 'defensive' innovator may well commit greater scientific and technical resources than the 'offensive' innovator. [. . .]

Both the 'offensive' and the 'defensive' innovator will be deeply concerned with long-range planning, whether or not they formalize this function within the firm. In many cases this may still often be the 'vision' of the entrepreneur and his immediate associates, but increasingly this function, too, is becoming professionalized and specialized, so that 'Product Planning' is a typical department for both 'offensive' and 'defensive' innovators. However, the more speculative type of 'technological forecasting' is more characteristic of the 'offensive' innovator, and still has considerable affinities to astrology or fortune-telling. It should probably still be regarded as a kind of sophisticated war dance to mobilize a faction in support of a particular project or strategy, but increasingly important serious techniques are being developed.

The 'defensive' innovator, then, like the 'offensive' innovator, will be a knowledge-intensive firm, employing a high proportion of scientific and technical manpower. Scientific and technical information services will be particularly important, and so will speed in decision-making, since survival and growth will depend to a considerable extent on timing. The defensive innovators can wait until they see how the market is going to develop and what mistakes the pioneers make, but they dare not wait too long or they may miss the boat altogether, or slip into a position of complete dependence in which they have lost their freedom of manoeuvre. R & D will be geared to speed and efficiency in development and design work, once management decides to take the plunge. Such firms will sometimes describe their R & D as 'advanced development' rather than 'research'. [. . .]

'Imitative' and 'dependent' strategies

The 'defensive' innovators do not normally aim to produce a 'carbon' copy imitation of the products introduced by early innovators. On the contrary, they hope to take advantage of early mistakes to improve upon the design, and they must have the technical strength to do so. [. . .]

The 'imitative' firm does not aspire to 'leap-frogging' or even to 'keeping up with the game'. It is content to follow way behind the leaders in established technologies, often a long way behind. The extent of the lag will vary, depending upon the particular circumstances of the industry, the country and the firm. If the lag is long then it may be unnecessary to take a licence, but it still may be useful to buy know-how. If the lag is short, formal and deliberate licensing and know-how acquisition will often be necessary. The imitative firm may take out a few secondary patents but these will be a by-product of its activity rather than a central part of its strategy. Similarly, the imitative firm may devote some resources to technical services and training but these will be far less important than for the innovating firms, as the imitators will rely on the pioneering work of others or on the socialization of these activities, through the national education system. [. . .]

The 'imitator' must enjoy certain advantages to enter the market in competition with the established innovating firms. These may vary from a 'captive' market to decisive cost advantages. The 'captive' market may be within the firm itself or its satellites. For example, a large user of synthetic rubber, such as a tyre company, may decide to go into production on its own account. Or it may be in a geographical area where the firm enjoys special advantages, varying from a politically privileged position to tariff protection. (This will be the typical situation in many developing countries.) Alternatively or additionally, the imitator may enjoy advantages in lower labour costs, plant investment costs, energy supplies or material costs. The former are more important in electrical equipment, the latter in the chemical industry. Lower material costs may be the result of a natural advantage or of other activities (e.g. oil refineries in the plastics industry). Finally, imitators may enjoy advantages in managerial efficiency and in much lower overhead costs, arising from the fact that they do not need to spend heavily on R & D, patents, training, and technical services, which loom so large for the innovating firm. The extent to which imitators are able to erode the position of the early innovators through these advantages will depend upon the continuing pace of technological change. The early innovators will try to maintain a sufficient flow of improvements and new 'generations' of equipment, so as to lose the 'imitators'. But if the technology settles down, and the industry becomes 'mature', they are vulnerable and may have to innovate elsewhere. [. . .]

Unless the 'imitators' enjoy significant market protection or privilege they must rely on lower unit costs of production to make headway. This will usually mean that in addition to lower overheads, they will also strive to be more efficient in the basic production process. They may attempt this by process improvements, but both static and dynamic economies of scale will usually be operating to their competitive disadvantage, so that good 'adaptive' R & D must be closely linked to manufacturing. Consequently, production engineering and design are two technical functions in which the imitators must be strong. [. . .]

A 'dependent' strategy involves the acceptance of an essentially satellite or subordinate role in relation to other stronger firms. The 'dependent' firm does not attempt to initiate or even imitate technical changes in its product, except as a result of specific requests from its customers or its parent. It will usually rely on its customers to supply the technical specifications for the new product, and technical advice in introducing it. Most large firms in industrialized countries have a number of such satellite firms around them supplying components, or doing contract fabrication and machining, or supplying a variety of services. The 'dependent' firm is often a sub-contractor or even a sub-sub-contractor. Typically, it has lost all initiative in product design and has no R & D facilities. The 'small' firms in capital-intensive industries are often in this category and hence account for hardly any innovations. [. . .]

'Traditional' and 'opportunist' strategies

The 'dependent' firm differs from the 'traditional' in the nature of its product. The product supplied by the 'traditional' firm changes little, if at all. The product supplied by the 'dependent' firm may change quite a lot, but in response to an initiative and a specification from outside. The 'traditional' firm sees no reason to change its product because the market does not demand a change, and the competition does not compel it to do so. Both lack the scientific and technical capacity to initiate product changes of a far-reaching character, but the 'traditional' firm may be able to cope with design changes which are essentially fashion rather than technique. Somtimes indeed, this is its greatest strength. [. . .] Their technology is often based on craft skills and their scientific inputs are minimal or non-existent. Demand for the products of such firms may often be very strong, to some extent just *because* of their traditional craft skills (handicrafts, restaurants and decorators). Such firms may have good survival power even in highly industrialized capitalist economies. But in many branches of industry they have proved vulnerable to exogenous technical change. Incapable of initiating technical innovation in their product line, or of defensive response to the technical changes introduced by others, they have been gradually driven out. These are the 'peasants' of industry. [. . .]

The efforts of firms to survive, to make profits and to grow have led them to adopt one or more of the strategies which have been discussed. But the variety of possible responses to changing circumstances is very great, and to allow for this element of variety one other category should be included, which may be described as an 'opportunist' or 'niche' strategy. There is always the possibility that entrepreneurs will identify some new opportunity in the rapidly changing market, which may not require any in-house R & D, or complex design, but will enable them to prosper by finding an important 'niche', and providing a product or service which consumers need, but nobody else has thought to provide. Imaginative entrepreneurship is still such a scarce resource that it will constantly find new opportunities, which may bear little relation to R & D, even in 'research-intensive' industries.

References

Hoffmann, W. D. (1976) Market structure and strategies of R & D behaviour in the data-processing market, *Research Policy*, vol. 5, pp. 334–53.

Mansfield, E. *et al.* (1971) *Research and Innovation in the Modern Corporation*, Norton.

Mansfield, E. *et al.* (1977) *The Production and Application of New Industrial Technology*, Norton.

Nelson, R. R., Peck, J. and Kalachek, E. (1967) *Technology, Economic Growth and Public Policy*, Brookings Institution.

Price, W. J. and Bass, L. W. (1969) Scientific research and the innovative process, *Science*, vol. 164, no. 3881, pp. 802–6.

Schott, K. (1976) Investment in private R & D in Britain, *Journal of Industrial Economics*, vol. 25, no. 2, pp. 81–99.

SPRU (1972) *Success and Failure in Industrial Innovation*, Centre for the Study of Industrial Innovation, London.

3.2

Japanese Management, Product Design and Corporate Strategy

Bill Evans

Most of us know of Japanese success through her products. There seem to be few consumer product areas which are not dominated by goods of Japanese origin (even if nowadays assembled elsewhere). So Japanese design is something we all know well, not particularly outstanding aesthetically, but always competitively priced and usually technologically advanced. What is it that enables Japan to manage her designs to rapidly meet changes in the market so cheaply, effectively and with such technological sophistication?

Even the most cursory viewer of current affairs will be aware of some of the contributory factors to Japan's success. The 'Art of Japanese management' is often mentioned, and has become something of a management fad in the UK. Quality circles and the principle of lifetime employment are often pointed to as good examples. In fact only 35–40 per cent of Japanese employees enjoy the benefits of lifetime employment, the rest of the economy taking up the slack. We can see in their products alone the level of investment, and hence confidence, they have in industry – and the massive investment in the technological infrastructure to support these products.

But we hear of the less acceptable side of Japanese success. Western commentators talk of workaholics and cramped social conditions. Japan is a high-pressure society and much of her social welfare towards the elderly and those not lucky enough to be in the lifetime employment system leave a lot to be desired by western standards. To reduce discussion to a list of transferable factors and resultant social problems, however, is to misunderstand the complex social, cultural and industrial interrelationship. We should not be so interested in copying Japan, but instead, by understanding her success and its consequences we should be searching for something more appropriate to our culture and expertise.

W. Evans, Japanese-style management, product design and corporate strategy, *Design Studies*, Vol. 6, No. 1, January 1985, extracts from pp. 25–33.
Bill Evans is Design Manager at Westra Environmental Ltd.

Economic Background

The economic build-up in Japan has happened primarily over the last 100 years, but much of its present high-technology success is based on plant less than seven years old. This investment by all the major companies, known as the Zaibatsu sector, is financed, first by the banks that each group has under its wing and, secondly, by the very high level of personal savings. [. . .]

These Zaibatsu companies, which rank among the largest in the world, overtaking the General Motors, Ford and IBMs of the world, have systematically worked their way through market sectors, establishing products first in their large domestic market (Japan has a population of 118 million), and then ruthlessly pursuing a proportion of the world market. The companies are often initially more concerned with market share than short-term profitability. But the oil shocks of the 1970s forcefully brought home to the Japanese their delicate trading position, with a heavy reliance on imported raw materials and energy. So Japan responded by gradually shifting towards the knowledge-intensive industries like electronics, telecommunications and robotics. This also coincided with a period during which her own R & D efforts were bearing fruit and she was less dependent on imported technology. Her investment in education, particularly technical education, also helped fuel the fires. For instance, in Japan there are four times as many engineers per capita as in the UK.

The Zaibatsu companies have always enjoyed a very close relationship with their suppliers, who are often smaller and not in the lifetime employment schemes. It is the smaller companies that bear the brunt of the fluctuations caused by the ups and downs of the world economy. This causes a distorted view of the viability of lifetime employment, which can often end up as permanent only for the industrial elite.

The industrial relations picture is very different, with workers encouraged by their Enterprise Union (single-company unions, rather than single-craft unions) to identify strongly with the aims of the company – 'What is good for the company is good for the workers' has an Orwellian ring, but the workers do reap the financial rewards of success. [. . .]

The bonus system is interesting because it is not a personal bonus for each worker, depending on output, but a bonus based on corporate success. So each worker's suggestion or increase in personal work speed is a contribution to everyone and requires a longer-term view by the individual. [. . .]

The workers' willingness to subjugate themselves to the goals of the company is the first thing that is visible on any factory tour: the uniforms, innumerable notices encouraging less waste, zero defects, extra safety and general fillips has an almost 'McDonald's' quality. The pace of work leaves the western visitor breathless; workers literally run around their work positions, methodically, but with almost dangerous haste assembling components onto subframes. One can stand and watch an operator load a machine for five wrist-wrenching minutes; he stands back to let a batch be processed; perhaps, one might think, a thirty-second pause; but no, he rushes off to a nearby machine for another series of tireless operations. The other major factor that the western visitor notices is that much of this fantastically high productivity is not happening on very sophisticated plant. There are obvious exceptions, but for instance, the car industry in the UK has some of the most modern plant (Metro and Sierra production), but the robot-dictated pace of work there seems almost pedestrian compared to the pace demanded and given by each Japanese worker's part in the factory machine.

Japanese Corporate Design Strategy

Design is central to the corporate policy of the Zaibatsu, more specifically design as part of product planning. But design is still, as in the UK, a service to management, who see their products essentially being pulled by the market. There are well documented exceptions to this, the Sony 'Walkman' (portable hi-fi cassette player) and 'Watchman' (pocket television) perhaps being the most famous, where a strong design concept forged its own market [see article 4.2 by Lorenz]. On the whole, a very careful and systematic product planning exercise is done on every new product, to the point where it becomes easy to see how many of the products look so similar. 'Market pull' also has a slightly different definition in Japan, where the strength of many companies' marketing position leaves a convenient blur between whether they 'pull' the market around themselves or whether it pulls them. There is much talk of 'needs and seeds', but often customers have a need satisfied that they did not even know existed. There is a subtle distinction here between giving people what they want and what they actually get. [. . .]

In the past, Japan was renowned for exploiting foreign technology to the full, but now she has concentrated on controlling and developing all the fundamental building blocks of her industries. Every Zaibatsu in the electronics sector has the most modern VLSI (very large-scale integrated) semi-conductor plant, helped considerably by a very pragmatic government approach to public spending on R & D. The government sees nothing contradictory about a vigorous private sector being made more vigorous by strategic public investment in new technologies. [. . .]

To some extent, a company's design strategy will depend on the product-cycle time both for development and life. Where cycle times are short, for instance consumer electronics, there is more emphasis on beating the competition to the market place and then staying one jump ahead. It is acknowledged by design managers that unscrupulous manufacturers can copy and bring to the market quite complex electronic products within six months. So innovation is the key. Even in apparently quite static areas such as radio, technological innovations are altering the product. If product cycles are longer, as with office automation (photocopiers take up to four years to develop), heavy patent protection and investment in process technology take over as the strategy.

The 'Human-ware' Age

Without fail, the major companies are developing a sophisticated analysis of the future which is becoming increasingly user-conscious, and they are looking to a period where advances in electronics are consolidating rather than rapidly advancing. As Ricoh [manufacturer of office and photographic equipment] puts it: we have moved from the Hardware Age (1965–75) through the Software Age (1975–85) towards the 'Human-ware' Age (see Figure 1). By this it is meant that the user's requirements will take over as the major dictator of a product's capability: 'We have the technology'. So the companies are now concentrating on the input of consumer life-style into the product, making the technology more intelligent, more flexible for users of different cultural backgrounds, and generally considering the social context of their products. This is a little more than rhetoric from a new Japan, sensitive to its public image. Many companies employ social scientists to work alongside their product designers in the product planning centres. They are not necessarily taken on to practise their specialization, but to let their education make a contribution to the complex and iterative product design process. [. . .]

The trends in production technology also mate with the Human-ware approach. FMSs (flexible manufacturing systems) are now common, for example, leading Ricoh to talk of producing such complex items as photocopiers in batches as small as 500 to meet the needs of their varied global markets. At present the changes between the models are confined to slightly different membrane keyboards with regionally oriented legends. But one is left with the impression that as they consolidate this approach, we will see greater regional variations. Sony already have a different TV for each country.

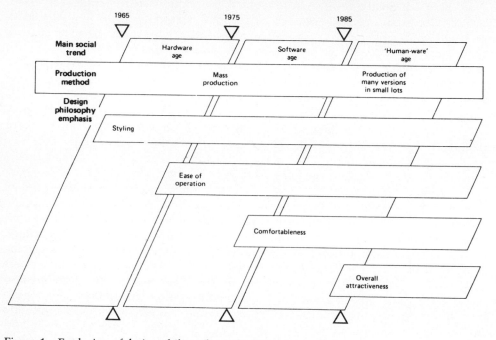

Figure 1   Evolution of design philosophy at Ricoh.

Group Technology

The availability of cheaper robotics for FMS has enabled the rapid development of the other major influence on design: group technology. Japanese electronic products have now become very systems and network-oriented, with interconnection and interdependence a vital feature. Group technology as an approach to design and production is an extension of this, and of course is something also practised in the UK on a smaller scale. It is based around the idea that despite the many variations of products made by a company, many similarities exist. By concentrating on the similarities, they can reduce the disadvantage of the variety. So products are divided into a limited number of groups where they share qualities and are considered technologically similar. So at Sharp, for example, the divisions are TV and video; audio; domestic appliances; solar systems; industrial instruments (calculators, computers and cash registers); and finally electronic components. As much of the plant that produces any of these groups as possible then becomes common or very similar. [. . .]

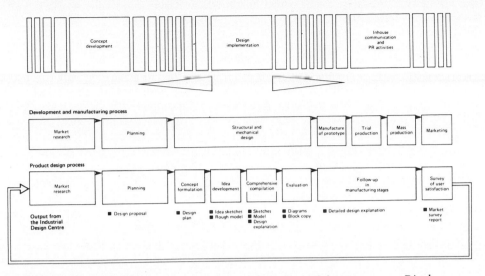

Figure 2   The design process and the work flow of the industrial design center at Ricoh.

Design Departments and the Product Planning Process

The product planning department is really at the centre of the major companies' structure. Design and engineering may or may not be in this section. Product planning (PP) might be a department of its own or a function of the corporate management. [. . .]

Basically, as the Ricoh design process diagram shows (Figure 2), Japanese companies adopt a very systematic linear flow of work from the initial concept, market research and planning to various stages of sketches, presentation models and pre-production prototypes, working closely with the engineering divisions to productionize the designs. But one cannot overstate the importance of the design-marketing link. With near-constant formal meetings to assess progress, many iterative loops and final exhaustive testing of the pre-production units, not much is left to chance. [. . .]

At all times the emphasis is on team work and, not unlike the UK, Japanese companies have their interdepartmental communication problems. But initiatives such as the design centre at Sharp where all the disciplines work together, and the near-maverick PP centre at Sony are obviously working at breaking down the barriers (along the lines of the UK interdepartmental teams like those found at Avery and Baker Perkins).

This area of corporate design methodology is most interesting in comparison with the UK. On paper, the process is similar in both countries and has been for many years. Yet the respective success is very different, as the UK's decimated industries prove. We have to look beneath the surface of the Japanese design process to gain a better understanding. Clearly, one of the major differences is the amount of real resources that are put into product development by companies directly and by governments indirectly. But despite this, one feels that Japanese companies have developed a methodology, perhaps linked to their thorough approach and acute awareness of national survival in a highly competitive globe. This enables them to better splice together marketing, design, engineering and production. The emphasis internally on co-operation rather than competition, and the confidence bolstered by designers' increasing status within the seniority system help. Japanese managers seem more at home with the contribution which design can make to industry. They view industrial design as a tool, rather than the icing on the cake. [. . .]

Conclusions

Japan has moved away from straightforward process innovation, which for so long had enabled her to keep making products more cheaply and efficiently. Design now plays a more central role as the glue that links research and the basic technological ability with engineering through to production techniques. Design is seen as a powerful visualizing and enabling tool by marketing to boost confidence in product ideas which will require massive corporate investment to realize an economic price. Design sometimes takes on the role of *agent provocateur* to stir a reluctant product planning or marketing department into a new unfathomed area. This contrasts against the UK's more analytical marketing techniques, which seem more concerned with what is here now (or imminent) than what 'could be'.

Japanese companies, in producing their designs so reliably, in their detailed attention to sophisticated labour management and investment policy, show the importance of taking the longer-term view. But one has to look beyond structural differences in the Japanese economy to explain her economic dynamism. It is not possible to separate this design management success from the culture in which it was born. For instance some UK companies have grasped at quality circles as a management technique. But often this is viewed as something for 'the production workers', rather than as a company-wide, well-considered and carefully trained-for event.

Japanese management does have a different attitude and more respect for its workforce, but it is still a society with clear social divisions. To the British, Japanese company paternalism looks pretty dominating. The lifetime employment and seniority system is under threat by a combination of new technology, an ageing working population and changing attitudes amongst the young about work. Our social and economic histories put us at very different stages in our development.

Japanese government–industrial links are also fundamental to her success. A 1984 OECD report *Science and Technology Indicators* looked at the investment in strategic industrial research. On the face of it, Britain ranks fourth in the world with government R & D as a percentage of GDP, but if this is considered as civilian R & D, we drop to seventh place. In fact Britain ranks second after the US in per-capita defence R & D; Japan, Germany and even France spend much less. Some 59 per cent of UK government R & D is spent on defence as opposed to 15 per cent in Japan. The much-vaunted spin-offs from defence R & D are just not happening in sufficient quantity to the proportion of spending. UK government support for design must be seen against this background. Indeed the Prime Minister, Margaret Thatcher, said in her acceptance message for an International Award for the Encouragement of Design given by Japan's Design Foundation in 1983, 'Design provides the means to achieve the efficient and competitive manufacturing industry on which our two countries depend'. But without adequate strategically placed resources that other countries' governments help to provide, Britain will not have such a vigorous technological base from which to innovate or as internationally competitive an industrial infrastructure to build on.

References

George, M. and Levie, H. (1984) *Japanese Competition and the British Workplace,* London: Centre for Alternative Industrial and Technological Systems, North London Polytechnic.

Kamata, S. (1973) *Japan in the Passing Lane*, Hemel Hempstead: Allen & Unwin.

Lorenz, C. (1983) Why Sony has given a new impetus to design, *Financial Times*, 9 September.

Magaziner, I. C. and Hout, T. M. (1980) *Japanese Industrial Policy*, Report no. 585, London: Policy Studies Institution.

NEDO (1982) *Transferable Factors in Japan's Economic Success*, London: National Economic Development Office.

Powell, D. (1982) Japan life-style: the human road ahead, *Design*, 407, November, pp. 38–40.

Wilkinson, E. (1983) *Japan versus Europe*, Harmondsworth: Penguin.

3.3

Technological Thresholds

Alan H. Cottrell

The Technological World Games

The goal of being second to none in most significant fields of science and technology is, of course, absolutely right for the United States. It could not possibly be otherwise for a country as large and vigorous as the United States, which contributes about one third of the entire gross product of the world and which spends on research and development several times more than any other country and more than twice as much as all the countries of Western Europe combined.

Would it make sense for any smaller country, however technically advanced, to set for itself this same goal? How advisable would it be for each of the major European countries, for example, to aim to be second to none in all science and technology? The goal may be heroic and inspiring, but is it realistic? For it to be scientifically and technologically realistic, each country would have to raise its research and development expenditure to about 25 per cent of its gross national product, so as to match the research and development effort of the United States. Where would it get the men and the money? What could it do with the results, when it was already spending on research and development alone more than twice as much as most countries now spend on all new industrial plants and machinery? In fact to keep things in balance – technologically, economically, and industrially – no country can afford to spend much more than about 2 or 3 per cent of its gross national product on research and development. But at this level of research and development expenditure the above goal then becomes quite unrealistic for a small country.

This, then, is the problem of the so-called 'technological gap'. For any single country with a gross national product of no more than, let us say, 5 per cent of the gross world product, it is either economically and industrially unrealistic, or scientifically and technically unrealistic, to aim at the goal of being second to none in all science and technology. What should the country do instead? Should it aim at excellence in a limited number of selected fields? Should it link up with other countries to form a larger technological and economic community? These questions require deliberate major national decisions. The need for such policies has first to be clearly recognized and accepted, and then they have to be

F. K. McCune (ed.), *The Process of Technological Innovation,* Symposium sponsored by the National Academy of Engineering, 24 April 1968, Washington, DC; National Academy of Sciences; extracts from pp. 48–59. Sir Alan Cottrell, former Chief Scientific Adviser to HM Government, is Master of Jesus College, Cambridge.

worked for energetically. The important point is that not making and pursuing such policies is, in effect, choosing the unrealistic goal. Every country has its enthusiasts for every branch of science and technology who, quite naturally, wish to attract as much support as they can for their particular projects. Every new scientific or technological advance in one country also raises fears in other countries that they may be missing some great opportunity for the future unless they add it to their own national lists of scientific and technological activities. And so we see that when there are no guiding policies, a host of repetitive projects can spring up. In the shadow of the great American and Russian space ventures, small-scale space projects have sprouted up everywhere. Several countries are individually striving to develop their own fast nuclear reactors. Much the same competitive fervor can be noted for computers, for desalination plants, and for many other products.

We have then, in effect, a scientific and technological 'Olympic Games' in which every technically advanced country can hardly resist the temptation to enter a candidate for every event. I am not arguing here against national competition, except perhaps in pure science where the true spirit of scientific inquiry may now be in some danger of being overridden by ambitions for prestigious success. In technology, particularly that of civil industry, competition provides a tremendous spur to progress. What I am arguing against is the feeble competition that is due to the frittering away of limited resources on too many independent projects that fail to reach the threshold of technological and commercial competitiveness.

Economics of Technological Innovation

The spearpoint of technological advance is *innovation* – the technical, industrial, and commercial steps that bring new products onto the market or new manufacturing processes and equipment into commercial use. Any manufacturing firm or organization in modern industry must bring out new or improved products and update its manufacturing processes regularly, if it is to withstand technological competition from rivals in its own field or from invaders in neighbouring fields. This competition is exerted through both the timing and the cost of innovations, and it is the close interplay of the time and cost factors that determines the threshold of competitiveness.

Let us look first at the question of timing. In a rapidly advancing technical field the prizes go largely to those who can get to the market first with a new product that is fully developed and tested and that can be produced on a large scale. The switch of market demand to a new product can be astonishingly rapid, as is shown by the history of civil aircraft, which has gone through piston engines, turboprops, pure jets, and now turbofan jumbo-jets. Traditionally it has taken some 20 to 50 years for a scientific advance to work its way through to full commercial exploitation, but under the pressure of technological competition this time has recently contracted considerably and is now sometimes no more than five to ten years. This time change has been noticeable, for example, in the introduction of new electronic devices. This is cutting the ground away from the traditional practice of starting very early on new projects that are kept a close secret until ready for launching onto an astonished market. The fast nuclear power plant, large desalination plant, or videotelephone system is not going to come like a bolt from the blue. Precisely when, where, and from whom they will come may still be open to question, but not the expectation of such innovations.

Systematic invention and innovation have become so elaborately organized and technological alertness so highly developed, in fact, that technical people everywhere are generally fully aware of what things in technology are ripe for development at any moment. What matters in this situation is the lead time necessary to exploit a product commercially, that is, the time required to take a new idea or invention through all the steps of project development, design, engineering, tooling-up, first production, testing, and marketing.

Given a field entirely to itself with no competition anywhere in sight, a firm could choose its own lead time to suit its resources and to phase the timing of its investments in accordance with its general financial policy. But when the field is full of competitors who are simultaneously setting out on rival projects, the lead time becomes a main instrument of technological competition. It is then essential for the firm in question either to make itself a leader (to exert the effort necessary to get the shortest possible lead time), or to be deliberately a follower in that particular project (to engage in it only by acquiring licenses from the leaders) and to channel its own major innovative resources instead into some other project. Either a large or a small innovative effort on a given project can be a sound policy. What is always unsound is a middling effort, which is too much effort for a follower and not enough for the leader in that particular technological enterprise.

Competitive values of lead times are usually well known. For small electronic goods such as marine radar sets and machine tool control equipment, for example, they are about three years. Larger projects, such as big computers and communications satellites take about four to five years.[1] The lead time of large aircraft projects goes up to about seven years.

The scale of research and development expenditure needed to develop a given innovation is also well known. [. . .] The total research and development cost and the competitive lead time for a given project then determine the minimum annual rate of research and development expenditure necessary to reach the threshold of technological competitiveness.[1] A computer project, for example, at $50 million total research and development cost and about a three-year lead time would demand an annual expenditure of about $15 million on research and development.

A high level of research and development alone is not sufficient to ensure a successful innovation, of course. There are many other requirements. For example, there must be a sufficiently large potential market to absorb the product. The sales price and marketing techniques for the product must be such as will gain a good share of the market. The manufacturing capacity must be big enough to meet the market demand.

We can continue our calculations, taking the computer project as an example, by either of two equivalent routes. In the first we recognize that, if it is to be economically competitive, a given product cannot carry more than a certain research and development cost in its sales price.[2] For advanced science-based products such as computers, this might amount to about 10 per cent of the sales price. In other words, at least $500 million worth of computers has to be sold from this one project. To produce this output, a manufacturing plant involving a capital outlay of some $200 million would be needed.

Thus, taking a given project and insisting upon meeting the threshold for technological competitiveness, and then adding in the market and production requirements to meet the further condition of cost competitiveness, we arrive at a minimum viable size of firm and market for the project. We see that a capital outlay of some $200 million and an annual output of some $200 million are necessary to balance the essential threshold research and development expenditure of some $15 million a year.

The second route considers the fact that research and development is generally only a small part of the total cost of a successful innovation. There is also the design and engineering work to be paid for, the tooling costs for first production, and the initial manufacturing and marketing expenses. A survey of American industrial innovations brought this point out strongly.[3] On the average, only about 5 to 10 per cent of the total launching cost of a successful new product goes into the research and development leading up to the basic invention, and only about 10 or 20 per cent goes into the engineering development and design of the product. All the rest is used in preparing for production and marketing. These general figures can be verified by many particular examples. For instance, the British textile material Terylene was invented in a research laboratory running at less than the equivalent of $60,000 a year. But when Imperial Chemical Industries obtained the United Kingdom commercial rights in the invention, it then spent the equivalent of some $11 million on a new plant for the first major commercial production.[4] [. . .]

The validity of the above ratio of research and development to total innovation costs depends on where the dividing line between development and first production is drawn. In aircraft projects, for example, it is usual to charge the first few production models and their expensive flight trials to development, so that research and development amounts to a large part of the total innovation costs. Furthermore, not all research and development leads to successful innovations. Some efforts fall by the wayside. Other research and development effort is used, not in the original innovation but in solving technical problems with existing products. Taking all these factors into account, a total innovative expenditure some five times larger than the total research and development expenditure seems a fair average, which brings us back again fairly close to the figures we estimated above for the computer firm. At the national level also, this is consistent with typical annual investments of about 10 per cent of the gross national product in new plants and machinery and about 2 per cent in research and development.[5]

This helps us understand why it is unrealistic for a small country to set itself the goal of being second to none in all science and technology. [. . .] People often speak of rapidly increasing research and development efforts and sometimes believe that these could, without much constraint, rise far above their present levels of 2 or 3 per cent of the gross national product. But as long as the main purpose of the major research and development expenditures is technological innovation for economic benefit, and as long as these research and development expenditure are geared to correspondingly larger innovative expenditures, the fact that these total innovative expenditures are a large fraction of the gross national product must always severely constrain the amount that can be profitably spent on research and development.

We have so far been examining the problem by starting from the research and development costs of particular projects and working out their larger consequences. But this still leaves open the question of choice of projects. To approach this we can put the problem the other way round and ask, 'What can be afforded?' Clearly, if a firm spends too little on new developments, it will become outdated and will lose its markets through the technological obsolescence of its products and processes. If it spends too much, it will not have the markets, productive capacity, or financial resources to reap the benefits of its investments in new technology. The sensible policies lie somewhere between these extremes.

Discounted cash flow

One way into this problem is through the concept of discounted cash flow. This concept has been applied by Hart[6] and Duckworth[7] to estimate research and development expenditure. I shall follow their method but apply it to total innovative expenditure.

Consider a manufacturing firm that is making a steady gross profit of, let us say, $10 million a year. Suppose, for a simple calculation, that its aim is merely to maintain its profit constant at this same level in the future. It thus expects to gain future profits of $10 million a year. These expected future profits have a present value to the firm. For example, if interest rates are 5 per cent a year, the present value of $10 million gained in one year's time is $9.5 million, since an investment of $9.5 million made today could increase to $10 million in one year. More generally, the present value of an amount S to be gained in n years' time, at an annual interest rate i is $S/(1+i)^n$. Summing this over all the values of n, from now to infinity, and taking an interest rate of 5 per cent, we find that the total present value of all the future $10 million annual profits is $205 million.

Consider next what would happen if the firm stopped innovating. Its product and processes would gradually become obsolete and its profits would eventually fall to zero. Estimating the future fall in profits due to obsolescence is not easy. To guide us in this, we might examine, for example, how many of the products that the firm was making 5, 10, 20 and 50 years ago could still be sold today if they had remained unchanged. Fortunately for the estimates, the discounting effect of large n values in the above formula means that the long-range forecasts of future profits need not be accurate. The returns over the next 10 years make the main contribution to the present value.

The method then is to predict the future fall in profits due to obsolescence and then to sum up the present values of each year's expected profits. Hart takes an example in which total obsolescence is reached over a period of 50 years. His figures, applied to our firm, give a total present value of $123 million. We then have two sets of figures, representing two different policies for the firm:

1. Innovation, to maintain profits permanently at $10 million a year. Present value of all future profits is $205 million.
2. No innovation. Profits fall to zero over 50 years. Present value of all future profits is $123 million.

The difference between these figures, $82 million, represents the present value of the additional profits gained from innovation and is equivalent to a steady annual profit of $4 million a year. In other words, $4 million a year, or 40 per cent of gross profits, is the most that it would be worth investing in innovation in order to maintain future profits at their present level. This is a maximum, of course. If competition is not severe, the firm could reasonably expect to hold its competitive position with a much smaller outlay on innovation.

This is a logical economic argument, but the figures must be treated with reserve for a number of reasons, as follows:

(a) The figures obviously depend on the value assumed for the interest rate.
(b) They are also sensitive to the obsolescence time. Obviously, if the field is one of rapid technical change and if the firm does not move with the times, its profits will drop rapidly and their present value will be small. [. . .] The time to obsolescence does, of course, vary enormously from one industry to another. It may range from as little as 5 years for electronic goods, to 50 or more years for traditional industries such as woollen goods pottery, and house building.
(c) The calculation assumes a steadily operating firm and, for this condition of constant profits, leads to a rate of investment in innovation proportional to these profits. This obviously does not hold for unsteady conditions. For example, if the firm starts to fall behind its competitors technologically, its correct response is not to reduce its expenditure on innovation in proportion to its shrinking profits, but to increase it. [. . .]

In spite of these reservations, the figures do provide some general indications. We would expect from them, for example, that over a long period of time it would be impossible for a large organization – a multiple-product corporation or a whole national economy – to be able to spend more than about one half of its gross trading profits on innovation. And if the figure fell well below one half, we might suspect that the organization was not keeping up-to-date. Since gross trading profits are typically about 20 per cent of the gross national product, this means that a reasonable expectation for expenditure on technological innovation is about 10 per cent of the gross national product, which is approximately what most countries spend on plant and machinery[5,8] and research development.[9]

Conclusion

The main fact emerging from these economic considerations is that, because investment in innovation cannot greatly exceed about one half of gross profits and because total innovative costs are often some five times larger than research and development costs, no organization whose gross profits are about one fifth of its output can, over a long period of time, economically afford to spend more than about 2 or 3 per cent of its output – or gross national product on a national basis – on the development of new products and processes. The problem for a medium or small firm or country, then, is to reconcile this with the fact that to secure competitive lead times in new technological projects, it has to achieve a certain absolute threshold of research and development expenditure in each project – a threshold that is independent of the size of the organization or its markets, since it is set by the levels of research and development effort being exerted by the largest organizations working in that field.

There are, however, many ways of dealing with this problem. Some countries such as Japan and Germany have prospered through a policy of importing information on advanced new technologies through foreign licenses. Since a country of medium size, however scientifically advanced, can hardly expect to create more than, at most, about one tenth of the world's new innovative efforts, it must import large amounts of information if it is to keep up with general technical progress.

Some countries have made a successful policy by selecting certain industrial fields and concentrating their resources heavily on them. Shipbuilding and small optical and electronic goods are well-known examples in Japan, as are chemical and metallurgical products in Europe. In Britain we have, of course, concentrated particularly strongly on nuclear power plants and aircraft engines. The difficult aspect of this policy of selective activity lies not in deciding what to support, but in deciding what *not* to support. Yet without these negative decisions to supplement the positive ones, any medium or small country faces the danger of becoming a jack-of-all-trades and master of none. [. . .]

References

1. C. Freeman, C. J. E. Harlow and J. K. Fuller, *National Institute Economic Review*, London, November 1965.
2. P. M. S. Blackett, *Technology, Industry and Economic Growth*, the Thirteenth Fawley Lecture, Southampton: University of Southampton, 1966.
3. US Department of Commerce, *Technological Innovation: Its Environment and Management*, Washington DC: US Government Printing Office, 1967.
4. J. Jewkes, D. Sawers and R. Stillerman, *The Sources of Invention*, London: Macmillan; and New York: St Martin's Press, 1958.
5. B. R. Williams, *Technology, Investment, and Growth*, London: Chapman & Hall, 1967.

6. A. Hart, *Symposium on Productivity in Research,* London: Institute of Chemical Engineers, 1963.
7. W. E. Duckworth, The determination of total research effort, *Operational Research Quarterly,* London, vol. 18, no. 4, December 1967, pp. 359–73.
8. T. P. Hill, Growth and investment according to international comparison, *Economic Journal,* Cambridge, England, vol. 74, June 1964.
9. *The Overall Level and Structure of R and D Efforts in OECD Member Countries,* Paris: Organization for Economic Cooperation and Development, 1967.

3.4

Innovation in the Public Sector

Michael Posner

It is sensible to think of the public sector in two parts – the trading sector and the public services. The public services, by definition, can be innovative largely in the way in which they conduct their own affairs, or in the purchases which they make from outside. The trading sector can, like any other part of the economy, innovate both in the products they sell and in the purchases they make.

Within the public services, the example of defence is well known, even though the details are complex and secret. The General Staff can buy freely on the world market, from a shop window filled by the armament industries of all our allies. A typical third or fourth-rate power will sensibly confine itself to choosing which innovations to pluck from someone else's tree. A first or second-rate power, however, can 'make money' in the sense of providing greater security per dollar of expenditure, by pursuing its own innovative paths in some particular lines of endeavour. It is already, in some lines, ahead of its competitors, and it would be denying itself opportunities for saving money if it were to refuse to follow the innovative chain where it leads.

But this cannot be a reason for insisting on innovation across the board. The power to manage the innovative process, and the capacity to finance the investment necessary in any particular group of products, means that in only some lines of defence expenditure is innovation the wise course to pursue.

More generally, in the public sector the choice between doing it at home and buying it abroad is most strikingly offered in the case of computers, and office hardware generally. Britain is in, perhaps, the awkward middle of the range position, without the dominance of the United States or Japan, and without the resigned acceptance of the innovative power of others which doubtless rules in pretty prosperous countries like Australia or Austria. Once again, it is a question of selecting, cautiously, but with entrepreneurial flair, which particular lines to choose.

So much for background, now a few words about the trading entities in the public sector – mainly the nationalized industries.

M. Posner, Innovation in the public sector, paper for a one-day symposium *Investing in Innovation*, British Association for the Advancement of Science, London, 2 May 1984.
Michael Posner is Economic Director-General of the National Economic Development Office.

Who should design the product we sell?

On the face of it, this is a straightforward question with only one answer — if we let our competitors or our suppliers design what we offer on the market, we will soon die. But this answer is not completely self-evident. In principle, the gas industry could market its product through American cooking stoves, control it through French electronic systems, transmit it through Japanese steel pipes. The electricity supply industry could wait for somebody else to design turbines and copy their success, and could have decided thirty years ago to follow the American lead in nuclear power stations. The railways could buy their locomotives and rolling stock from Europe or North America, and concentrate on running their trains on time and keeping them clean.

But the analogy of defence suggests that to follow the market in this way is not always the best policy. There are features of the British market or British geography, or British population distribution, which make our business different from those of the foreigners. There are particular lines of activity in which we are, at any particular date, ahead of the competition and where we would be surrendering opportunities for profit, if we were to fail to press further our original competitive advantage. It would be stupid for the railways to design their own typewriters; it might be very sensible for them to design some at least of their own civil engineering equipment, and some at least of their own commuter rolling stock.

The choice of how much innovation to do at any particular time is a matter in part of how much to do in-house rather than buying from outside; and in part a question of management — there is a limit to the amount of management effort which can be usefully applied at any one time to the innovative process.

Managing Innovation

It is, alas, easier to give striking examples of how not to do this job, rather than examples of how successful we have been. Horror stories abound, from delays in the commissioning of large power stations, unanticipated 'plant outage' both for power station equipment and for railway locomotives, computer installations whose teething problems seem to run well into middle age. All this experience leads to some very familiar negative rules:

1. Do not innovate in any one product in too many different ways simultaneously. There is almost certainly a mathematical rule here suggesting a very substantial increase both in the number of required prototypes and in the amount of testing time required, as the number of features which are 'new' increases only modestly. Railway troubles with the Advanced Passenger Train, or successive difficulties in getting the AGR reactors to work, are good examples to avoid.
2. Innovate only where there is a clear commercial market for the product which your engineers believe they can develop. It remains remarkably difficult to find an appropriate marriage between the marketing and the development of a new product, and I regret to report that experience teaches that the market man must be in the lead. I am much encouraged by the way in which this lesson seems to have been learned in several public industries, particularly at the moment the railways.
3. Try to avoid putting all your money on a single large development, no bit of which can be replaced by a component drawn from elsewhere. There was a time at which British

Telecom seemed to be caught in this trap with its System X development. There are well-known and highly successful counter-examples, e.g. float glass.

4. Do not despise numerous small innovations in processes or in minor products, even though their public relations aspects may not be as good as the big bangs themselves. Public relations can often be satisfied by a lick of paint and a new dustpan and brush.

5. Finally, pay enormous attention to following through innovations. Anything really big in most public industries will take ten or fifteen years to follow through, and can be accomplished most successfully if the young research engineer in his thirties, who started the thing off, is allowed to follow it through himself until he is at the verge of becoming Managing Director of his whole large company. He should be positively encouraged and, perhaps, indeed required, to use the innovative process as an escalator for his own career.

Conclusions

I fear I have offered a somewhat negative and jaundiced view of the innovative process in this short talk. I make no apology for that. We need a rather greater explicit reliance on borrowing from elsewhere most of what we want, and of selecting rather brutally on a narrow front those areas where innovation is really worthwhile.

COMPANY ORGANIZATION
AND
PROJECT MANAGEMENT

3.5

Generating Effective Corporate Innovation

Edward B. Roberts

Effective corporate innovation requires the planned integration of staffing, structure, and strategy. This view arises from 15 years of research at MIT and elsewhere on the problems of managing industrial research and development and the technological innovation process. [. . .]

There are four critical areas to note:

1. The *staffing* of technical organizations must provide for the several key functions necessary to achieve successful innovation.
2. The organization must be *structured* to enhance the flow of technical and market information into research and development.
3. The organization's structure must also assure strong links with *marketing*, to assure that innovations effectively move forward into commercial success.
4. The company must adopt strategic *planning* methods that improve integration of top management's technical plans with other dimensions of overall corporate strategy.

Five different key staff roles must be fulfilled if innovative ideas are to be generated, developed, enhanced, commercialized, and moved forward in the organization:

1. The *creative scientist or engineer,* the source of creativity within the organization about whom so much – perhaps too much – has been written.
2. The *entrepreneur* who pushes the technical idea (it may be his or someone else's) forward in the organization toward the point of commercialization.
3. The *project manager,* who can focus upon the specifics of the new development and indicate which aspects will go forward, which can be economically supported, and which must be deferred and who can coordinate the needed efforts.
4. The *sponsor,* the in-house senior individual who provides coaching, back-up, and large skirts behind which entrepreneurs and creative scientists can hide. His role is that of protector and advocate – and sometimes bootlegger of funds – so that innovative technical ideas survive past the birth stage to gain the confidence of the technical organization.

E. B. Roberts, Generating effective corporate innovation, *Technology Review,* October/November 1977; extracts from pp. 27–33.
Edward B. Roberts is David Sarnoff Professor of Management at Massachusetts Institute of Technology.

5. The *gate-keeper,* who brings essential information into the technical organization. Gate-keepers come in two varieties: the technical gate-keeper and the market gate-keeper; both of them account disproportionately for the information that is used in developing innovative ideas and moving the resulting processes and products forwards into manufacturing and the marketplace.

In studies of many research and development organizations over the last 15 years, we have observed deficiencies primarily in all but one of these key roles needed for organizational effectiveness. The role of creative scientist seems to be over-emphasized; organizations tend to assume that having creative people on the payroll guarantees effective development of new products, new processes, and product improvements. This assumption is far from correct, and it has tended to cause systematic neglect of the other functions necessary for effective innovation.

This observation is important in the light of our conclusion that each of the several roles required for effective technical innovation presents unique challenges and must be filled with very different types of people, each type to be recruited, managed, and supported differently, offered different sets of incentives, and supervised with different types of measures and controls. Most technical organizations seem not to have grasped this concept, with the result that all technical people tend to be recruited, hired, supervised, monitored, evaluated, and encouraged as if their principal roles were those of creative scientists. But only a few of these people in fact have the personal and technical qualifications for scientific inventiveness; a creative scientist or engineer is a special bird who needs to be singled out and cultivated and managed in a special way. He is probably a strong, innovative, technically well-educated individual who enjoys working on advanced problems, often as a 'loner'. In an industrial laboratory, he is likely to be among the minority of scientists and engineers with doctorates, but education itself is by no means the criterion for creativity.

The entrepreneur is a special person, too – creative in his own way, but his is an aggressive form of creativity appropriate for selling an idea or a product. The entrepreneur's drives may be less rational, more emotional than those of the creative scientist; he is committed to achieve, and less concerned about how to do so. He is as likely to pick up and successfully champion someone else's original idea as to push something of his own creation. Such an entrepreneur may well have a broad range of interests and activities, and he must be recruited, hired, managed, and stimulated very differently from the way a creative scientist is treated in the organization.

The project manager is a still different kind of person – an organized individual, sensitive to the needs of the several different people he's trying to coordinate, and an effective planner; the latter is especially important if long lead time, expensive materials, and major support are involved in developing the ideas that he's moving forward in the organization.

The sponsor may in fact be a more experienced, older project manager or former entrepreneur who now has matured to have a softer touch than when he was first in the organization; as a senior person he can coach and help subordinates in the organization and speak on their behalf to top management, allowing things to move forward in an effective, organized fashion. Many organizations totally ignore the sponsor role, yet our studies of

CREATIVE SCIENTIST

ENTREPRENEUR

PROJECT MANAGER

industrial research and development suggest that many projects would not have been successful were it not for the subtle and often unrecognized assistance of such senior people acting in the role of sponsors. Indeed, organizations are most successful when chief engineers or laboratory directors take on this sponsor role as part of their natural behaviour.

Finally, there is the information gate-keeper, the communicative individual who, in fact, is the exception to the truism that engineers do not read – especially that they do not read technical journals. If you're looking for a flow of technical information in a research and development organization to enhance new product development or process improvement, you have to look to these gate-keepers.

But those who do research and development need market information as well as technical information. What do customers seem to want? What are competitors providing? How might regulatory shifts impact the firm's present or contemplated products or processes? For answers to questions such as these research and development people need people I call the 'market gate-keepers', engineers or scientists, or possibly marketing people with technical background who focus on market-related information and communicate effectively to their technical colleagues. Such a person reads trade journals, talks to vendors, goes to trade shows, and is sensitive to competitive information. Without him, many research and development projects and laboratories become misdirected with respect to market trends and needs. [. . .]

The Social Aspects of Technology

The structure of an organization also affects the success of its creative efforts. The need, of course, is for an interrelationship which enhances the flow of the right kind of information into and through the technical organization, assures its appropriate use there, and encourages the flow of results of technical programs from the research and development group to the other parts of the organization where they can be made to count.

No research and development organization produces a profit. At best such organizations can produce the technical bases that will permit the firm's marketing and manufacturing activities to produce the profit. Thus the search for effective, profitable innovation must embrace the interface relationships that bring information into research and development and move its results forward to other parts of the firm.

My colleague Professor Thomas J. Allen is responsible for some of the best studies in the country on the factors that affect technical information flow in an organization.[1,2,3] He has found, for example, that if you separate two technical people by 60 or 70 feet, you've suppressed the likelihood of technical communication by two-thirds; separate them by another 70 feet and you've essentially eliminated 90 per cent of the possibility of technical communication between them; furthermore, he finds no difference in the impediment to communication between 3,000 miles and 3,000 feet.

He has also found that the social relationships between technical people are critically related to the technical relationships between the same people. The person with whom you

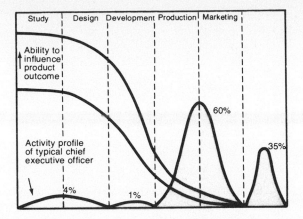

Figure 1   The expertise of a chief executive can most influence any new technology-based product development program in the program's early stages – during preliminary study, design, and development. But current research suggests that chief executive officers actually devote only trivial amounts of their time and attention to these early stages of such new-product programs. Instead, they typically have significant involvement only during production and marketing – when it's too late to do anything that can influence the outcome. (This illustrative figure comes from related work by Foster and Gluck of McKinsey and Co.)

go to lunch or dinner is also the person with whom you'll talk about new technical ideas; the sources of technical problem-solving ideas within the firm correlate strongly with the sources of information about the Sunday afternoon football game.

Professor Allen's approach emphasizes the social aspects of managing a technical organization, an area that technical managers have seldom considered. If technology is to be useful in product improvement, new products, and new business, we must take a broader, more cultural view of what in fact takes place in the creation and enhancement of technical information flows. [. . .]

At the very beginning of a new technology-based product development activity, a senior manager is able to influence the direction of the project in almost any way he chooses; he can stop it, enlarge it, accelerate it, redirect it, make it go for one piece of the market or another. The top manager's opportunity to be the bold inventor or the aggressive obsoleter of technology is greatest at that earliest stage. As the product moves forward into development, production, and finally marketing, the ability of the executive to influence the outcome goes down – eventually very close to zero.

But if you look at how managers typically spend their time, you find a very different pattern (Figure 1). Studies of this subject seem to indicate that chief executive officers spend trivial amounts of their time on the study and design stages of major new projects, the redirection projects of the firm. Instead the typical chief executive is primarily involved during the production and marketing stages of a project, when it's too late to do anything that can influence the outcome. The lesson from this is simple: if you want to affect the future of your firm, you need not only the right kind of staffing in your technical organizations and the right kinds of structures to enhance information flow in and transfer out. You need a strategic posture which displays critical points for paying attention to certain dimensions of product and technology, and you need to have an allocation of managerial time that brings the best talents of the company to bear on these focal points at the critical time.

References

1. T. J. Allen, Communications in the research and development laboratory, *Technology Review*, October/November 1967.
2. T. J. Allen and A. R. Fusfeld, Design for communication in the research & development laboratory, *Technology Review*, May 1976.
3. T. J. Allen *Managing the Flow of Technology*, Cambridge: MIT Press, 1977.

3.6(a)

Organising Design

Mark Oakley

Introduction

Prominent amongst factors which may influence the outcome of design projects is the question of organising design activities. Senior management must consider very carefully the special features of design work and design departments; it is then necessary to decide how best to structure such departments and how they may work with and relate to other parts of the company.

Several different structures may be encountered in practice. Some companies organise themselves with separate design units headed by a senior manager to whom is delegated (or abandoned) part or all of the responsibility for product design and strategy decisions. At the other extreme are companies whose design activities are diffused throughout the organisation. In this case, top management may or may not set policy guidelines or seek to ensure some degree of conformity in design results.

Some companies identify a role for 'product champions' to push new product projects through any organisational or procedural barriers which may be encountered. In such companies, a 'task force' approach is quite common with small groups of engineers, designers and others assigned to specific projects. Other companies may prefer to use a more broadly based method, perhaps built on one or more large units within which there are a number of projects. Individuals move in or out of these projects depending upon the need for their particular skills. It is clear from the available literature that there is little agreement about the precise circumstances in which each method should be applied, although it is possible to derive guidance in broad terms about selecting a suitable approach.

Whatever organisational structure is adopted, leadership and management will be required. Some experts see a need for a special kind of 'design manager' who, amongst other functions, is able to bridge the knowledge and language gap that frequently exists between ordinary managers and designers.[1,2]

Hence, the purpose of this article is to examine these organisational questions and to try to give an idea of the advantages and disadvantages of the alternative structures that may

M. Oakley, *Managing Product Design*, London: Weidenfeld & Nicolson, 1984; extracts from Chapter 4.
Mark Oakley is a lecturer in operations management at the University of Aston Management Centre.

be adopted. The special demands placed upon those who manage design are compared with the demands placed upon managers in other functions.

Concepts of Change

All managers are confronted by change – if no change ever occurred then there would be no need to have managers at all. If markets were static or totally predictable, no-one would need to take decisions about production quotas. If customers' tastes and loyalties never changed then no new products would be needed. But of course, in the real world change is an ever-present fact of life. For millions of different reasons – some natural, some man-made, some accidental, some deliberate – all companies must continually adjust their activities to compensate for changes which are taking place. In the same way that changes in the market dictate a need for product design, the act of designing new products, or improving old ones, gives rise to some degree of change within the company. This change may be slight, as when modifications are made to minor components, or it may be major, as in the case where a factory must be totally re-organised and re-equipped to accommodate a new product.

Regardless of its extent, however, many organisations will resist change right up to the point where survival of the organisation is seen to be at stake. In the normal course of events, preoccupation with current activities may leave little scope for any real innovation. Many firms demonstrate this attitude by paying lip-service to their product design activities whilst proclaiming official doctrines of innovation. Such firms may, for example, encourage new product ideas, only to find consistently that none of them meets the stringent criteria laid down in advance.[3] Similarly, many firms effectively eliminate change by oscillating between support and resistance – an 'on-again, off-again' approach to design and innovation.

Not infrequently, a myopic concentration on the manufacture of existing products means that new products are never successfully developed. [. . .]

According to Twiss,[4] resistance to change is more often a feature of older companies than younger ones. He has found that as companies reach maturity, strategies become more defensive and fewer projects lead to new products which depart substantially from current practice. He believes that the reason for this often lies with the 'Chief Technologist' who is normally found to be the creative force in a young company, but is less effective as the company evolves and a management team is built up around him. This means that in the mature company it becomes necessary to design some formal approach in order to cope with change.

An American study by Lynton[5] has examined large and small organisations and found that each may face circumstances in which they can no longer deal with change by intuitive means. It may be thought that an organisation's size is the primary factor in handling change. But Lynton has found that a decision to formally cope with change is not so much related to the size of the organisation, as to the degrees of uncertainty in technology, markets and the environment. The necessary redesign of the organisation into one which can accommodate higher degrees of uncertainty is invariably more difficult in older companies, but it is not markedly so in larger companies as opposed to smaller ones. New designs always involve some change and for this reason they may be resisted. After all, whilst the modern company is often built around the production process which is (usually) rational and standardised, design is not always seen as rational and can be disruptive to those affected by it. Bright[6] offers 12 reasons (Table 1) why an innovative change such as a new design be resisted by the employees of a company. [. . .]

Table 1   Reasons for resisting an innovative change

1   To protect social status.
2   To protect an existing way of life.
3   To prevent devaluation of capital invested in an existing facility.
4   To prevent a reduction of livelihood because the change would devalue the knowledge or skills presently required.
5   To prevent the elimination of a job or profession.
6   To avoid expenditure such as the cost of replacing existing equipment.
7   Because the change opposes social customs, fashions and tastes and the habits of everyday life.
8   Because the change conflicts with existing laws.
9   Because of rigidity inherent in large or bureaucratic organisations.
10 Because of personality, habit, fear, equilibrium between individuals or institutions, status and similar social and psychological considerations.
11 Because of tendency of organised groups to force conformity.
12 Because of the reluctance of an individual or group to disturb the equilibrium of society or the business atmosphere.

Not only is change itself important, so is the rate of change in a company. A firm has to change at a sufficient rate and in an ordered manner to meet the conditions imposed from outside. A rate which is too high leads to chaos; one which is too low may end in bankruptcy. Also, in order to achieve the goal of prosperity which is the reason for the existence of most companies, the effects of competition demand that each subsequent design change has to be done a little better and a little more profitably than the preceding one. It is disturbing that so few organisations seem prepared for this challenge. Many are engulfed in systems which perpetuate conformity, precedent and procedure, and reaction to crisis continues to be the primary model of adjusting to change.

The problem is huge (and in its entirety is quite outside the scope of this discussion) involving issues of governmental policies, the attitudes and activities of trade unions, division of wealth and power and the availability of resources such as fuel and materials. However, as already noted, one major factor is the style of management which is practised and encouraged in design and other innovative activities. Efficiency in recognising the need for new products and the careful screening and selecting of ideas will all be in vain if the management of the company is not suited to the special requirements of new product work.

Table 2   Comparison of organisational features of production with those of product design

Features of production	*Features of design*
1 Rational, standardised, predictable	1 Irrational, novel, unpredictable
2 Operations accurately timed	2 Accurate timing of activities usually impossible
3 Long runs of identical products	3 Activities frequently changing
4 Creativity and initiative not developed in workforce	4 Highly creative personnel essential
5 Work closely controlled – essential for profitability. Risk eliminated	5 Profitability related to skill, change, judgment, intuition, risk taking, etc.

The nature of these requirements can be better understood if the special features of design departments are compared with those of other departments, particularly production with which design is often closely linked, since both are often considered to be the 'technical' parts of the business and to have similar methods of operation. Table 2 summarises the main organisational features which distinguish design and production.

Comparison with other functions such as marketing or finance will also show differences of emphasis; for example, marketing's need for short lead times on new products compared with design's concern to spend as long as possible on projects to achieve the best results. In the light of these differences, we need to consider both the organisation of the design unit itself, and its relationship with the rest of the company.

Organisation of Product Design Units

A number of years ago Burns and Stalker[7] analysed the organisational aspects of a sample of firms involved with the design of new products; their work remains important today. They observed that within these companies there were organisational styles ranging from what they termed 'mechanistic', which were very formal, hierarchical, bureaucratic and inflexible, to styles which they termed 'organic', which were informal, based on teams and tended to 'shape' themselves to the problems being tackled.

They concluded that mechanistic systems work satisfactorily only where conditions are relatively stable – flow line production departments for example, or other situations where close control of highly specialised work is essential. Mechanistic forms of organisation are not likely to prove satisfactory when applied to design units, which need flexibility in many respects. Here organic systems are more appropriate and, as Burns and Stalker observed, such systems improve the prospects of success for new products. Table 3 lists some of the features that may be found in organic systems. Managers responsible for product design departments need to consider how they can promote 'organic features'.

Table 3   Typical features of organic systems

1   Unifying theme is the 'common task' – each individual contributes special knowledge and skills – individual's tasks are constantly re-defined as the total situation changes.
2   Hierarchy does not predominate – problems are not referred up or down, but are tackled on a team basis.
3   Flexibility – jobs not precisely defined.
4   Control is through the 'common goal' rather than by institutions, rules and regulations.
5   Expertise and knowledge located throughout the organisation, not just at the top.
6   Communications consist of information and advice rather than instructions and decisions.

Promoting such features may be a delicate matter, especially within those firms which are otherwise organised along precise and inflexible lines. Even in situations where they do not have the opportunity to develop ideal systems, managers should be aware of the actions they can take to assist creative work so that organisationally desirable features predominate. [. . .]

It is important to understand that mechanistic and organic styles are categories of organisation which are unlikely, in practice, to be found in 'pure' forms. There are many intermediate stages between these categories and most design units will exhibit features of both styles, with the 'right balance' depending on the nature of the company, the industry and the projects being undertaken. In all cases, the best policy for the design manager is to be vigilant for the evolving of mechanistic characteristics and then take steps to prevent them becoming dominant within the design operation. The kind of features that need to be checked include:

 – Increase in the number of levels of supervision.
 – Control by detailed inspection of work methods rather than by evaluation of results.
 – Communication consisting of instructions and decisions rather than exchange of advice, consultation or information.
 – Confrontation of win-lose nature rather than collaboration.
 –Insular attitude of top management with sense of commitment to past decisions. [. . .]

Location of Design within the Firm

It is not always easy to decide where design activities should be located within the firm. As already stated, product design is often considered to be a technical activity of a kind similar to production. However, because of the fundamentally different natures of the two functions, giving control of design to production may result in failure. This may happen either because organisational conditions inappropriate to design are imposed or simply because production has resistance to new products (because of the disruption involved). This resistance may take the form of constant rejections of new designs, refusal to supply information and help, or just general obstruction – all while paying lip-service to the need for new products. These attitudes may be particularly acute in long-established firms where design work has been limited previously to improvements and modifications. Nevertheless, in a great many companies design departments are to be found as part of the production set-up, under the control of a production or technical manager, as in Figure 1.

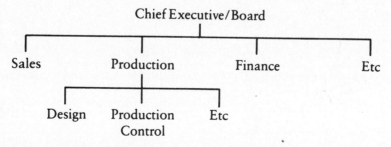
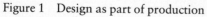

Figure 1   Design as part of production

It will be noted that design is effectively isolated from the highest level decision maker, the chief executive. Information and directives may be 'filtered' by production to the extent that design work is reduced to a mundane level. Modifications to existing designs and extensions to current ranges of products may well be achieved, but more ambitious projects are likely to be stifled. Any work that is undertaken will be subject to pressure to ensure that production considerations are given paramount attention – at the expense of

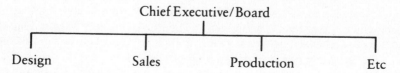

Figure 2   Design as an independent function

other requirements including those of the customer. Very similar problems may arise if design is organised as a part of one of the other functions such as marketing.

Some companies seek to avoid these problems by establishing design as an independent department equal in status to the other major functions, as shown in Figure 2.

This may well lead to improved design performance because, with direct linkage to the highest decision level, projects can be monitored and encouraged. Radical projects may be attempted and some success anticipated. However, some drawbacks will remain. The design manager (or design director) may have to bargain with other departmental bosses for cash and resources. Because design is a long-term activity it may seem to senior managers in other functions that it is just a wasteful consumer of the income which they create. Boardroom politics may work to the disadvantage of the design function; power is often related to the size of financial budgets. Sales directors or finance directors whose scale of operation may be many times greater than that of design, might acquire or assume correspondingly greater influence in decision-making. A further disadvantage with an independent design unit is that it may well be viewed with suspicion by the rest of the firm. The fact that it *is* separate from other functions may cause speculation about its relevance to company objectives. This usually manifests itself when a new product is ready to be handed over at the end of a design exercise. The product may be rejected because it is seen as an intrusion from outside – sometimes referred to as the 'not invented here' syndrome.

Some companies try to overcome this problem by directing new product operations through a steering committee which represents all major functions. Others appoint 'project champions' whose job it is to push the new product through all barriers and solve any problems which may arise. Unfortunately, project champions are not often given the authority which is necessary for them to be effective. Companies sometimes delude themselves into believing that a bright young manager will be able to achieve the progress that has been denied to others simply by force of personality. Frequently, the problem is not one of leadership at all, but of availability and allocation of resources. For example, if the design unit needs to use a particular manufacturing process for test purposes and that process is fully occupied with production work designated as top priority by the Board, then no amount of dynamic design management will ensure success.

Despite these reservations, the steering committee approach as summarised in Figure 3 does offer a number of advantages. The most important is the potential ability of the committee to ensure both the relevance of new design activities and the acceptability of design results.

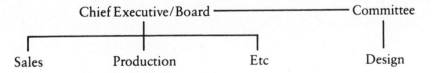

Figure 3   Design Guided by Steering Committee

The intimate involvement of the committee members, drawn from all major areas of the company, should promote well-integrated design, in theory at least. In practice, the method may be less successful. As well as the usual problems associated with management by committee – compromise, indecision and procrastination – individual managers may still succeed in undermining projects which they find undesirable. These managers may find that conflict arises out of their dual roles (as departmental managers and as steering committee members) and that this causes the taking of inappropriate decisions.[8] The use of a 'project champion' in place of a committee may avoid many of these problems, although selecting the right person for the job assumes crucial importance. He must be able to understand the technical aspects of the project and do his best to deliver an acceptable new product within the deadlines set. Whilst satisfying colleagues elsewhere in the firm, he must always see the completion of the project as the main priority and he must inspire his team towards this goal.

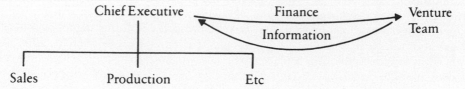

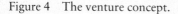

Figure 4   The venture concept.

In order to encourage this approach, some firms move one step further still and, except for a 'financing link', effectively set free the design project so that it can grow and mature independently of the main organisation. This concept of organising for new products is sometimes referred to as the 'venture method' (Figure 4). It is especially attractive for traditionally organised companies which find it difficult to manage design within its normal boundaries. By setting up design projects as 'mini-businesses', a high degree of success may be achieved.

References

1 M. Farr, *Design Management*, Hodder & Stoughton, 1966.
2 A. Topalian, Designers as directors, *Designer*, February 1980.
3 D. A. Schon, The fear of innovation, in R. M. Hainer *et al.*, (eds.), *Uncertainty in Research Management and New Product Development*, Reinhold, 1967.
4 B. C. Twiss, *Managing Technological Innovation*, 2nd edn, Longmans, 1980.
5 R. P. Lynton, Linking an innovation sub-system into the system, *Administrative Science Quarterly*, September 1969, pp. 398–415.
6 J. R. Bright, *Research Development and Technological Innovation*, Irwin, 1964.
7 T. Burns and G. M. Stalker, *The Management of Innovation*, Tavistock Publications, 1961.
8 R. K. Mueller, *The Innovation Ethic*, American Management Association, 1971.

3.6(b)

Managing Design: Practical Issues

Mark Oakley

Product Specifications and Project Briefs

The starting point for a new product should always be some kind of 'specification'. Depending on the complexity or scale of the project, the specification may be little more than a rough sketch and a few notes.

However, for a new product such as a motor vehicle, the specification may run to many volumes and describe in precise terms the multitude of parameters which must be observed. Such a document might take several months, or even years, to prepare and will be based on extensive market investigations and internal policy deliberations. Generally though, specifications should be as simple as possible and should be presented in such a way that designers are allowed maximum discretion. There is little point in employing creative design staff if freedom of action is then restricted by excessive direction.

The essential information which ought to be provided by the specification may be summarised as follows:

- Exact type of product to be designed – described in terms of existing alternative productions if available, or in terms of the functions required if the product is a totally new concept.
- Major technical requirements – such as speed of operation, maximum and minimum dimensions that can be allowed, performance levels required, etc.
- Styling requirements – general nature of appearance, shapes and colours preferred, carrying capacity, arrangements of major features, etc.
- Operational requirements – size of controls, forces required to operate, safety features, compatibility with operator's dimensions or other items of equipment, etc.
- Cost constraints – target selling prices or production costs of the product, requirements regarding maintenance and service cost levels if relevant, and operating costs, etc.
- Special requirements – for example, 'need for product to be safe for use by small children'.

M. Oakley, *Managing Product Design*, London: Weidenfeld & Nicolson, 1984; extracts from Chapter 6.
Mark Oakley is a lecturer in operations management at the University of Aston Management Centre.

[. . .] The product specification usually forms a part of a larger document, the project brief (or 'design brief'). Success in design work is not dependent upon simply achieving a final version of the new product which satisfies the specification. It is also important that the work is completed within a period of time and at a cost which will enable the product to stand a chance of becoming a commercial success. Information about the time and cost constraints is a major element of the project brief.

The duration of a design project may be important in marketing terms – if a project lasts too long, competitors may become firmly established in the market and overall prospects for the new product will be greatly diminished. Another consequence of a late project is that total design costs will increase which, coupled with a delayed product launch (and hence delayed sales income), may push the whole project into irretrievable financial deficit [see Figure 6 in article 1.6 by Caldecote].

If design costs are running at a high level, a very short delay indeed may transform the prospects for the new product from good to doubtful. Worse still, if the increased debt caused by the delay cannot be covered by the company then bankruptcy may be the inevitable result. In recent years, spectacular failures in the aircraft industry have received much attention but many other companies of more modest scale have suffered similar disastrous consequences as a result of their inabilities to sustain the losses caused by out-of-control design projects.

To avoid these problems, great care must go into planning design projects and then preparing briefs which give clear guidance on all constraints, but particularly on time and cost. This is simply stated but in practice is not so easy to achieve. There have been attempts to analyse the discrepancies which arise in many projects between original cost estimates and the eventual outcome. These have led to various rules-of-thumb for calculating the actual final cost of a project, such as multiplying the first estimate:

By $\pi/2$ if a new design is similar to other work done by the firm.
By π if it is similar to work done by another firm.
By 2π if no-one has done anything like it before.[1]

Preparing design briefs requires considerable experience and judgement if problems such as these are to be overcome. Companies that lack such expertise should consider engaging a design consultant – or better, a design management consultant – to advise on preparing the brief. To press forward without a brief at all is very dangerous and generally an outside designer will not start work on a project unless a brief is provided. However, internal design staff often have little choice in the matter and it is unfortunate that the briefings they are given are often unclear, incomplete or even non-existent.

The importance of the brief or specification is not often stressed by writers on design matters so it is encouraging to see high priority given to the topic by Corfield[2] in his report on product design. [. . .]

The Development Operation

It has been stated already that the designer's task is to seek solutions which satisfy the requirements of the specification. Usually a final design cannot be achieved without a good deal of testing of ideas and experimental comparisons of alternative solutions. Hence, design units are often supported by some kind of 'development' facility – basically a workshop or laboratory where test-pieces can be made and subjected to whatever examination may be required.

Design and development proceed hand in hand. During the early stages of the design programme, the demand on development may be for information about material strengths or corrosion resistance, for example. Later, when this information has been used and the design of the product has proceeded to the point where a general layout has been achieved, individual components may be built by the development section and tested to assess properties or performance. In the light of the results obtained, the design can be changed as necessary to meet the standards demanded by the specification. Eventually, a final version of the new product will be built and tested, typically to examine features like reliability, durability, safety and general performance.

Depending on the scale of the product being designed, the development effort may be concerned with scale-models or full-size models. Where possible, full-size tests should always be encouraged because 'scale-up' results usually involve some error; obviously this is not always feasible when dealing with products like aircraft, ships or similar objects. Similarly, it should be remembered that full-sized models (sometimes called 'prototypes') which have been made using 'one-off' techniques rather than mass production methods may exhibit features different from those of the finished product. Consequently, the final activity of the design operation should be to examine the product as soon as the production process has been set up. Some unexpected features may be discovered which must be rectified before large scale production commences. Sometimes quite major problems are encountered at this stage as a result of lack of co-operation between production and design during earlier stages of the project. In article 3.6(a) it was noted that relationships between these departments may be difficult because of the differences in organisational styles and objectives. Nevertheless, it is essential that production is involved in design decisions as they are made.[3] Failure to do so is a recipe for disaster – years of design effort may be wasted if assumptions are made about production matters without checking the true facts. [. . .]

Designer's Responsibility for Quality, Time and Cost of Production

As far as quality is concerned, the designer must aim to achieve the standards demanded by the specification, but at the same time not exceed the capabilities of the production department. This may not be an easy task because the determinants of quality are frequently difficult to identify. The specification may be explicit enough in the terms in which quality of the new product is described – for example, in terms of minimum acceptable working life, adequacy of performance and other aspects. The designer must decide how these are affected by features of the production process, as well as by his choice of materials, particular design solutions, etc. If he knows that a certain manufacturing operation can only achieve, say, very poor dimensional tolerances, he must decide whether such an operation can be used for the new product. If it cannot, he must devise an alternative solution.

Clearly the skill of the designer can have substantial impact on production times and costs. It is a skill which is certainly influenced by aptitude, training and experience, but it depends also upon close co-operation between production staff and designers and upon a 'sense of value' within the firm as a whole. [. . .]

Need for New Products to be Compatible with Existing Production

Designing to appropriate standards of quality and minimum levels of cost and time are certainly major responsibilities of designers. Where new production facilities are being created specifically for the new product, these may be the only responsibilities. But where existing production systems are to be used, there are other factors which must be considered before the design is completed to ensure that the new product will be compatible.[4] This is a topic which is covered in few texts yet, in this writer's experience, is perhaps the greatest cause of friction between production managers and designers.

Compatibility must be achieved with the production process and with existing products. This involves ensuring that design features are appropriate to existing methods and that they do not cause disruptions. An extreme example of non-compatibility might be where a product was designed to be built using equipment not present in the factory. Another example could be where manufacture of the new product would cause overloading of a particular process.

The key to the problem lies in the designer's knowledge of the production set-up. Many designers have practical experience of production and fully understand the limitations and capabilities that they must work within. Unfortunately, there are also many who do not have this experience and, quite simply, do not appreciate the systems that they are supposed to be designing for. When this situation arises, the responsible production manager will adopt a constructive attitude and ensure that the design department is provided with a detailed picture of the true nature of his department. This information should cover:

– Type of production system (e.g. batch, line, etc).
– Processes available (e.g. casting, welding, etc).
– Handling and storage facilities.
– Nature of the workforce and skills available.
– List of preferred sub-contractors and their skills.
– Breakdown of current products.
– Utilisation of processes and machines.
– Quality limits and inspection procedures.
– Materials used and stockholding facilities.

Only with this basic knowledge can designers hope to achieve compatibility; the more complete the knowledge, the better the chances of a smooth start-up of production.

In many, perhaps most, companies in the United Kingdom there is an assumption that production departments should adapt to whatever demands are made. 'It is the job of Design to dream up the ideas and of Production to make them work' sums up a common attitude. In fact, what frequently happens is that the ideas do not work and a great deal of wasted energy is expended. Both production managers and design managers are jealous of their own specialisms and may see co-operation as a threat to autonomy. This is an attitude that can be seen in other departments too, and is reflected in the highly segmented way in which many companies organise themselves. The message for top management is that successful innovation must be a major objective and that it is the responsibility not only of designers, but of all functions in the organisation.

References

1. R. R. White, Development costs – a law of overspend? in *Engineering Progress through Development*, London: Mechanical Engineering Publications, 1978.
2. NEDO, *Product Design* (The Corfield Report), London: National Economic Development Office, 1979.
3. C. Flurschiem, *Engineering Design Interfaces*, London: Design Council, 1977.
4. H. A. Harding, *Production Management* (3rd edn), London: Macdonald and Evans, 1978.

3.7

Large and Small Firms in the Emergence of New Technologies

Roy Rothwell

Introduction

It is clear from recent policy statements on technological and economic change that, generally speaking, governments in the advanced market economies increasingly have laid greater emphasis on measures to support small and medium sized manufacturing firms* (Rothwell and Zegveld, 1981a). This is based on the belief that small and medium sized firms (SMFs) are a potent vehicle for the creation of new jobs, for regional economic regeneration and for enhancing national rates of technological innovation.

The debate concerning firm size and innovation is of long standing, some commentators arguing that large size and monopoly power are prerequisites for economic progress via technological change, others that because of behavioural and organisational factors small firms are better adapted to the creation of major innovations. Some of the advantages and disadvantages variously ascribed to large and small firms in innovation are listed in Table 1, which suggests, *a priori,* that comparative advantage in innovation is unequivocally associated neither with large nor small scale.

Table 1   Advantages and disadvantages† of small and large firms in innovation

	Small firms	Large firms
Marketing	Ability to react quickly to keep abreast of fast changing market requirements. (Market start-up abroad can be prohibitively costly.)	Comprehensive distribution and servicing facilities. High degree of market power with existing products.

---

* In Europe, for the purposes of government policy generally taken as firms with employment between 1 and 499.

R. Rothwell, The role of small firms in the emergence of new technologies, *Omega*, vol. 12, no. 1, 1984, extracts from pp. 19–22, 27.
Roy Rothwell is a Senior Research Fellow at the Science Policy Research Unit, Sussex University.

Table 1 (continued)

	Small firms	Large firms
Management	Lack of bureaucracy. Dynamic entrepreneurial managers react quickly to take advantage of new opportunities and are willing to accept risk.	Professional managers able to control complex organizations and establish corporate strategies. (Can suffer an excess of bureaucracy. Often controlled by accountants who can be risk-averse. Managers can become mere 'administrators' who lack dynamism with respect to new long-term opportunities.)
Internal Communication	Efficient and informal internal communication networks. Affords a fast response to internal problem solving; provides ability to reorganize rapidly to adapt to change in the external environment.	(Internal communications often cumbersome; this can lead to slow reaction to external threats and opportunities.)
Qualified Technical Manpower	(Often lack suitably qualified technical specialists. Often unable to support a formal R & D effort on an appreciable scale.)	Ability to attract highly skilled technical specialists. Can support the establishment of a large R & D laboratory.
External Communication	(Often lack the time or resources to identify and use important external sources of scientific and technological expertise.)	Able to 'plug-in' to external sources of scientific and technological expertise. Can afford library and information services. Can subcontract R & D to specialist centres of expertise. Can buy crucial technical information and technology.
Finance	(Can experience great difficulty in attracting capital, especially risk capital. Innovation can represent a disproportionately large financial risk. Inability to spread risk over a portfolio of projects.)	Ability to borrow on capital market. Ability to spread risk over a portfolio of projects. Better able to fund diversification into new technologies and new markets.
Economics of Scale and the Systems Approach	(In some areas scale economies form a substantial entry barrier to small firms. Inability to offer integrated product lines or systems.)	Ability to gain scale economies in R & D, production and marketing. Ability to offer a range of complementary products. Ability to bid for large turnkey projects.
Growth	(Can experience difficulty in acquiring external capital necessary for rapid growth. Entrepreneurial managers sometimes unable to cope with increasingly complex organizations.)	Ability to finance expansion of production base. Ability to fund growth via diversification and acquisition.
Patents	(Can experience problems in coping with the patent system. Cannot afford time or costs involved in patent litigation.)	Ability to employ patent specialists. Can afford to litigate to defend patents against infringement.
Government Regulations	(Often cannot cope with complex regulations. Unit costs of compliance for small firms often high.)	Ability to fund legal services to cope with complex regulatory requirements. Can spread regulatory costs. Able to fund R & D necessary for compliance.

† The statements in brackets represent areas of potential *disadvantage*. Abstracted from Rothwell and Zegveld (1981b).
Source: Rothwell (1983).

Data from the UK on some 2,300 important innovations introduced by – though not necessarily developed by – British companies during the period 1945 to 1980 have thrown some light on this issue (Townsend *et al.*, 1981). These data (on some 35 sectors of industry) showed that, at an aggregate level, SMFs' share of innovations in the UK has, during the period covered, consistently averaged about 20% of the total (Table 2). At the same time the share enjoyed by firms in the largest size category (greater than 10,000 employees) increased progressively from 36% during 1945–9 to 59% during 1975–80. Moreover, the data also showed that the larger firms increasingly have innovated via smaller units and that independent firms increasingly were displaced by subsidiaries of larger firms as the major source of innovations. [. . .]

Table 2   Percentage of innovations in each firm size category for each five-year period

Number of employees	1945–9 (%)	1950–4 (%)	1955–9 (%)	1960–4 (%)	1965–9 (%)	1970–4 (%)	1975–80 (%)	Total (%)
1–199	16.0	12.0	11.0	11.0	13.0	15.0	17.0	14.0
200–499	9.0	6.0	8.0	6.0	7.0	9.0	7.0	7.0
500–999	3.0	2.0	7.0	5.0	5.0	4.0	3.0	4.0
1,000–9,999	36.0	36.0	25.0	27.0	23.0	17.0	14.0	23.0
10,000 and over	36.0	44.0	50.0	51.0	52.0	55.1	59.0	52.0
Total	100.0	100.0	100.0	100.0	100.0	100.0	100.0	100.0
Number of Innovations	94	191	274	405	467	401	461	2293

Source: Wyatt (1982).

In terms of share in total sectoral innovations, in some sectors, e.g. pharmaceuticals, SMFs played a very small or zero role; in other sectors, e.g. scientific instruments, SMFs played a consistently significant role. Not surprisingly where R & D requirements are very large and capital costs very high, high entry costs prohibit the participation of small firms; where technical, capital and marketing start-up costs are relatively low, entry by small firms is entirely possible.

An interesting case is electronic computers. Between 1945 and 1969 innovative activity (and output) was dominated by large firms producing predominantly mainframe computers. This involved high capital costs, a large R & D effort and the establishment of comprehensive production and servicing facilities. During the period 1970–80, however, SMFs have emerged as a significant force and accounted for 40% of all important innovations introduced in the UK.

This reflects the introduction of high density integrated circuits and the microprocessor which made possible the entry of new small firms producing mini- and micro-computers. These are skill-intensive, require considerably less capital investment than previous models and have opened up a large variety of new market niches suitable for exploitation by technical entrepreneurs. Thus, while one type of technological change, i.e. that requiring high development costs and large investment for commercial realisation, can pose a barrier to entry by small firms, other types of technological change can provide them with many new opportunities.

Two important points emerge from the above discussion. The first is that the debate concerning firm size and innovation should proceed on a sector-by-sector basis. The second is that a *dynamic* approach clearly is necessary: the relative contribution of firms of different sizes to innovation in a particular industry might depend on the *age* of that industry; the *type* of innovation typically produced by large and small sized firms at different stages in the industry cycle might vary also, i.e. product or process innovations. [. . .]

If this pattern of evolution is valid, then while the initial small entrepreneurial firms are concerned primarily with new product innovation and major product improvement, the subsequent large established firms become increasingly involved in process innovation and minor product improvements. A final point is that the debate so far has focused largely on the issue of small firms *or* large firms and generally has failed to recognise that the two will often be related. In other words, we should be concerned with the *dynamic complementarities* that can exist between small and large firms during the industry cycle. [. . .]

Discussion

It is clear that it is indeed necessary to consider the interactions between small and large firms if we are fully to understand the evolutionary dynamic (Rothwell and Zegveld, 1981b; Kaplinsky, 1982). In the evolution of the electronic computer industry, existing large corporations played the major initial role in invention, producing new devices largely for in-house use only. The major role in the initially rapid market diffusion of these new devices, however, was played by new, small but fast growing companies founded by technological entrepreneurs. Moreover, the technical know-how, the venture capital and the entrepreneurs themselves very often derived from the established corporations, as well as, in the case of the latter two, from major companies operating in other areas. Thus we see – at least in the US – a system of dynamic complementarity between the large and the small: both had their unique contribution to make; both were necessary, the former to the initiation of the new technological paradigm, the latter to rapid market diffusion and general commercial exploitation.

What our discussions do suggest is that established technology-based large corporations can be extremely effective in creating new technological possibilities; they are highly inventive. While they are adept at utilising the results of their inventiveness in-house (new technology for existing applications), they are less well adapted to the rapid exploitation of their inventions in new markets (new technology for new applications). It appears that new firms, initially, are better adapted to exploit new techno/market combinations; they are unconstrained by existing techno/market regimes within which established corporations, for historical cultural and institutional reasons, might be rather strongly bound. Referring back to Table 1, it appears that during the early phases in the evolution of a new industry the behavioural advantages of small scale are crucial; as the industry evolves, technological possibilities become better defined and market needs become increasingly well specified, the advantages of large scale begin to dominate. Comparative advantage shifts to the larger firms and the industry develops towards a mature oligopoly, a situation characteristic of the semiconductor and CAD industries today.

References

Kaplinsky, R. (1982) *The Impact of Technical Change on the International Division of Labour: the illustrative case of CAD,* London: Frances Pinter.

Rothwell, R. (1983) Firm size and innovation: a case of dynamic complementarity, *Journal of General Management,* Spring.

Rothwell, R. and Zegveld, W. (1981a) *Industrial Innovation and Public Policy,* London: Frances Pinter.

Rothwell, R. and Zegveld, W. (1981b) *Innovation and the Small and Medium-sized Firm*, London: Frances Pinter.

Townsend, J., Henwood, F., Thomas, G., Pavitt, K. and Wyatt, S. (1981) *Innovations in Britain since 1945*, Occasional Paper Series no. 16, Science Policy Research Unit, Sussex University, Brighton.

Wyatt, S. (1982) The role of small firms in innovative activity: some new evidence, *mimeo*, Science Policy Research Unit, Sussex University, Brighton.

3.8

Can Technological Projects be Controlled?

David Fishlock

Around the mid-1960s scientists at the (then) Ministry of Aviation began to chart in an ingenious new way the research costs of the TSR.2, a replacement for the Canberra bomber, by then costing about £4 million a month. What they learned was highly disturbing. The new 'cost slip' chart seemed to show that, after six years of research and development effort on TSR.2, there was still no end in sight to the escalating costs of this project. What in fact was happening was that the Service chiefs were trying, bit by bit, to turn a project for Canberra replacement into a fully-fledged supersonic bomber – a project the government had killed years before.

Early in the 1970s Dr Ieuan Maddock and his colleagues at the Department of Trade and Industry used the same technique to chart the costs of two civil projects whose research costs were becoming very worrisome: Concorde and the hovertrain. The shape of their charts followed closely the pattern of TSR.2 until it was axed in 1965. The hovertrain project was abandoned early in 1973. It had cost a total of £5 million – 2.5 times the original estimate for demonstrating the feasibility of such a transport system. And it was poised to leap into a far more costly phase, with no trace of a possible customer in sight.

As with TSR.2, the decision to stop the hovertrain project brought forth a torrent of public protest from those financially or emotionally attached to the project. The protesters even included a Parliamentary Select Committee which, with several years' experience of scrutinising technical projects in Britain, might reasonably have been expected to take a more detached view.[1] As I wrote when it published its report vehemently criticising the government's decision, this committee might have written a very different report had it spent less time listening to the grievances of a handful of scientists and engineers and paid more heed to the larger socio-economic questions involved. 'As it is, it has fallen right into the trap of supporting the view that technology should dictate the pace.'[2]

First Labour then Tory governments in Britain struggled hard to get off that particular hook. It was a line of thinking that led to a host of spectacular ventures – Concorde,

D. Fishlock, *The Business of Science: The Risks and Rewards of Research and Development*, Associated Business Programmes, 1975; extracts from Chapter 6.
David Fishlock is science editor of the *Financial Times*.

hovercraft, vertical take-off airliners, sea-bed tractors, novel electricity generating systems – of highly dubious commercial value. The adventurous engineer's instincts are always to see how far the technology will stretch.

The hovertrain project

Tracked Hovercraft was such a venture. Soon after the first hovercraft was unveiled in 1959, enthusiasts began to discuss a tracked version that might break right away from the 'steel-wheel-on-rail' systems. Speeds of several hundred m.p.h. were contemplated of hovertrains free from such restraints as adhesion and a rigid suspension. The original research on the hovertrain was done within a wholly-owned subsidiary of the National Research Development Corporation, called Hovercraft Development, but by the mid-1960s the research team was eager to launch a project on a scale large enough to demonstrate its claims. After some heart-searching, chiefly because no commercial interest showed the slightest sign of investing, the Labour government gave its approval in September 1967. A new wholly-owned NRDC subsidiary was formed, with Mr Tom Fellows as chief executive and initial funds of £2 million.

Figure 1   Full-scale Tracked Hovercraft Research Test Vehicle (RTV) 31

Fellows' brief was to explore on a large scale two inventions in which NRDC had proprietorial interest. One was Christopher Cockerell's air cushion suspension and the other was university research by Professor Eric Laithwaite of Imperial College, London, into linear induction methods of propulsion. A test track 3 miles in length was planned at a site near Cambridge, capable of extending to a twenty-mile track. Two further allocations of funds were approved by the government – albeit very reluctantly the second time – bringing the total to £5.25 million. For this sum the research team succeeded in running an unmanned test vehicle at speeds up to 107 m.p.h., and thus demonstrating that the skills were there to take the project much further [Figure 1].

By 1971, however, a new government was asking much tougher questions than before. Here was a project all set to leap into the tens of millions of pounds bracket yet arousing no more enthusiasm from potential backers at home or abroad than had been the case four years before. Other nations were showing great interest in the project – as was to be expected since they were pursuing similar projects – but none wished to join forces with Tracked Hovercraft. At no time during the life of the project – which included a spell of considerable publicity as it neared the end – did anyone offer to contribute significant sums to the work.

One important distinction between the hovertrain project and similar projects overseas was that all the others, although heavily government-supported, also involved private funds. For example, the three West German organisations in the field, Krauss-Maffei, Messerschmitt-Bolkow-Blohm and a consortium composed of Siemens, Brown Boveri and AEG, between them had chipped in about one-third of the £10 million or so spent by 1973 on such projects, with the government finding the rest. Locomotive makers in Britain such as Brush (Hawker Siddeley) and GEC showed no such enthusiasm to invest their own funds. One reason – the main one perhaps – was that British Rail showed no glimmer of interest in the hovertrain project for it could make no use of existing BR track. Past experience has taught British industry that untried systems which have found no market at home rarely find one abroad.

Another important distinction between the hovertrain and its rivals overseas was the inflexibility of the project. It was a project for a very high speed train, suitable only for long straight stretches of track, which would not adapt to the sinuous situations of urban transport requirements. Mr Dennis Lyons, director-general of research at the Department of Environment, told the Select Committee bluntly that Tracked Hovercraft was 'on too fixed a direction and directed to too definite a project'. It was a view from the main potential customer for such a project that the Select Committee chose to ignore completely.

The APT project

Mr Lyon's department is the one responsible for British Rail. In the mid-1960s British Rail had embarked on its own very ambitious project to develop a new and much faster train – yet one that would work with the existing track and signalling system. Track and signalling represent a capital investment of about £459 million (1972), an investment no one would abandon lightly. By exploring the limitations of the existing systems at a more fundamental level than ever before, British Rail's scientists at Derby produced a train concept tailored to the twists and bends of the track. For £5 million – the same sum as Tracked Hovercraft spent in developing a robot and one mile of new track – they had developed an experimental advanced passenger train (APT–E) replete with electronic controls and capable of running at 150 m.p.h. on existing routes [Figure 2]*. What is more, they believed that the new principles on which it was based could be stretched to greater speeds, perhaps 200 or 250 m.p.h. Such speeds might require some of the propulsion technology that was being developed for Tracked Hovercraft. But this possibility was scarcely justification – as the Select Committee seemed to be arguing – for also pursuing the alternative hovertrain project.

In a highly dubious attempt to justify their own project the hovertrain enthusiasts tried hard to cast doubt on the APT concept. The report quoted the 'distinguished academic witness' (unidentified) who assured the committee that at 150 m.p.h. British Rail would be

[* For full details see Potter, S. with Roy, R.(1986) Research & Development: British Rail's fast trains, T362 *Design and Innovation* (Block 3), Open University Press.]

Figure 2 The experimental Advanced Passenger Train (APT–E) demonstrates its high curving speeds during trials.
Source: British Rail.

'banging their heads against a limit'. The report argued that the government would be unwise to commit its future transport policy to a vehicle that had still to complete a technical proving programme. Yet that conclusion was reached at a time when the hovertrain was in so primitive a stage of development that not a single passenger had been able to travel upon it. The German, French, American and Japanese engineers who were pursuing very high-speed surface transport projects had all taken care to design test vehicles that could at least offer their patrons an exhilarating ride. Neither had British Rail, with its unique APT approach, neglected this important psychological point.

The Select Committee reached the conclusion that the government's decision to close down the hovertrain project was 'both premature and unwise'. Its case rested mainly on the fact that several other nations continued to put a major effort into high-speed surface transport.* Let us for the moment put to one side the argument – dubious though it is – that Britain should pursue a direction of research not because it is needed but because it is fashionable. Let us concentrate instead on one nation, West Germany, of similar size and ambitions to Britain, which is pursuing no less than three similar projects.

The German scene is worth studying more closely than the Select Committee – which took no evidence from the overseas projects – had bothered to do. First, government officials have told me frankly that they are very interested in British Rail's APT. This makes good sense, for Germany has a very highly developed system of existing railways, and plans to spend DM11,000 million (nearly £2,000 million) on their development by 1985, yet has no comparable project for extending performance. Second, the German programme which began in 1968 – after Tracked Hovercraft was formed – rapidly became convinced that one of the two innovations Britain was trying to exploit was a non-starter. The much-vaunted air-cushion suspension was – to quote Krauss-Maffei, which claims to

---

* I once asked a group of West German Science Ministry officials in Bonn what 'magic formula' they used to pick the project to back. The most senior man present said it was to look at what other countries like Britain were backing.

have built a test vehicle of this type more advanced than either Britain or France* – too noisy and too unstable. All three German projects have opted for magnetic levitation ('maglev') of various kinds. No more German money was to be spent on the air cushion.

Tracked Hovercraft's interest in 'maglev' had been confined to making paper studies once it became clear that the US, Japan and West Germany were all moving this way. Only after the axe fell in 1973 was any serious interest asserted in Eric Laithwaite's latest schemes for dual-purpose propulsion and levitation systems, using his so-called 'river of magnetism'. Bonn science officials had another point of importance to make: that they were supporting three industrial 'maglev' projects in the expectation that soon they would be able to choose the one of greatest promise and discontinue support for the other two.

One might wish it had come from a more dispassionate source, but Richard Marsh, chairman of British Rail, made a very pertinent comment to the Select Committee when he said it was 'one of the unique occasions when the Government has cancelled a project before it had wasted a great deal of public money'. Regrettably this remark probably accounts in part for the peevish tone of the report and the subtle efforts it makes to discredit British Rail's APT project – ironically on grounds that seem to argue simultaneously that it is too ambitious and not ambitious enough.

The 'eureka' factor

How on earth is the businessman, the civil servant, the Parliamentary Select Committee, to evaluate the advice – particularly the estimates of cost – he is given by scientists or engineers? To say: 'Pick a research manager you can trust' is no answer. How does a research manager evaluate the cries of 'eureka' from his own staff? [. . .]

If any one person can be held responsible for steering the Select Committee and, by his television appearances, a great many members of the general public, to a false conclusion about the hovertrain it is probably Professor Eric Laithwaite. Science has comparatively few really colourful fellows of high academic repute and the capacity to grip a lay audience with excitement at the ideas they propound. Laithwaite is certainly one of the few. A big, boisterous fellow with a broad north country accent and a penchant for the right biblical quotation, he takes great delight in surprising or astonishing an audience.

No modest man, Laithwaite lets it be know that he is the world's greatest authority on a type of electrical propulsion, little used yet but likely to find large markets in the future, known as the linear induction motor (LIM). Where most engines are designed to drive a shaft round and round, the LIM – like the jet engine – propels in a straight line. In Laithwaite's adroit hands it can be made to eject a shell from a cannon or to simultaneously levitate and propel a model train along a track. His audience is treated to an exhilarating verbal and visual display in which objectivity towards either his own ideas and inventions or those of his rivals plays a secondary part.

Of his work on the LIM, Laithwaite has been claiming – he was even quoted in the scientific press as claiming – 'we're six years ahead of everybody and we intend to stay there'.[3] Such a claim is probably unwarranted for *any* new idea or invention of recognisable commercial or military value today.

However, in a more reflective moment during the month his claim was published, Laithwaite admitted to me that the hovertrain project, to which he was consultant, promised him opportunities to try out his latest ideas for large-scale linear motor systems that he could not afford to explore in his laboratory at Imperial College, London, and which (as

---

* The French Aerotrain (hovertrain) project was abandoned in July 1974.

a part-time businessman himself) he would expect either GEC or Linear Motors, the companies with which he is associated, to finance themselves. With disarming frankness he admitted that between the four centres of interest in linear motors 'we had it made' – until the Government wielded its axe. Nevertheless, he stood by his claim of a six-year lead.

Some research managers claim to know their man well enough to be able to apply a factor – it might be called the 'optimistic factor' or the 'eureka factor' – to each individual in a position to put forth substantial requests for funds. This is the factor by which the boss mentally multiplies the sum requested before deciding whether to fund the project. For one man the 'eureka factor' might be say 1.2, for another it might be two, three or some still greater figure. [. . .]

Had the politicians and civil servants concerned with aerospace projects in Britain since the war had their 'eureka factor' to apply to proposals, Britain might have indulged in fewer costly failures. A retrospective appraisal has shown that British aerospace projects since the Second World War pretty consistently underestimated their costs by a factor of more than three. Mr R. Nicholson, programme director of Rolls-Royce's RB.211 aero-engine project at the time of the crash, when asked by the government's inspectors what went wrong with his project, admitted that he couldn't improve on the statement: 'We failed to appreciate it was going to cost us so much.' Confessed Nicholson to his inquisitors: 'We ought to have known, there is no question. We just have to look at the history of projects through the 1960s.'[4] The £65 million project estimate in 1968 finally turned out to be a bill for £195 million – a factor of three.

The factor of about three had, in fact, been suggested some years before. In a lecture on 'The development of inventions' in 1970 T. A. Coombs remarked that the ratio of actual time and cost to estimated time and cost of development programmes was not two, as had been believed during the 1950s, but π.[5] But even when experience had shown what could happen, he said, people maintained an outrageous optimism. 'I was recently told, with a use of words which would have delighted Sir Ernest Gowers, "We do not anticipate any difficulties." The unfortunate thing is that almost everything unexpected that happens to a development project is bad.' From this Mr Coombs concluded, however, that 'it is therefore just as well that one does not know too much otherwise one would never start anything new'.

Most persistent defect

The reluctance of the engineer to take costs into account was possibly his most persistent defect, Dr Ieuan Maddock, Chief Scientist at the Department of Industry, told engineers gathered in London in October 1973 during his presidential address to the Institution of Electronic and Radio Engineers. He found this reluctance to face the full significance of costs in nearly every project – 'a fair number by now' – with which he had come in contact. He confessed, moreover, that he himself had been guilty of the same defect when advancing projects of his own in the past.[6]

Maddock isolated a number of reasons why very few projects in advanced technology stayed within their cost forecasts. One was that the man with the idea failed to appreciate that for each unit spent on research and development, the project would require ten units to bring it to the market place – and might need as many as a hundred units to fully exploit a market (see Figure 3). This fourth phase of exploitation included expanded or modernised production plant, stockpiling of parts, creation of agencies, provision of leasing finance and extended credit – all factors that tended to be completely forgotten in the early stages of a project. Another reason for escalating costs was that the difficulties bound to

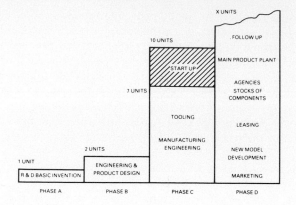

Figure 3   Four phases in the innovation process – each more expensive than its predecessor.
[*Source:* (7)   See also critical comment on these figures in article 3.1 by Freeman.]

arise as projects are expanded from the conceptual to the practical scale are consistently underestimated. The error is particularly great when disciplines outside the person's own knowledge and experience are required, as is increasingly the case with high technology projects today. These other disciplines the engineer tended to dismiss as 'mere engineering detail', said Maddock. A third reason for escalating costs was deliberate underestimating by the engineer, who even argued sometimes that if their estimate were to include realistic margins for contingency the project would never be approved at all. Such an argument ignores the annoyance subsequent revelations are likely to cause the sponsor. [. . .] Maddock described the Concorde project as 'one of the most publicised and best documented cases of cost escalation', not because it was special but because it was typical of so many projects in advanced technologies.

References

1. House of Commons (1973) *Tracked Hovercraft Ltd, Third Report of the Select Committee on Science and Technology,* HMSO.
2. D. Fishlock, (1973) Why the hovertrain had to go, *Financial Times,* 7 September.
3. E. Laithwaite, (1973) *Physics Bulletin,* September.
4. Department of Trade and Industry (1973) *Rolls Royce Ltd, Report by R. A. MacCrindle and P. Godfrey,* HMSO.
5. T. A. Coombs, (1970) The development of inventions, *Lecture to Institution of Mechanical Engineers,* 11 November.
6. D. Fishlock, (1973) How to read the warning signals, *Financial Times,* 6 November.
7. United States Department of Commerce (1967) *Technological Innovation: its Environment and Management,* US Government Printing Office.

Section 4

Consumers and Users

Introduction: Design, Marketing and the Diffusion of Innovations

Robin Roy

This section is divided into two parts. The first is concerned with the factors that stimulate the creation of new products and technical innovations that enter the market place; the second with the factors influencing how such products and innovations spread through the market into widespread use.

Design, marketing and the consumer

As Christopher Freeman has already pointed out in article 1.3, one of the longest running issues in innovation theory is whether successful innovations originate as a result of 'technology push' or of 'market pull' (Freeman uses the terms 'science push' and 'demand pull'). In other words, do successful innovators create a new product, device or system, and then try to sell it? Or do they first try to discover what prospective customers need or want, and then develop an innovation to meet those demands?

Many empirical studies of new products and technical innovations have tended towards the conclusion that 'market pull' innovation is a more certain route to commercial success than 'technology push'. Various studies have suggested that between two-thirds to three-quarters of all successful technical innovations were stimulated by market demand (see Table 1).

James Pilditch in article 4.1, 'How Britain can compete through marketing', argues strongly that the recipe for success is 'quite simply summed up in the six words – find a need, *then* fill it'. He points to the tendency in Britain of concentrating far too much talent and effort on inventing and developing 'technically brilliant failures' like Concorde, at the expense of designing more mundane products to satisfy market and user requirements. He gives examples of market-oriented products ranging from the Boeing 747 'jumbo jet' to a redesigned battery lamp. This view is echoed by Rothwell, Schott and Gardiner (1985) in their booklet *Design and the Economy* (pp. 19–22).

Several of the articles chosen for this sub-section, however, point to the weakness of relying on a simplistic 'market-pull' approach to design and innovation. In article 4.3, Bennett and Cooper argue that the wholesale adoption by American companies of the market-

Table 1 A comparison of studies of the proportions of innovations stimulated by market needs and technological opportunities.

Author	Proportion from market, mission, or production needs (%)	Proportion from technical opportunities (%)	Sample size
Baker *et al.* (1967)	77	23	303*
Carter and Williams (1957)	73	27	137
Langrish *et al.* (1972)	66	34	84
Myers and Marquis (1969)	78	22	439

*ideas for new products and processes.   *Source:* Adapted from Utterback (1974).

ing approach has resulted in a preoccupation with minor incremental changes to existing products, restyling exercises and trivial innovations (from vaginal deodorants to the 'pet rock'). Breakthrough innovations, like the telephone, electric lighting, lasers and instant photography, they say, must be the result of 'technology push'. First, the market cannot signal demand for really novel products and processes. Secondly, as consumers and society often resist major innovations when they first appear, it is almost impossible for market researchers to distinguish between distrust of the new and a genuine lack of need or demand.

In article 4.2, Christopher Lorenz suggests two approaches to the problem of assessing the potential market for major innovations and radical new design concepts. The first is to rely on the 'gut feel' of far-sighted individuals like Sir Clive Sinclair, who sensed the potential demand for a low-cost home computer, or Sony's chairman Akio Morita, who likewise believed, contrary to the market-research evidence, that there would be a demand for the 'Walkman' personal stereo. The second approach is to engage in a type of market research different from the traditional quantitative survey methods – namely, 'qualitative' research or 'creative marketing' (Alexander, 1985). Creative marketing is based partly on attempts to anticipate long-term trends in life-styles and technology, and partly on open-ended discussions between users, designers and marketing staff to identify problems with existing products and to isolate emerging user wants. A combination of these approaches has enabled Japanese companies to be highly successful in developing products, like the home video-recorder, which have created new markets that British and American companies completely failed to anticipate (see also article 3.2 by Evans). Nevertheless, radical innovation remains a risky business, and it is easy to find examples of novel products that failed when first launched onto the market (e.g. the Bell videotelephone; British Telecom's Prestel videotex service (Bruce, 1986); quadrophonic hi-fi; 'Corfam' synthetic leather; and the Wankel rotary engine).

Pannenborg (a board member of the multi-national company Philips) shows in article 4.4 that the relative importance of 'technology push' and 'market pull' depends on the type of customer and the nature of the product. He distinguishes between 'professional users' and 'non-professional users', either of whom may be 'experts' in the product. For instance, an expert professional user might be a research laboratory purchasing scientific instruments or an airline purchasing aircraft. In such cases, it makes sense for the innovator to *involve* the user in the specification of the new product, and perhaps even in its design and development. Von Hippel (1978) has shown that a high proportion of scientific instrument innovations were originated by their *users,* not their manufacturers. This point is also made by Rothwell, Schott and Gardiner (1985, pp. 23–5) in *Design and the Economy.*

In common with several of the authors in this section, Pannenborg concludes that successful innovation combines technology push with market pull. This is in line with Freeman's conclusions in article 1.3:

> Whilst there are instances in which one or the other may appear to predominate . . . any satisfactory theory must take into account *both* elements . . . the crucial contribution of the entrepreneur is to *link* the novel ideas and the market . . . The vast majority of innovations . . . involve some imaginative combination of new technical possibilities and market possibilities.

Diffusion of innovations and new product marketing

Regardless of whether a particular new product or technical innovation had its origins in a new technological possibility or a market demand, it is not of much commercial, economic or social relevance until it has been adopted by a significant proportion of its potential users – that is, until it has undergone *diffusion*.

There is an interesting polarity of views regarding innovation diffusion. On the one hand, we find writers like Alvin Toffler (1970) warning us of 'futureshock', due to over-rapid diffusion of new technologies. On the other, Donald Schon (1971) argues that contemporary organizations and social systems are 'dynamically conservative' – that is, they 'fight like mad to stay the same' when faced with new technologies.

In article 4.5, taken from the classic text on the subject of diffusion, Rogers agrees that getting an innovation widely adopted is often very difficult and in many cases proves impossible, even if the innovation has major advantages (he gives the example of an improved typewriter keyboard). In article 4.6, Freeman also notes that diffusion is often slow and usually much more difficult than inventors and designers expect. Nevertheless certain innovations, like microelectronic products and processes, have diffused more rapidly than many ever envisaged.

Diffusion theory is concerned with the factors that influence the rate and extent that individuals and organizations adopt innovations. An innovation in this context need not objectively be 'new' (in the sense of a patented invention), but only needs to be perceived as new by the adopter. Nevertheless, Freeman points out that the factors involved may differ, depending on whether the innovation is totally novel or merely new to the adopter, and on whether the adopter is a private individual or an organization. For instance, a firm deciding whether to adopt a novel manufacturing process will tend to use different criteria to a private consumer deciding whether to buy a new model of car.

Rogers and Freeman adopt somewhat different approaches, although both are really concerned with the diffusion of significant innovations, rather than with the marketing of new products. Rogers sets out the basic tenets of diffusion theory from a mainly sociological perspective, whereas Freeman is concerned with wider socio-economic, technical and political factors.

Rogers establishes that the rate and extent of adoption of an innovation depends on two crucial factors:

1. The characteristics of the *innovation* itself.
2. The characteristics of the potential *adopters* of the innovation.

Figure 1 shows how the total sales of an innovation are typically distributed between different categories of adopter.

Freeman brings in the wider factors that influence diffusion, such as an innovation's environmental acceptability, the scale of investment required in order to adopt the innova-

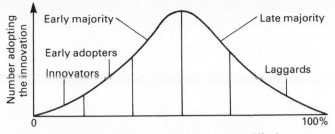

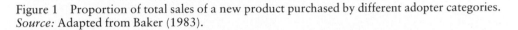

Figure 1   Proportion of total sales of a new product purchased by different adopter categories. *Source:* Adapted from Baker (1983).

tion, and the legal and regulatory constraints and stimuli surrounding it. Freeman uses these factors to discuss the specific case of the diffusion of microelectronic technology. Microelectronics is a particularly interesting technology because its diffusion involves individuals adopting novel consumer goods such as home computers, and organizations adopting capital goods such as process control equipment and automatic machinery. Freeman shows that microelectronics has diffused very rapidly compared to other techni-cal innovations, such as nuclear power and supersonic air travel, because the commercial, political and social factors have been especially (and perhaps uniquely) favourable.

A key area which Rogers and Freeman only comment upon very briefly is the relevance of diffusion theory to the sales, marketing and advertising of new products. Baker (1983) has used Rogers' concepts to produce guidelines relevant to those concerned with sales and marketing. Baker notes that new products are often best targeted initially towards the 'in-novators' and 'early adopters' in the population of potential customers (Figure 1). These are the individuals and organizations who are most receptive to new ideas and who tend to share certain characteristics, for example, they are generally wealthier, younger, of higher status, more willing to take risks, and better informed than the other categories of adopter. Innovators and early adopters also tend to rely more on *personal communica-tions* for information (Rogers calls these 'inter-personal channels'). Personal communica-tion involves direct person-to-person contact: this may be 'seller-initiated', as when a sales representative calls, or 'buyer-initiated', as when someone asks a friend for an opinion of a particular product. By contrast, *non-personal communication* (Rogers calls this 'mass media channels') involves one of the media – print, television, etc. – and is the major pro-vince of advertising.

Rogers notes that adoption of an innovation normally follows five stages: knowledge; persuasion; decision; implementation; confirmation. Baker (1983) observes:

> Non-personal media are usually most effective in establishing awareness and interest, while personal influence is necessary to move the members of an audience up the hierar-chy of effects through desire to action . . .
>   However, while the emphasis may be on one or other sort of communication, it is usual to find both employed, together . . . impersonal channels are almost invariably affected by the mediation of personal sources. This is usually referred to as 'the two-step flow of communication'. According to the two-step model, mass media communica-tions are picked up first by only a small proportion of the intended audience, who, in turn, pass on the message to other members of the audience. These intermediaries are designated *opinion leaders.*

Despite the difficulty in identifying and persuading opinion leaders to adopt an innova-tion, Baker believes their role to be crucial to diffusion. Rogers confirms this view when he observes the importance of the network of social contacts in diffusion: 'At the heart of the diffusion process is the modelling and imitation by potential adopters of their network partners who have adopted previously'.

References

Alexander, M. (1985) Creative marketing and innovative consumer product design, *Design Studies*, vol. 6, no. 1, January, pp. 41–50.
Baker, M. J. (1983) *Market Development: A Comprehensive Survey*, Harmondsworth: Penguin.
Baker, N. R. *et al.* (1967) *IEE Transactions on Engineering Management*, EM–14, 156, December.
Bruce, M. (1986) Marketing: British Telecom's Prestel, T362 *Design and Innovation* (Block 2), Open University Press.
Carter, C. F. and Williams, B. R. (1957) *Industry and Technical Progress*, Oxford, Oxford University Press.
Langrish, J. *et al.* (1972) *Wealth from Knowledge: Studies of Innovation in Industry*, London: Macmillan.
Myers, S. and Marquis, D. G. (1969) *Successful Industrial Innovations*, NSF69–17, Washington, DC: National Science Foundation.
Rothwell, R., Schott, K. and Gardiner, J. P. (1985) *Design and the Economy: The Role of Design and Innovation in the Prosperity of Industrial Companies*, 3rd edn, London: Design Council.
Schon, D. A. (1971) *Beyond the Stable State: Public and Private Learning in a Changing Society*, London: Temple Smith.
Toffler, A. (1970) *Futureshock*, London: Bodley Head.
Utterback, J. M. (1974) Innovation in industry and the diffusion of technology, *Science*, vol. 183, 15 February, pp. 658–662.
Von Hippel, E. (1978) Users as innovators, *Technology Review*, vol. 80, no. 3, January, pp. 30–4.

DESIGN, MARKETING AND
THE CONSUMER

4.1

How Britain Can Compete through Marketing

James Pilditch

[. . .] Total imports of finished manufactured goods cost us £8853m in 1976 – more than the National Health Service (£6169m), and more than all education (£7000m). As a matter of fact, these imports – largely of goods we ought to be making ourselves – cost every household in Britain the equivalent of more than £9 a week. What is even more disturbing, as Figure 1 shows, is that the trend is worsening.

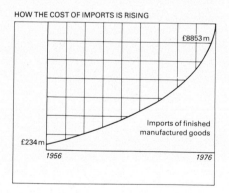

Figure 1   How the cost of imports is rising.

The priority for action, therefore, is clear. UK industry must provide products people in this country *want* to buy and *can* buy in preference to any other. (They want to buy Land Rovers, but cannot, so production is a factor, too.) And what we can profitably do is to examine those firms which, among the general decline, still manage to grow inexorably, year after year. What is it they have, these great growth companies? Singing the company song every morning, as they do in Japan, may help, but we can be sure of

J. Pilditch, How Britain can compete, *Marketing*, December 1978; extracts from pp. 34–8.
James Pilditch is Chairman of AIDCOM International and of the Business and Technician Council's Design Board.

one thing: wherever they are – in Japan, Germany, the US, and in Britain too – growth companies possesses no magic ingredients.

Find a need, then fill it

What they do is quite simply summed up in the six words 'Find a need, then fill it'. Everyone knows that is what is meant by the marketing concept. But those six words serve in a number of ways.

1. They state the need in a nutshell.
2. They get the priorities right (find the need first, *then* fill it).
3. They show how to focus our business (outward not inward).
4. They describe the key difference between companies that prosper and those that do not.
5. Finally, they show why most R & D expenditure is wasted.

Those six words sum up the difference between *invention* (at which we excel) and *innovation* (at which we need to). The distinction is important because the figures suggest that Britain is not being inventive enough. If we were more inventive, after all, wouldn't we be more successful in our home market? That is what a lot of people think. Michael Fores at the Department of Industry has been studying this. He says 'part of our manufacturing malaise is due to the propagation by scientists and economists of the false idea that innovation in manufacture arises from the application of new scientific knowledge.'

Are invention and innovation the same? And, does it matter? It was considered important enough fifteen years ago for the US Department of Commerce to set up a working party to study the subject. It found that companies that had 'committed them- selves to innovation as a way of life', grew about seven times as fast as the national aver- age – and did so over a twenty-year period. The definition of innovation the working party came to is quite long. Paraphrased, it is 'the *whole* process by which new ideas or processes are translated into the economy'.

If invention were the same as innovation Britain should be well off, that is if we accept Nobel prizes as a sign of inventiveness. For a population a quarter the size of the US, the British achievement in science is remarkable. As Figure 2 shows, since 1901 US scientists have won 65 Nobel prizes, British scientists have won 48, Germans 42, French 15, Swiss 10, Netherlands 7, and the Japanese have won only three.

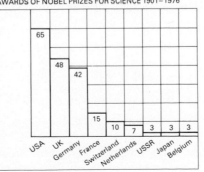

Figure 2   Awards of Nobel prizes for science 1901–76.
Source: Encyclopedia Britannica.

Brilliant failure

Is this what we need to create the market-winning new products on which our future depends? Well, Concorde was a good invention, and Boeing 747 was perhaps less brilliant and less inventive, but commercially the position is reversed. Only nine Concordes have been sold (with five more on option), whereas 406 Boeing 747s have been sold. One has earned about £270 million; the other at least £8120 million. One moved a new idea successfully into the economy, the other has not. One satisfied the six-word formula, the other did not.

The truth is that while Britain wins Nobel prizes it loses markets. So, if we say we urgently need more innovation we must ensure that we mean more of the *whole* process, not just more R & D. There are two reasons for this.

First, the academic aura that surrounds so much R & D actually reduces the chance of innovation. As Lord Wilfred Brown wrote recently, 'if the university tradition' (which he described as 'the untrammelled search for knowledge') 'holds sway in the minds of those accountable for a company's product design the effect can be disastrous.' R & D, like anything else, needs to be given practical goals and must produce measurable results.

Second, too great a reliance on R & D, or whoever the inventors are, makes us fail to recognize, respect or harness other skills at least as important to successful innovation. Indeed, of the many disciplines necessary in 'the *whole* process' I put marketing and design highest. It is a marriage of the two, of rational analysis and practical imagination, that leads to dynamic new products.

There are many reasons why design is this important, but four are paramount:

1. People buy things; not reports, not analyses, not financial plans. Everything must be designed.
2. Though we must analyse markets far more than most of us presently do, others can analyse too. The more rationally everyone analyses a market the more similar their response and therefore their products will become. Design is a great way to create or show the vital difference between one product and another.
3. Finding a need is not easy in rich, saturated markets, and increasingly the need is an emotional one – a fact that applies as much to industrial equipment as it does to consumer goods. Designers act as the interpreters of manufacturing into things that please people and satisfy their needs and wants, emotional as well as rational.
4. Good design is the way to turn the marvels of incomprehensible modern technology into simple objects people feel safe with, can use and like and want to possess.

By design, I should say, I do not mean styling. My definition is: finding the right problem, defining it simply, then creating an effective and, ideally, an elegant solution (elegant, that is, in the mathematical sense of economy and harmony). Finding the right problem: that is the key, and the link to marketing. An example shows how.

A solid square object

Ever-Ready, now called Berec, used to make a lamp for motorists. It was a solid, square object designed originally to take a big square battery (Figure 3).

The designers could certainly have restyled that, by changing the colour, altering the handle and so on. But the real problem was different. The big, square battery. It was quite

Figure 3   Old (top) and new versions of a motorists' battery lamp made by Berec (formerly Ever-Ready). It was redesigned by Pilditch's consultancy firm, AIDCOM, and shows the importance of market-oriented design.

likely that the would-be lamp owner could go to several shops before finding one, particularly outside the UK. And, to tell the truth, the batteries had often been in stock so long they had lost some of their power.

A second problem was that because the lamp was so bulky it was difficult to keep. If it was kept in the boot it bounced around and it was not surprising if connections worked loose and failed. And finally foreign competitive products were cheaper.

What the designers did was first to redesign the lamp to take ordinary batteries available everywhere. Second, at the expense of the swivelling beam, they made the lamp flat, so that now it can be kept in a glove compartment. And as an added attraction the handle has been made to move to form variable legs. Third, they simplified the manufacturing process to attack costs. The old lamp had 72 parts and 49 assembly operations. The new lamp has half the number of parts (36) and only 28 assembly operations. It costs 25% less to make and even with an improved profit margin it meets the price of comparable imports.

In the new lamp's first year domestic sales have more than doubled and, according to the company, are definitely eating into imports, which shows that even in an export-dominated market it is possible to hit back. Ever-Ready is beating the opposition because it has given people what they want. The interesting thing is that the company knew that before it made the lamp. It had analysed the market, and it had the product redesigned to meet a clear brief. Then it used models to test the new proposition, and it even value-engineered the product before tooling. So when the time came for investment Ever-Ready knew it was on the right lines. Sadly, this process is unusual among engineering firms, whether of industrial goods or consumer products. Generally they expect technical staff to develop new products which manufacturing makes and salesmen sell. (See Figure 4).

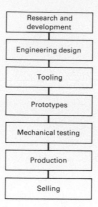

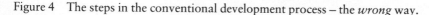

Figure 4   The steps in the conventional development process – the *wrong* way.

The right development method

Figure 5 demonstrates the right way to develop new products, and it can be seen from the left-hand side of *Stage 1* that the starting point is the gathering of market data. This involves the usual forms of market research, but it is worth stressing the particular value of discussions with market experts and opinion formers who can often make clear things that it might otherwise take months to discover. An example occurred last year when a company was considering diversifying into the DIY market. Eight editors of DIY magazines were invited to lunch. Among questions they were asked was one about letters they received from readers. Each was asked to write the ten most common problems these revealed. Not only were their lists identical, they were (with trivial exceptions) in the same order of priority.

It should go without saying that a close examination of the competition must be carried out, but sufficient attention is rarely given to this. In a one-day seminar in London called 'Meeting the competitive challenge', McKinsey described the lengths Japanese companies go to to understand how their competitors' products are made, how much each component costs, and from this what must be done to beat them. It was argued: 'In saturated markets when competitors fight for the same customers, the competition sets standards – of performance, price and delivery.' Few companies study their competitors in such depth.

It is important to assess strengths and weaknesses, and it is surprising how often companies undervalue their strengths or are hesitant to identify shortcomings. If a company is to build on strengths it must know what they are.

From an assessment of the market and of the company's capacity it is possible to write a list of screening criteria by which any new idea should be judged. Each screen is written to suit the particular circumstance. As an example, the one used in the DIY case cited above specified that each product concept should:

– Be distinctive.
– Exploit the company's main strengths.
– Have a compatible value-added (e.g. cost of goods 26, overheads 6, distribution 2, selling 6, marketing 15, profit 10).
– Provide the company with a 65% share of the retail selling price and the trade with a 35% share.

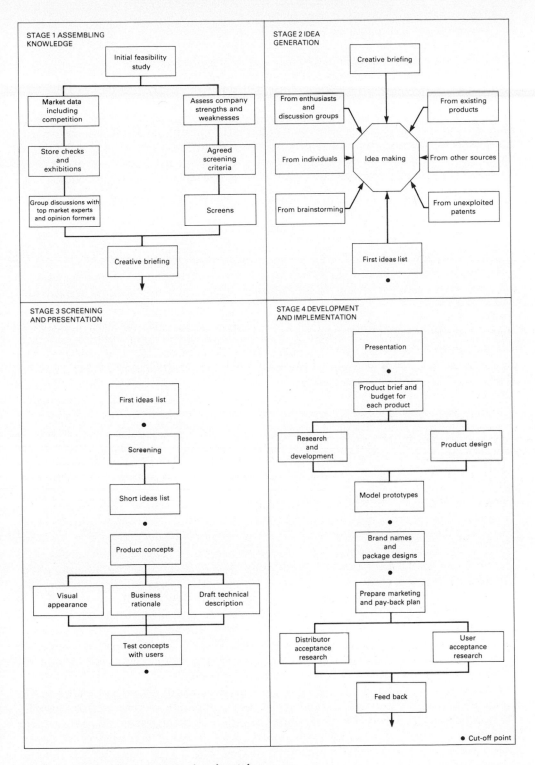

Figure 5   The marketing approach – the *right* way.

- Have a compatible life cycle (e.g. not be a fashion item).
- Be simple (no technology).
- Be compact in size.
- Have a minimum turnover of $1m per annum per product.
- Fit existing warehousing/physical distribution facilities.
- Have a ratio of turnover to fixed capital of 3 : 1.
- Not be 'cash hungry'.

Stage 2 is concerned with idea generation and it is important that this should be separate from analysis, because the mental process is different. Ideas are encouraged from all sources, by searching, from enthusiasts and from creative people, either alone or brought together for brainstorming sessions. Typically over 100 ideas would be generated.

Cut-off points are vital

Notice the dots throughout the programme. They are cut-off points, which are vital for three reasons. First, to ensure sustained executive control; second, to avoid dissipation of time and money on ideas unlikely to prove fruitful; third, to ensure the focus remains only on the best ideas.

Ideas in the first list when approved pass through the marketing screen, as shown in *Stage 3*. This screen winkles out poor ideas, ideas which may be good, but for which there is too small a market, others which would take too long to develop, and still others that do not fit the company's strengths. Typically perhaps 15 to 30 ideas pass this screen, but at this stage they remain words on paper.

Next, while marketing people define a business rationale for each idea, designers (using both their imagination and practical skills) turn them into three-dimensional mock-ups or models. If appropriate, packaging is designed, the product is named. These designs or packages or names need not be definitive at this stage. The point is to turn words on paper into tangible objects quickly. (A criticism of many development programmes is that they remain too abstract too long.) Such mock-ups make it easier for management to judge the value of ideas.

It may not always be possible, but if one can derive a reliable guide from consumers at this stage a great deal will have been achieved. An example from Switzerland illustrates the point. Following the process described so far, some twelve product concepts were tested among defined consumer groups. The client company and market research agency carefully devised a scale rising from one to eight. For example, over six points was thought to demonstrate positive consumer acceptance. Seven of the concepts appeared broadly acceptable to consumers, and of these, three were outstandingly popular.

This sort of testing means that in, say, six months (the time from start) with no R & D and tooling costs, the company knows that any further investment will be in the right place. It is not until the next stage, *Stage 4,* that R & D comes in. Of course, this sequence – of finding out how to make something after consumers have said they want it – is not universally possible; but it is certainly better than its opposite.

Of the other actions shown in the figure, it is perhaps worth drawing attention to the requirement to carry out distributor and acceptance research. In a recent test, for example, supermarket operators said they liked new products put to them, but would not buy them because the packaging was not sufficiently pilfer-proof. It would have been a costly error to launch the products in question without that knowledge.

The process I have outlined is the opposite of the usual approach and is not only much faster and cheaper, because heavy investment is not made until a need for the product has been determined. More important, the eventual product is more likely to produce customer-satisfying results. But it does mean managing skills in new ways and bringing in other often undervalued skills. Britain does not lack the skills or power to win markets but it uses them in the wrong way to solve the wrong problems.

4.2

'Gut Feel' is Market Research, Too

Christopher Lorenz

If Clive Sinclair and Akio Morita are to be believed, market research is the enemy of successful design and innovation. Sinclair says he did no market research whatever before ordering 100,000 sets of parts for his first line of ZX personal computers: he simply had a hunch that the time had come when the public would buy really cheap, simple computers. The ZX series has gone on to become the world's top seller [Figure 1].

At almost every executive level within the remarkably innovative Sony organisation, the story is the same. 'Whenever we come out with a new product, people say it won't sell,' says chairman Morita. 'I don't believe in market research – it doesn't help us develop new products,' declares his design chief, Yasuo Kuroki, pointing with delight to the thumbs-down which early market research gave to the Walkman cassette player. As the world now knows, the Walkman went on to become a phenomenal success.

The point is made even more forcibly by Mitsuru Inaba, who as head of Sony's New Jersey design centre is Kuroki's eyes and ears in the USA: 'Designing American-style market research into a product won't make it competitive – you're only getting today's and yesterday's information, nothing about the future.'

Sony certainly has a successful history of getting places by ignoring the results of market research (as most of us know it). When the company launched its first small-screen monochrome TV at the all-important Chicago trade show in 1960, US General Electric – then one of the giants of US consumer electronics – had just concluded from a series of research studies that there was no market for such a product. Yet Sony's eight-inch set was an instant success, thanks to a highly imaginative advertising campaign and Morita's decision to by-pass the sceptical trade by selling direct to department stores and other major retailers. Where Sony led, the Japanese followed. Within two years, Japanese TV imports to the USA had soared from zero to 120,000; in 1965 they hit a million, and by the mid-seventies they had all but devastated the domestic industry.

With such forceful evidence from two of the world's most innovative companies, and

Extracts from C. Lorenz (ed.), Market research: a fear of feedback? *Design*, December 1983; pp. 31–41.
Christopher Lorenz is management editor of the *Financial Times*.

Figure 1   In 1979 when Sinclair Research was developing the first computer in its ZX range it didn't commission specialists to test the market. It would have been little use asking people if they could find a use for a personal computer. You'd be lucky to find anyone who could even visualise the product, never mind use it. So, solely on the basis of the market awareness of its entrepreneurial founder, the firm launched the ZX80 and went on quickly to sell more computers than anyone else in the world with the ZX81 and the ZX Spectrum.

with plenty of support from leading designers on both sides of the Atlantic, the case against market research seems proven.

But wait. Why then does Nick Butler, whose BIB consultancy has probably done more work for major Japanese companies than any other top British design group, report that Minolta, Daihatsu, National Panasonic (Matsushita) and his other clients all rely intensely on market research? Minolta, which with Canon has swamped the European camera industry, has a 'phenomenally large market research operation for a company of its size', says Butler. His point is borne out by Britain's National Economic Development Office, which has repeatedly pointed to the reliance placed by Japanese industry on market research. And James Pilditch, who has reinforced his original AID design consultancy with extensive market research offshoots, is emphatic that 'you must find out what the customer wants before you make it. Supposing people don't like your Concorde, your Allegro, your calculator, cooker or kettle?' [see article 4.1]. In essence, the answer revolves around your reply to the question: 'What *sort* of market research, and for what particular purpose?'

The problem with General Electric's research in 1960 – and with many studies even now – is that they simply measure what the consumer knows he or she wants. This may be fine when a company is deciding whether to change the colour of an existing product, to improve its reliability, or to add a new feature. But when the firm is dealing with an entirely new product concept, it is a completely inappropriate form of research, except insofar as it can sometimes help one spot a gap in the market. But such cases are no longer all that frequent.

What Sony did instead was to observe the increasing penetration of TV sets into the American lounge, notice the growing proliferation of television channels, and marry this with its own technological developments. Its small-screen TV was neither market-driven nor technology-driven, it was both. In effect, the company looked beyond consumers' expressed needs and led the market by the careful stimulation of a new 'want'. It has since done precisely the same with the video-cassette recorder, the Walkman, and now the 'Watchman' flat-screen TV [Figure 2].

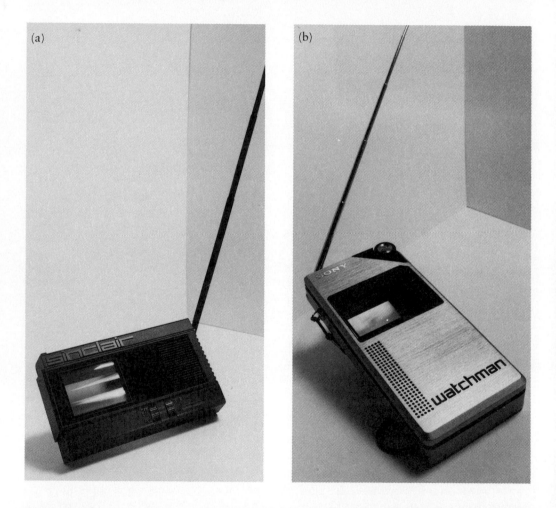

Figures 2a, 2b   Sony's latest, smallest TV, the flat-screen Watchman (2b), is proving to be a much bigger success in the market than research suggested it would. It has recently been re-engineered for higher volume production, smaller size, and lower price. The sales figures of the Watchman are all the research statistics Sinclair Research needs to know that its newly launched, smaller and cheaper flat-screen TV (2a) has a huge market. Market-aware companies can innovate without hiring research consultants.

In other words, whether through formal research techniques or the informal observation and intuition of the likes of Morita and Kuroki, Sony actually does make heavy use of market research. Not the conventional kind, but what is known among the marketing fraternity as 'qualitative research': an amalgam of fact-finding and social forecasting. Having mounted a small test marketing exercise in Japan, it rapidly gears up production and rushes the product out onto the world market.

But test marketing is a luxury that highly capital-intensive industries cannot afford. Which is why some companies take a far greater risk than Sony when they revolutionise their use of market research in the quest for more innovative products.

Just such a transformation lay behind Ford of Europe's development of the Cargo truck and the controversial Sierra car. For decades, in common with hallowed Detroit practice, Ford had placed extremely heavy reliance on the results of product 'clinics' – sessions in which carefully disguised competitive vehicles, plus several existing and potential Ford models, are compared and assessed by samples of dealers and consumers.

Participants were asked to rate all sorts of details about the various models, right down to bumpers, rear lights, armrests, and so on. Not surprisingly, the guinea pigs tended to judge everything against their current vehicles. The result? This classic market research exercise, much aped by other industries, tended to produce support for only slight design improvements, and to stiffen resistance to dramatic change.

The new approach is for the questions in the clinic to be 'much more about people's perceptions of a vehicle five years hence,' says Uwe Bahnsen, Ford of Europe's vice-president for design. By various means, 'we try to get them to forget about what they've just been driving'.

Most important of all, the results of Ford's clinics are no longer seen by its marketing men as the holy grail. To coin a phrase which is being used more and more by designers and enlightened general managers in all sorts of companies, 'market research should never be allowed to become a decision substitute. The best it can do is to provide decision support.' In the final analysis, the key factor must be adventurous managerial judgement.

4.3

Beyond the Marketing Concept

Roger C. Bennett and Robert G. Cooper

The marketing concept was pioneered by the General Electric Company after World War II, and Theodore Levitt's famous 1960 article, 'Marketing Myopia', cemented its acceptance as a corporate philosophy for business planning.[1] In North America, the marketing concept has been a way of life for larger consumer goods firms, at least since the early sixties.

Thirty years of experience provide ample evidence to subject the maturing concept to a rigorous critique. The marketing concept, with its simple and intuitive appeal, has been heralded by practitioners and academics alike as the most advanced philosophy of business. For decades, authors have extolled its virtues. But can any business philosophy be so perfect? Can it really be a case of all roses and no thorns? Surely not! Perhaps it is time to take a second look, with a particular focus on the negative aspects of this business philosophy. [. . .]

Product Innovation

New products are the life blood and hope for the future of many firms. And it is probably in the new product arena that a religious adherence to the marketing concept has its most detrimental effects.

Market Pull vs Technology Push

The marketing concept suggests that in their new product efforts, firms must be customer oriented, and that new products should be conceived and introduced to meet the needs and wants of buyers. In practice, this means that buyers' needs and wants should be identified, qualified, and quantified as part of product idea conception and prior to product development. And marketing research is often cited as the tool for need identification. In short, a 'market-pull' approach to product innovation is the logical outcome of implementing the marketing concept in product development.

But the evidence over the years suggests otherwise. In fact, many of the great product

R. C. Bennett and R. G. Cooper, Beyond the marketing concept, *Business Horizons,* June 1979, extracts from pp. 76–83.
Robert C. Bennett is Assistant Professor of Management; Roger G. Cooper is Associate Dean of the Management Faculty at McGill University, Montreal.

innovations throughout history have been the result of a technological breakthrough, a laboratory discovery, or an invention, with only a vague notion of a market need in mind. Often these 'great ideas' originated from men and women far removed from customers, from market needs and wants, and from the industry itself. Inventors, scientists, engineers, and academics, in the normal pursuit of scientific knowledge, gave the world the telephone, the phonograph, the electric light, and in more recent times, the laser, xerography, instant photography, and the transistor. In contrast, worshippers of the marketing concept have bestowed upon mankind such products as new-fangled potato chips, feminine hygiene deodorant, and the pet rock. Extreme cases perhaps, but ones which were deliberately chosen to make a point. True product innovation depends to a large extent on scientific discovery, which often must proceed in the absence of a clear and definable customer need or want.

There should always be a place for pure or scientific research unshackled by the immediate demands of the marketplace. Reflect for a moment on what might have occurred had creative inventors, such as Edison or Bell, been forced into a rigid market-oriented approach. 'Technology push' is the antithesis of 'market pull' in product innovation. Yet history shows us that the major breakthroughs, the truly innovative products, are often the result of 'technology push'.

An executive at a large electrical company summed up the problem: 'We used to produce many major innovations. But since the scientists have become marketing oriented, the major breakthroughs have not occurred here. We spent a lot of effort explaining to them that their ideas should meet market needs, but the main effect seems to have been that they have left the forefront of technology.'

Defining Needs

The logical rebuttal is that all this is history, that tomorrow's innovations will be the result of a strong market orientation. But here, too, it can be shown that the marketing concept, backed by marketing research, effectively discourages the development of product innovations. Marketing research is cited as the answer to the operational problem of defining market needs and wants for generating new product ideas, but market needs and wants are invariably described by users in terms of the familiar – existing products and existing product classes. Ask the customer what his needs are for urban transportation, and he will give you a list of suggested improvements to his bus or subway service, or to his commuter car. As one senior executive in a multinational agricultural equipment firm noted, 'Whenever we do market research to generate new product ideas, all we get back are descriptions of minor improvements, or worse yet, our competitors' features. But the bold new ideas come from in house.' The inability of the typical buyer to raise himself above the level of the familiar means that any end-user market research is likely to identify 'new' needs and wants in a very limited perspective. The end result is a preoccupation with 'me, too' products and minor modifications, while true innovative efforts take a back seat.

Gauging Product Acceptance

Any market-oriented, new product program will almost certainly result in killing off any genuinely new products. Implementation of the marketing concept leads logically to customer tests of the product concept at some point during the development process. In the case of product modifications, such research results prove invaluable. Because the potential user is familiar with the product class, he can make constructive comments about desired features and may even indicate an intent to purchase. But the situation changes dramatically when innovative product concepts are tested. Here, the result is likely to be quite negative.

Picture the would-be market researcher eighty years ago attempting to gauge market reaction to a proposed new product, the automobile. Respondents to any questionnaire would have assured the market-oriented innovator that cars would frighten horses, make too much noise, run too fast, and be generally unreliable. The competition of that time, the horse, would be judged just too strong for a successful market entry. The product would be labeled as a 'bad product' with no future. Had it not been for tough-minded innovators and entrepreneurs who were driven by a vision of what they believed and who persisted in spite of negative market reaction, the automobile might never have come of age as quickly as it did.

The problem of gauging market acceptance of innovative products is complex and has serious consequences. Operationally, it is virtually impossible to distinguish between a normal distrust for something new and the genuine lack of a market need or want for the innovation. Three-dimensional movies, for example, were not a success, but multichannel sound and instant photography brought fortunes to their developers. Although we can identify *post facto* reasons for these different results, it is not clear that the most careful consideration of consumers' needs would have resulted in a correct forecast thirty years ago. And the more innovative the product, the worse the problem. Edwin Land has been quoted as saying that Polaroid's products are so innovative, marketing research just does not work. Implementing the marketing concept for innovative products by researching market acceptance is certain to spell disaster for most of them.

Small Changes

The marketing concept goes further than merely discouraging product innovation; it actually fosters a preoccupation with minor product changes. A customer emphasis has led not only to the development of sophisticated marketing research tools but to the emergence of such tools as major strategic weapons. Take, for example, multidimensional scaling with all its applications in the field of product-space segmentation and product positioning. In the days before such methods were available, the market strategist, wishing to attack a new market, went to the product development group and requested a genuine new product. But with the marvels of modern marketing, this is no longer necessary. Rather, a product positioning study is commissioned. And behold, an existing product is 'repositioned' by tinkering with the elements of the marketing mix, so 'new and improved' products reach the supermarket shelves daily – products which really are not that new or improved. But through shrewd use of media and packaging, they can be positioned in the consumers' mind to be 'new'. Strategic tools such as product positioning backed up by various supporting elements of the marketing mix in some industries have all but superseded the traditional notion of better products through product development.

This tendency to discourage true innovation and focus resources on product modification may not be an entirely bad situation from the firm's perspective. Certainly, product modification is less costly and less risky than the innovation game, at least in the short run. Considerable profits have been earned by focusing on reformulations and repackaging, particularly in crowded markets. But as a long term strategy, a strictly market orientation does have its pitfalls. In the first place, those opportunities which require innovative products backed by basic and applied research may simply never be exploited by the firm. If the market-oriented firm, content with responding to market wants, does not reply with genuinely new products, then eventually other firms will.

Second, society at large is the loser; society is cheated out of the talent and resources that are channeled away from pioneering product work into the nitty-gritty world of product modification. Today, facing more problems than ever, society requires bold and innovative technological breakthroughs as answers to crises in transportation, housing, energy, food, and the like. Many observers have wondered at the lack of genuine innovation in these major areas. Some of the fault may be ascribed to those who have slavishly accepted the marketing concept. Instead of major breakthroughs, they have given America thirty thousand new supermarket products each year. [. . .]

Staying abreast of technology

Many of the major inventions of the last century have been commercialised by firms formed to exploit the invention or by small companies whose nature was rapidly changed by the invention. Ironically, the leading firms of those days had little to do with many of these great breakthroughs. Today's market giants in countless industries have emerged and prospered on the basis of technology which they invented or were quick to adopt. AT & T grew with the telephone, which its founder had patented. General Motors sprang from a two-wheeled cart company, a thriving concern but hardly world renowned, to become the automotive giant in America. The aircraft manufacturers tend to be without ancestors, except General Electric. The radio manufacturing companies were also new, although most have since grown into television. Xerography, a significant technological discovery, paved the way for the small Haloid Company to become Xerox Corporation, the leader in the plain paper copying business, while Edwin Land formed the Polaroid Corporation to exploit instant photography.

These new growth businesses were not the product of large corporations, nor did any major firm of that time exploit them fully. Yet all of these inventions had dramatic effects on existing products that the new inventions paralleled or replaced. Since history repeats itself, it is quite reasonable to predict that these new companies, born through inventions, could themselves fall prey to technological progress.

The marketing concept suggests that the way to achieve organizational goals of profit and growth is to focus on the marketplace and to concentrate on market needs. But such an orientation would have been of little help to the firms victimized by the innovations above. Levitt was right when he wrote, 'The history of American business is littered with the skeletons of companies who mistakenly thought their products or services were immune from the ravages of time and competition.' And he was right in saying that many firms had suffered because they had concentrated solely on the product they produced. For any firm which narrowly concentrates on one particular strategic element, without giving others due consideration, is likely to be hurt. So Levitt's remedy – to concentrate on the needs of the market – may be equally as deadly as the illness it was intended to cure. In each of the cases mentioned, a concentration on staying abreast of technological developments was perhaps more important than being market oriented.

In no way are we suggesting the marketing concept is not a useful tool. It is, and dozens of successful examples will illustrate the point. Our arguments are merely that the marketing concept is only one approach. A blind acceptance of this concept as the 'one and only' guide could be just as dangerous as failing to heed the proponents of the philosophy.

Before deciding which orientation is most suitable, a firm must analyze the elements required for success. Concurrently an assessment of the firm's own strengths and weaknesses and an identification of its distinctive competence are required. The blending of these elements will target the firm's strategic and new product efforts. The resulting path may well be a strong market orientation with a high degree of market synergy. But often a technology or production orientation could be the route to follow.

The marketing concept has been with us for three decades and will probably be alive for many more. But if it is to be an effective tool in the years ahead, we must be careful not to 'oversell' it. Rather, in the very tradition of the marketing concept, we should be need-oriented ourselves; we should see that the strategic philosophy adopted by a corporation is the one which really suits that firm and not one, although popular, which may be totally inappropriate for its needs and circumstances.

Reference

1. Levitt, T. (1960) Marketing myopia, *Harvard Business Review*, July/August, pp. 45–56.

4.4

Technology Push versus Market Pull – The Designer's Dilemma

A. E. Pannenborg

It seems proper before embarking on the main theme of this article to give some description, if not definition, of the terms used in the title. If we talk about the task of the designer in the design of a product, it seems worthwhile to try to describe the essence of this occupation. The task of the designer might be described as the production of manufacturing instructions that allow 'the factory' to manufacture the products according to agreed specifications. The latter comprise, of course, a long list which, however, may be broken down into performance specifications and cost/price specifications. In addition a timetable and a budget will be specified within which the development has to be completed. It is the first aspect, i.e. the performance, on which this article will concentrate and more specifically on the methods used to arrive at the proper answers.

Performance of a product can be analysed in terms of functions which the product has to fulfil for the user. In general these can be split into a primary function and a collection of secondary functions, which enhance the primary function to the user through comfort engineering (ergonomics), pleasing outward appearance and measures to ensure correct (primary) functioning under varying circumstances. When only products without certain desirable secondary functions are available, they will still be bought (e.g. black and white TV before the advent of colour TV). Clearly the competition innate in a free-market economy furthers the development of secondary functions.

In most cases the primary function is quite obvious and fairly lasting. The introduction of new primary functions is very rare. The telephone is a case. Never before had man been able to engage in a conversation with a partner well beyond shouting distance. Radio broadcasting also was a new function, although it might be argued that the town crier was, functionally speaking, its predecessor. Somewhat more frequent are basically new solutions for existing functions, with the characteristic that either they open up the possibility

A. E. Pannenborg, Technology push versus market pull – the designer's dilemma, *Electronics and Power,* vol. 21, pt 69, 1975, pp. 563–6.
A. E. Pannenborg is Vice-President and Vice-Chairman of N. V. Philips Gloeilampenfabrieken, Eindhoven.

of significantly reduced prices or open the way to vastly improved secondary functions. An example is given in Figure 1.

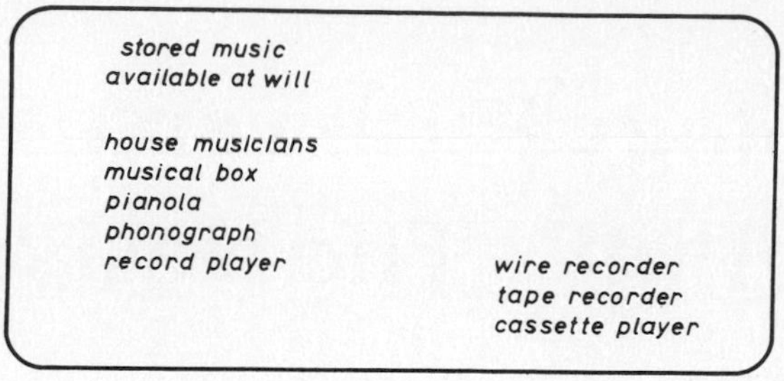

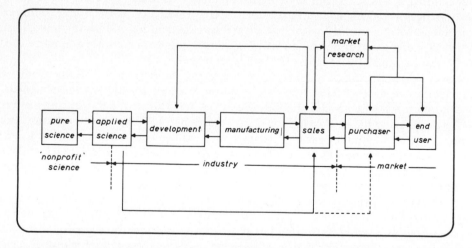

Figure 1   Although the record player once seemed to bring something really new, in the functional sense, it is just one station in a chain.

Innovation chain

Over the last 15 years, much study has been expended on the analysis of the process of new product design. One of the models proposed for this mechanism is the so-called 'innovation chain' (see Figure 2). From this model, the notions of technology push and market pull have been derived. They represent two aspects of the mechanism, the reality consisting always of a mixture of the two. Technology push denotes the situation where the initiative for a new possibility or a new solution for an existing problem comes entirely out of the realm of science and technology. Market pull (sometimes also denoted as society pull or demand pull) describes the situation where a specific request originates from the user to which science and technology respond and provide an answer.

Figure 2   Innovation chain with some of the feedback loops.

The main criticism of the innovation chain model is that innovation is not a linear process. This criticism is basically correct because, for a proper description of reality, the chain has to be extended with a number of feedback loops. Innovation is not a linear phenomenon but a multiple-loop feedback system.

Shift from technology push to market pull

It can be claimed that, for the first 20 years after the Second World War, the major force in innovation was technology push. This is plausible in view of the overwhelming demand for material goods in the postwar years and also in view of the impetus given to science and technology and their application during the war years. The many new possibilities opened up for electronics by solid state research found application in many new devices and, through these, in many new products. This meant basically new technical solutions for recognised functions which paved the way for a much wider application of these facilities than was ever dreamt of before.

In the same period, an unparalleled extension of R & D facilities in the electronics industry (and in other industries and institutions) took place. This gave rise to some overproduction of scientific research and corresponding saturation in these fields. The frontier research in a discipline like physics in that period also branched out into areas which for the time being gave no link with applied physics and technology. In short, the arsenal of technological solutions available for electronics has become relatively so complete that the emphasis has shifted more to the use of these possibilities in new combinations for hitherto unfulfilled functions for society.

In the terms used above, this means a shift from technology push towards market pull and therefore a good understanding of market pull becomes a prime requirement. Here much is still open, and many methods currently used seem haphazard and hit-and-miss.

Market pull

The expression 'market pull' with its keyword 'market' has such a vague connotation that it seems useful to start with a rough breakdown of the various forms of market (see Figure 3). A generalisation like this cannot cover every situation but a useful breakdown seems to be:

(a) *The professional user*
   Case 1. The user's profession makes him knowledgeable and expert in the technology of the product he requires.
   Case 2. The user's profession is essentially divorced from technology; the user has no technological knowledge to allow him to judge the product, except by its external performance.
(b) *The nonprofessional user*
   Case 1. The consumer who chooses and decides on a product by himself.
   Case 2. A community of consumers delegates the task of catering for its requirements to an institution.

A further analysis of these four categories shows that the degree of difficulty and the mechanisms for a proper investigation of the need are quite different.

Consumer empowers an authority to act

The traditional examples of case 2 in category (b) are the services which a modern nation requires for its infrastructure, such as the electricity supply, the telephone and telegraph administrations, broadcasting, public transport, traffic control and the like. In all these cases, in Western European countries, the public has delegated the responsibility for providing these facilities to a centralised institution, usually publicly owned. All these fields have the aspect of systems or networks with the inherent need for standardisation. The organisational concentration achieved by the delegation of responsibility leads to a level of expertise within the organisation which transforms this category, as seen from the supplier, into a professional market (case 1, category (*a*)).

An example of the forces involved are the PTT [Post Telegraph and Telecommunications] administrations which generally have established for the formulation of future requirements laboratories that are available as expert discussion partners for the supplier. In this dialogue between parties who are both abreast of the possibilities and limitations set by technology, concepts for the extension of service to the user are created.

Two remarks arise at this point. First, because of the centralisation on the one hand and the specialised nature of the subject on the other, the decision-making mechanism sometimes becomes dictatorial to the end user. Secondly, there is the real danger that market pull is neglected in favour of technology push with the possibility of developments which do not correspond to a real social need. A pertinent case was the introduction of the Picturephone by Bell, which was launched in a form and at a time when the American public did not feel a real need for it.

A special class within category (*b*), case 2, has recently been found in rapidly developing countries where the central government decides on the installation of a public service, does not have the necessary expertise and accordingly orders a turnkey project from Western industry. In these cases, it is by and large left to the imagination of the supplying industry to decide on the performance specifications in detail. Usually it leads to a copy of systems already supplied to the highly developed world, with the danger, because of insufficient analysis of the demand, that the system supplied does not fit the local needs in the best way.

Serving the expert customer

Examples within category (*a*), case 1, not derived from the previous category can be found in products supplied by one industry to another, e.g. production and manufacturing equipment and tools. Outside electronics, shipbuilding is a good example. The mechanism often used by the supplying industry to develop better solutions is close co-operation with a user who is considered to be advanced in this trade and therefore representative for that whole sector of the user world.

One profession serving another

Examples in the market category of one profession serving another are the medical profession, the world of education, and, in the broadest sense, the office with its requirements for computers, paging and intercommunication equipment. Already we encounter here that difficulty which will come out most vividly in the next category, namely, that a dialogue between supplier and user can only really start after hardware, possibly experimental, has been supplied. There is no real pull emanating from the user. A difficult communication pattern between users, semi-experts on behalf of the user, technical experts working out in the field for the supplier and the design department within the supplying industry has to be built up. A further aggravating factor is that the majority of products involved are very complicated and often of a systems nature with consequent pressures towards economies of scale within the supplying industry, which inevitably lead to the design of products which can be marketed the world over.

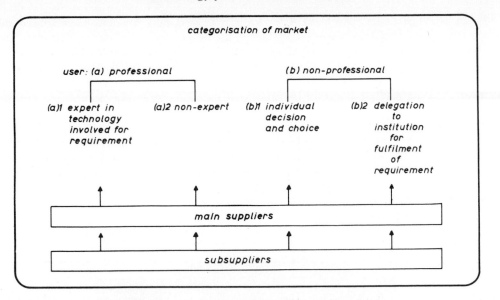

Figure 3 Market divided into several segments, each of which presents its own type of interaction with the new product designer.

Thus the supplying industry must build up a deep insight into the requirements of the user and must become almost as expert in the functions and procedures to be fulfilled as the user himself. A continuous evaluation must go on whether the prime requirement of the user world is in the direction of unchanged performance, but at a lower price, or whether the emphasis rests on increased performance more or less regardless of price, as applies in the military and the medical profession, at least up till now.

General-purpose computers are a special case within this category. One can see that technology push has been predominant until now in the development of such machines. The user world has had to adapt itself, its organisation and its procedures to a significant extent in order to make full use of this new tool. The user world has been rather subservient in this respect. People have even taken the trouble to learn new languages to be able to make use of computers. Perhaps the most significant influence of demand pull on the computer industry has been towards a long mean time between failures. On the other hand, it must be admitted that technology push in computers has been impressively successful in bringing the price/performance ratio down.

Another illustrative case within this category is X ray diagnostic equipment. Here the primary function of making two-dimensional images of nontransparent objects at the required resolution and contrast at a minimum dose for the patient has reigned unchanged for some 70 years. The great improvements, especially over the last 30 years, have covered mainly the secondary functions, such as minimising the dose incurred by doctors and nurses, increasing the comfort of those operating the installations through automation, and significant progress in systems for storing the images taken.

Private consumer

We now turn to that large sector which is usually termed consumer electronics. Here we encounter the classical economic mechanism of supply and demand in its purest form. Market research is considered to be the discipline which provides guidance to industry where to go. The classical mechanism, however, is still basically one of action and reaction, i.e. a new product is launched into the market, the degree of success in the market is carefully observed and conclusions for changes in design are derived from this. No mechanisms devised have as yet been truly effective for real anticipation. For products of relatively small value and in a phase in which only slight modifications of design are called for, this is quite acceptable. The situation becomes really dangerous in the case of one-product industries, such as an automobile manufacturer, where one wrong design can bring about the death of the company.

As stated in the introduction to this article the arrival of new functions is rather rare. Accordingly the secondary functions and qualities of a product determine to a large extent the success of a product. Here many often imponderable aspects come in, such as fashion and comfort engineering. Again and again it is seen that the average user is out for comfort, or, to put it somewhat unkindly, is extremely lazy and impatient; so that to the engineer seemingly trivial additions are often decisive for the success or failure of a product [see Figure 4].

Figure 4   An example of engineer's delight: the 'monoknob' radio receiver of the mid 1930s. All controls were combined in one joystick. The user, however, needed unusual dexterity to tune the set to his liking.

New aspects

If we exchange the expression market pull for the expression society pull, we can observe that, recently, society pull has received additional content. This content is not connected with further additions to performance, but is related to restrictions which can be expected to be imposed increasingly on the Western economy and technology. In view of these new restrictions in the fields of energy consumption, pollution and material use, we should remember that the design engineer has always lived and worked within the boundaries of a large package of restrictions (Figure 5).

These restrictions, as shown in the figure, can be related to the various locations within

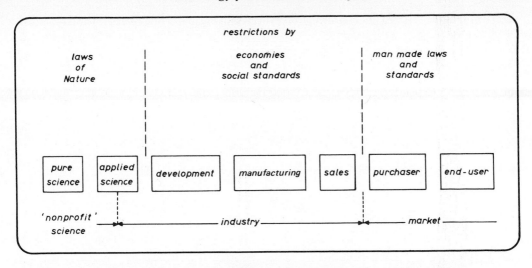

Figure 5   The product designer has to reckon with many restrictive boundary conditions.

the innovation chain. Technology has always had to reckon with the principal restrictions set up by the laws of nature. Industry has always had to work within the rules imposed by economy and the social standards of the national community, and finally the market at large has rightly decided to regulate its daily life by a large volume of legislation including laws for standards and safety. New restrictions do not mean a qualitative change as compared with the past. Personally I am convinced that the design engineer will find as much challenge in working within the new restrictions as ever before.

Finally it seems wise for us Europeans to watch closely the mechanisms which prove successful in other societies. Specifically, an analysis of the Japanese situation gives much food for thought. One striking aspect of that community is the degree of co-operation between government and industry, which in our eyes sometimes almost appears as collusion. Many examples can be found in Japan where the Government has, by law, created requirements on the home market at an early time, offering in that way a field of trial to Japanese industry which has placed the latter in an advantageous position for exporting their products somewhat later to other countries.

A completely different aspect of the Japanese approach seems to be the practice on a much larger scale than we are accustomed to of what I call the hit-and-miss method. Often we have classified the Japanese somewhat sneeringly as 'gadgeteers'. Undoubtedly, there is much waste in industrial developments in Japan, where new-product propositions in the form of prototypes are continually tried out. Even with a modest percentage of success, this mechanism places the Japanese in front of many European industries. This seemingly wasteful strategy is probably fostered by the relatively low cost of the Japanese professional designer, whose income hitherto has been only slightly higher than that of the skilled worker.

DIFFUSION OF INNOVATIONS

4.5

Elements of Diffusion

Everett M. Rogers

One reason why there is so much interest in the diffusion of innovations is because getting a new idea adopted, even when it has obvious advantages, is often very difficult. There is a wide gap in many fields, between what is known and what is actually put into use. Many innovations require a lengthy period, often of some years, from the time when they become available to the time when they are widely adopted. Therefore, a common problem for many individuals and organizations is how to speed up the rate of diffusion of an innovation.

The following case illustration [see the Box] provides insight into some common difficulties facing programs of diffusion. [. . .]

Nondiffusion of the Dvorak Keyboard

Most of us who use a typewriter – and this includes about 18 million individuals who earn a living as typists – don't even know that our fingers tap out words on a keyboard that is called 'QWERTY', named after the first six keys in the upper row of letters. Even fewer of us know just how inefficient the QWERTY keyboard is. For example, this typewriter keyboard takes twice as long to learn as it should, requires twice as long to use as it should, and makes us work about twenty times harder than we should. But QWERTY has persisted out of inertia since 1873, and today unsuspecting individuals are still being taught to use the QWERTY keyboard, unaware that a much more efficient typewriter keyboard is available.

Where did QWERTY come from, and why does it continue to be used, instead of much more efficient alternative keyboard designs? QWERTY was invented by Christopher Latham Sholes in 1873, who designed this keyboard to slow the typist down. In that day, the type bars on a typewriter hung down in a sort of basket, and pivoted up to strike the paper; then they fell back in place by gravity. When two adjoining keys were struck rapidly in succession, they often jammed. Sholes rearranged the keys on a typewriter keyboard to minimize such jamming; he 'anti-engineered' the arrangement to make the most commonly used letter sequences awkward and slow to use. By thus making it difficult for a typist to operate the machine, and slowing down typing speed, Sholes' QWERTY keyboard allowed these old typewriters to

---

E. M. Rogers, *Diffusion of Innovations*, 3rd edn, New York: Free Press, 1983; extracts from Chapter 1. Everett M. Rogers is Professor of International Communication at Stanford University.

operate satisfactorily. His design was used in the manufacture of all typewriters.

Prior to about 1900, most typists used the two-finger, hunt-and-peck system. But thereafter, as touch typing became popular, dissatisfaction with the QWERTY keyboard began to grow. Typewriters became mechanically more efficient, and the QWERTY keyboard design was no longer necessary to prevent key jamming. The search for an improved design was led by Professor August Dvorak at the University of Washington, who in 1932 used time-and-motion studies to create a much more efficient keyboard arrangement. The Dvorak keyboard has the letters A, O, E, U, I, D, H, T, N, and S across the home row of the typewriter. Less frequently used letters were placed on the upper and lower rows of keys. About 70 per cent of the typing is done on the home row, 22 per cent on the upper row, and about 8 per cent on the lower row. On the Dvorak keyboard, the amount of work assigned to each finger is proportionate to its skill and strength. Further, Professor Dvorak engineered his keyboard so that successive keystrokes fell on alternate hands; thus, while a finger on one hand is stroking a key, a finger on the other hand can be moving into position to hit the next key. Typing rhythm is thus facilitated; this hand alternation was achieved by putting the vowels (which represent 40 per cent of all letters typed) on the left-hand side, and placing the major consonants that usually accompany these vowels on the right-hand side of the keyboard.

Professor Dvorak was able to avoid many of the typing inefficiencies of the QWERTY keyboard. For instance, QWERTY overloads the left hand; it must type 57 per cent of ordinary copy. The Dvorak keyboard shifts this emphasis to 56 per cent on the stronger right hand and 44 per cent on the weaker left hand. Only 32 per cent of typing is done on the home row with the QWERTY System, compared to 70 per cent with the Dvorak keyboard. And the newer arrangement requires less jumping back and forth from row to row; with the QWERTY keyboard, a good typist's fingertips travel more than twelve miles a day, jumping from row to row. These unnecessary intricate movements cause mental tension, typist fatigue, and lead to more typographical errors.

One might expect, on the basis of its overwhelming advantages, that the Dvorak keyboard would have completely replaced the inferior QWERTY keyboard by now. On the contrary, after more than 40 years, almost all typists are still using the inefficient QWERTY keyboard. Even though the American National Standards Institute and the Computer and Business Equipment Manufacturers Association have approved the Dvorak keyboard as an alternative design, it is still almost impossible to find a typewriter (or a computer) keyboard that is arranged in the more efficient layout. Vested interest are involved in hewing to the old design: manufacturers, sales outlets, typing teachers, and typists themselves.

No, technological innovations are not always diffused and adopted rapidly. Even when the innovation has obvious and proven advantages.

As the reader may have guessed by now, the present pages were typed on a QWERTY keyboard.

Four Main Elements in the Diffusion of Innovations

We can define *diffusion* as the process by which (1) an *innovation* (2) is *communicated* through certain *channels* (3) over *time* (4) among the members of a *social system*. The four main elements are the innovation, communication channels, time, and the social system (Figure 1). They are identifiable in every diffusion research study, and in every diffusion campaign or program. The following description of these four elements in diffusion constitutes an overview of the main concepts and viewpoints.

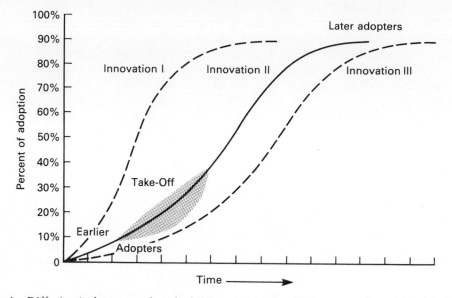

Figure 1   Diffusion is the process by which (1) an *innovation* (2) is *communicated* through certain *channels* (3) over *time* (4) among the members of a *social system*.

1.   The Innovation

An *innovation* is an idea, practice, or object that is perceived as new by an individual or other unit of adoption. It matters little, so far as human behaviour is concerned, whether or not an idea is 'objectively' new as measured by the lapse of time since its first use or discovery. The perceived newness of the idea for the individual determines his or her reaction to it. If the idea seems new to the individual, it is an innovation.

Newness in an innovation need not just involve new knowledge. Someone may have known about an innovation for some time but not yet developed a favorable or unfavorable attitude toward it, nor have adopted or rejected it. The 'newness' aspect of an innovation may be expressed in terms of knowledge, persuasion, or a decision to adopt. [. . .]

It should not be assumed that the diffusion and adoption of all innovations are necessarily desirable. In fact, there are some studies of harmful and uneconomical innovations that are generally not desirable for either the individual or his or her social system. Further, the same innovation may be desirable for one adopter in one situation but undesirable for another potential adopter in another situation. For example, mechanical tomato pickers have been adopted rapidly by large commercial farmers in California, but these machines were too expensive for small-size tomato growers, and thousands have thus been forced out of tomato production. [. . .]

Characteristics of Innovations
It should not be assumed, as sometimes has been done in the past, that all innovations

are equivalent units of analysis. This is a gross over-simplification. While it may take consumer innovations like blue jeans or pocket calculators only five or six years to reach widespread adoption in the United States, other new ideas such as the metric system or using seat belts in cars may require several decades to reach complete use. The characteristics of innovations, as perceived by individuals, help to explain their different rate of adoption.

1.   *Relative advantage* is the degree to which an innovation is perceived as better than the idea it supersedes.The degree of relative advantage may be measured in economic terms, but social-prestige factors, convenience, and satisfaction are also often important components. It does not matter so much whether an innovation has a great deal of 'objective' advantage. What does matter is whether an individual perceives the innovation as advantageous. The greater the perceived relative advantage of an innovation, the more rapid its rate of adoption is going to be.

2.   *Compatibility* is the degree to which an innovation is perceived as being consistent with the existing values, past experiences, and needs of potential adopters. An idea that is not compatible with the prevalent values and norms of a social system will not be adopted as rapidly as an innovation that is compatible. The adoption of an incompatible innovation often requires the prior adoption of a new value system. An example of an incompatible innovation is the use of contraception in countries where religious beliefs discourage use of birth-control techniques, as in Moslem and Catholic nations.

3.   *Complexity* is the degree to which an innovation is perceived as difficult to understand and use. Some innovations are readily understood by most members of a social system; others are more complicated and will be adopted more slowly. For example, the villagers in Los Molinos, Peru, did not understand germ theory, which the health worker tried to explain to them as a reason for boiling their drinking water. In general, new ideas that are simpler to understand will be adopted more rapidly than innovations that require the adopter to develop new skills and understandings.

4.   *Trialability* is the degree to which an innovation may be experimented with on a limited basis. New ideas that can be tried on the installment plan will generally be adopted more quickly than innovations that arc not divisible. Ryan and Gross (1943) found that every one of their Iowa farmer respondents adopted hybrid-seed corn by first trying it on a partial basis. If the new seed could not have been sampled experimentally, its rate of adoption would have been much slower. An innovation that is trialable represents less uncertainty to the individual who is considering it for adoption, as it is possible to learn by doing.

5.   *Observability* is the degree to which the results of an innovation are visible to others. The easier it is for individuals to see the results of an innovation, the more likely they are to adopt. Such visibility stimulates peer discussion of a new idea, as friends and neighbours of an adopter ask him or her for innovation–evaluation information about it. Solar panels on a household's roof are highly observable, and a California survey found that the typical solar adopter showed his equipment to about six of his peers (Rogers *et al.*, 1979).

About one-quarter of all California homeowners know someone who has adopted solar equipment (even though only about 2.5 per cent of the state's homeowners had adopted by 1979), and about two-thirds of this quarter (15 per cent of all homeowners) have seen their friend's solar panels. Solar adopters often are found in spatial clusters in California, with three or four adopters located on the same block. [. . .]

In general, innovations that are perceived by receivers as having greater relative advantage, compatibility, trialability, observability, and less complexity will be adopted more rapidly than other innovations. These are not the only qualities that affect adoption rates, but past research indicates that they are the most important characteristics of innovations in explaining rate of adoption.

Re-Invention

Until about the mid-1970s, it was assumed that an innovation was an invariant quality that was not changed as it diffused. I remember interviewing an Iowa farmer in 1954 about his adoption of 2,4-D weed spray. In answer to my question about whether he used this innovation, the farmer described in some detail the particular and unusual ways in which he used the weed spray on his farm. At the end of his remarks, I simply checked 'adopter' on my questionnaire. The concept of re-invention was not in my theoretical repertoire, so I condensed his experience into one of my existing categories.

Then, in the 1970s, diffusion scholars began to pay more attention to the concept of *re-invention*, defined as the degree to which an innovation is changed or modified by a user in the process of its adoption and implementation. Some researchers measured re-invention as the degree to which an individual's use of a new idea departed from the 'mainline' version of the innovation that was promoted by a change agency (Eveland *et al.*, 1977). Once scholars became aware of the concept of re-invention and began to work out measures for it, they began to find that a fair degree of re-invention occurred for some innovations. Other innovations are more difficult or impossible to re-invent; for example, hybrid seed corn does not allow a farmer much freedom to re-invent, as the hybrid vigor is genetically locked into seed (for the first generation) in ways that are too complicated for a farmer to change. But certain other innovations are more flexible in nature, and they are re-invented by many adopters who implement them in different ways.

We should remember, therefore, that an innovation is not necessarily invariant during the process of its diffusion, and adopting an innovation is not necessarily a passive role of just implementing a standard template of the new idea.

Given that an innovation exists, communication must take place if the innovation is to spread beyond its inventor. Now we turn our attention to this second element in diffusion.

2.   Communication Channels

Communication is the process by which participants create and share information with one another in order to reach a mutual understanding. Diffusion is a particular type of communication in which the information that is exchanged is concerned with new ideas. The essence of the diffusion process is the information exchange by which one individual communicates a new idea to one or several others. At its most elementary form, the process involves: (1) an innovation, (2) an individual or other unit of adoption that has knowledge of, or experience with using, the innovation, (3) another individual or other unit that does not yet have knowledge of the innovation, and (4) a communication channel connecting the two units. A *communication channel* is the means by which messages get from one individual to another. The nature of the information-exchange relationship between the pair of individuals determines the conditions under which a source will or will not transmit the innovation to the receiver, and the effect of the transfer.

For example, mass media channels are often the most rapid and efficient means to inform an audience of potential adopters about the existence of an innovation, that is, to create awareness-knowledge. *Mass media channels* are all those means of transmitting messages that involve a mass medium, such as radio, television, newspapers, and so on, which enable a source of one or a few individuals to reach an audience of many. On the other hand, interpersonal channels are more effective in persuading an individual to adopt a new idea, especially if the interpersonal channel links two or more individuals who are near-peers. *Interpersonal channels* involve a face-to-face exchange between two or more individuals.

The results of various diffusion investigations show that most individuals do not evaluate an innovation on the basis of scientific studies of its consequences, although such objective evaluations are not entirely irrelevant, especially to the very first individuals who adopt. Instead, most people depend mainly upon a subjective evaluation of an innovation that is conveyed to them from other individuals like themselves who have previously adopted the innovation. This dependence on the communicated experience of near-peers suggests that the heart of the diffusion process is the modelling and imitation by potential adopters of their network partners who have adopted previously. [. . .]

3. Time dimension

The time dimension is involved in diffusion (1) in the innovation decision process by which an individual passes from first knowledge of an innovation through its adoption or rejection, (2) in the innovativeness of an individual or other unit of adoption – that is, the relative earliness/lateness with which an innovation is adopted – compared with other members of a system, and (3) in an innovation's rate of adoption in a system, usually measured as the number of members of the system that adopt the innovation in a given time period.

The Innovation-Decision Process

The *innovation-decision process* is the process through which an individual (or other decision-making unit) passes from first knowledge of an innovation to forming an attitude toward the innovation, to a decision to adopt or reject, to implementation of the new idea, and to confirmation of this decision. We conceptualize five main steps in the process: (1) knowledge, (2) persuasion, (3) decision, (4) implementation, and (5) confirmation. *Knowledge* occurs when an individual (or other decision-making unit) is exposed to the innovation's existence and gains some understanding of how it functions. *Persuasion* occurs when an individual (or other decision-making unit) forms a favorable or unfavorable atitude toward the innovation. *Decision* occurs when an individual (or other decision-making unit) engages in activities that lead to a choice to adopt or reject the innovation. *Implementation* occurs when an individual (or other decision-making unit) puts an innovation into use. Re-invention is especially likely to occur at the implementation stage. *Confirmation* occurs when an individual (or other decision-making unit) seeks reinforcement of an innovation decision that has already been made, but he or she may reverse this previous decision if exposed to conflicting messages about the innovation. [. . .]

For purposes of simplicity, we have restricted our present discussion of the innovation-decision process mainly to a single individual, and thus to the case of individual-optional innovation-decisions. But many innovation-decisions are made by organizations or other types of adopting units, rather than by individuals. For example, an organization may decide to purchase word-processing equipment on the basis of a staff decision or by an official's authority decision; the individual office worker in the organization may have little or no say in the innovation-decision. When an innovation-decision is made by a system, rather than by an individual, the decision process is usually much more complicated.

Innovativeness and Adopter Categories

Innovativeness is the degree to which an individual or other unit of adoption is relatively earlier in adopting new ideas than the other members of a system. Rather than describing an individual as 'less innovative than the average member of a social system', it is handier and more efficient to refer to the individual as being in the 'late majority' or some other adopter category. This short-hand notation saves words and contributes to clearer understanding, for diffusion research shows that members of each of the adopter categories have a great deal in common. If the individual is like most others in the late majority category, he is low in social status, makes little use of mass-media channels, and secures most of his new ideas from peers via interpersonal channels. *Adopter categories* are the classifications of members of a social system on the basis of innovativeness. The five adopter categories are: (1) innovators, (2) early adopters, (3) early majority, (4) late majority and (5) laggards [see Figure 1 in the Introduction to this section].

Innovators are active information seekers about new ideas. They have a high degree of mass media exposure and their interpersonal networks extend over a wide area, usually outside of their local system. Innovators are able to cope with higher levels of uncertainty about an innovation than are other adopter categories. As the first to adopt a new idea in their system, they cannot depend upon the subjective evaluations of the innovation from other members of their system.

Obviously, the measure of innovativeness and the classification of the system's members into adopter categories are based upon the relative time at which an innovation is adopted.

Rate of Adoption

There is a third specific way in which the time dimension is involved in the diffusion of innovations. *Rate of adoption* is the relative speed with which an innovation is adopted by members of a social system. When the number of individuals adopting a new idea is plotted on a cumulative frequency basis over time, the resulting distribution is an s-shaped curve. At first, only a few individuals adopt the innovation in each time period (such as a year or a month, for example); these are the innovators. But soon the diffusion curve begins to climb, as more and more individuals adopt. Then the trajectory of the rate of adoption begins to level off, as fewer and fewer individuals remain who have not yet adopted. Finally, the s-shaped curve reaches its asymptote, and the diffusion process is finished.

Most innovations have an s-shaped rate of adoption. But there is variation in the slope of the 's' from innovation to innovation; some new ideas diffuse relatively rapidly and the s-curve is quite steep. Another innovation may have a slower rate of adoption, and its s-curve will be more gradual, with a slope that is relatively lazy. One issue addressed by diffusion research is why some innovations have a rapid rate of adoption, and why others are adopted more slowly (Figure 1).

The rate of adoption is usually measured by the length of time required for a certain percentage of the members of a system to adopt an innovation. Therefore, we see that rate of adoption is measured using an innovation or a system, rather than an individual, as the unit of analysis. Innovations that are perceived by individuals as possessing greater relative advantage, compatibility, and the like, have a more rapid rate of adoption.

There are also differences in the rate of adoption for the same innovation in different social systems. Clearly, there are aspects of diffusion that cannot be explained only by the nature of individual behavior. The system has a direct effect on diffusion, and also an indirect influence through its individual members. What is a social system?

4. A Social System

A *social system* is defined as a set of interrelated units that are engaged in joint problem solving to accomplish a common goal. The members or units of a social system may be individuals, informal groups, organizations, and/or subsystems. The system analyzed in a diffusion study may consist of all the peasants in an Asian village, high schools in Wisconsin, medical doctors in a hospital, or all the consumers in the United States. Each unit in a social system can be distinguished from other units. All members cooperate at least to the extent of seeking to solve a common problem in order to reach a mutual goal. This sharing of a common objective binds the system together.

It is important to remember that diffusion occurs within a social system, because the social structure of the system affects the innovation's diffusion in several ways. [. . .]

Social Structure and Diffusion

Generally, the fastest rate of adoption of innovations results from authority decisions (depending, of course, on how innovative the authorities are). Optional decisions can usually be made more rapidly than collective decisions. Although made more rapidly, authority decisions are often circumvented during their implementation.

The type of innovation-decision for a given idea may change or be changed over time. Automobile seat belts, during the early years of their use, were installed in autos as optional decisions by the car's owner, who had to pay the cost of installation. Then, in 1968, a federal law was passed requiring that seat belts be included in all new cars in the United States. An optional innovation-decision thus became a collective decision. But the decision by the auto driver or passengers to fasten the belts when in the car was still an optional decision – that is, except for 1974 model cars, which a federal law required to be equipped with a seat belt–ignition interlock system that prevented the driver from starting the engine until everyone in the auto's front seat had fastened their seat belt. So for one year, the fastening of seat belts became a collective authority-decision. But the public reaction to this draconian approach was so negative that the US legislature reversed this law, and the fastening of auto seat belts again became an individual-optional decision. [. . .]

Consequences of innovation

A social system is involved in an innovation's consequences because certain of these changes occur at the system level, in addition to those that affect the individual.

Consequences are the changes that occur to an individual or to a social system as a result of the adoption or rejection of an innovation. There are at least three classifications of consequences:

1. *Desirable* versus *undesirable* consequences, depending on whether the effects of an innovation in a social system are functional or dysfunctional.
2. *Direct* versus *indirect* consequences, depending on whether the changes to an individual or to a social system occur in immediate response to an innovation or as a second-order result of the direct consequences of an innovation.
3. *Anticipated* versus *unanticipated* consequences, depending on whether the changes are recognized and intended by the members of a social system or not.

Change agents usually introduce innovations into a client system that they expect will be desirable, direct, and anticipated. But often such innovations result in at least some unanticipated consequences that are indirect and undesirable for the system's members. An illustration is the case of the steel ax as introduced by missionaries to an Australian aborigine tribe (Sharp, 1952). The change agents intended that the new tool should raise levels of living and material comfort for the tribe. But the new technology also led to a breakdown of the family structure, the rise of prostitution, and 'misuse' of the innovation itself. Change agents can often anticipate and predict the innovation's form, the directly observable physical appearance of the innovation, and perhaps its function, the contribution of the idea to the way of life of the system's members. But seldom are change agents able to predict another aspect of an innovation's consequences, its meaning, the subjective perception of the innovation by the clients. [. . .]

References

Eveland, J. D. *et al.* (1977) The innovation process in public organizations, *mimeo,* Department of Journalism, University of Michigan, Ann Arbor, Michigan.

Rogers, E. M. *et al.* (1979) *Solar Diffusion in California: A pilot study,* Report, Institute of Communication Research, Stanford University, Stanford, California.

Ryan, B. and Gross, N. C. (1943) The diffusion of hybrid seed corn in two Iowa communities, *Rural Sociology,* vol. 8, pp. 15–24.

Sharp, L. (1952) Steel axes for stone age Australians, in E. H. Spicer, (ed.), *Human Problems in Technological Change,* New York: Russell Sage Foundation.

4.6

The Diffusion of Innovations – Microelectronic Technology

Christopher Freeman

The Diffusion of Innovations and New Technology

It is a common experience that new products and processes often diffuse much more slowly than their inventors had imagined or hoped. This disappointment arises partly because technologists may often overlook or minimise the social, economic and cultural barriers and constraints to the acceptance of new ideas and products.

These are especially important in the case of a totally new product which is not simply substituting for an inferior product already in use. Simple unfamiliarity may not be as great a barrier as is often supposed since novelty has its own attractions, but uncertainty about future performance, maintenance or social and environmental acceptability may be very great inhibiting factors. In the case of 'substitution' diffusion, these barriers are likely to be less severe and both the innovators and the adopters approach the decision with greater confidence. A need is known to exist and a measurable group of potential 'adopters' can be identified and systematically canvassed. However, even in this case innovators are often disappointed at the slowness of response which they frequently ascribe to ignorance or conservatism on the part of potential adopters.

They are often right in believing that these influences are quite strong but research on diffusion has shown that other factors are frequently more important at least for the main group of adopters. Sociologists have classified potential adopter populations into 'early' or 'pioneer' adopters, 'average' or 'normal' adopters and 'laggard' or late adopters. Further refinements can be made in this sub-division but this simple classification brings out the main point that the approach to the adoption decision varies and that this variation may be related to economic, educational, social and psychological characteristics. 'Pioneer' adopters may have low risk aversion, high incomes, high educational qualifications, extrovert personalities and experience of foreign travel whilst 'laggards' may have the

C. Freeman, Some economic implications of microelectronics, in B. Lundvall and P. R. Christensen (eds.), *Technology and Employment,* Industrial Development Research Series, no. 20, Ålborg, Denmark: Ålborg University Press, 1981; extracts from pp. 18–35.

Professor Freeman is the former Director and currently Deputy Director of the Science Policy Research Unit, Sussex University.

reverse characteristics. For the main body of adopters the experience of the pioneers and attitudes of opinion leaders is important but so too are fairly straightforward 'rational' calculations of economic advantage. These sort of explanations fit the logistic curves which characterise many (but by no means all) diffusion processes. The slow start reflects the hesitations and uncertainties associated with the first trials, and the long flat tail reflects the number of incorrigible laggards, whilst the slope of the steep part of the curve reflects the advantages perceived by the main body of adopters and the speed of their response.

Work on the diffusion of innovations was originally pioneered mainly by sociologists and in 1961 Rogers[1] could reasonably complain that industrial economics had made virtually no contribution. The sociologists had concentrated their effort chiefly in the areas of agriculture, education and medicine, and almost the only industrial study which Rogers could cite was Bruce Williams' work on the tunnel oven in the pottery industry. In the 1960s and 1970s, however, a great deal more work has been done by economists, starting with Mansfield's[2] research in the early 1960s and including the major international project of the NIESR,[3] IFO and the Swedish Industrial Institute and several doctoral dissertations.[4]

As a result of all this empirical work and the associated theoretical generalisations, it is now possible to synthesise some of the main conclusions deriving both from the sociological and from the economics research. Predictably the economists put the main emphasis on profitability, scale of investment, and relative costs, while the sociological work stressed characteristics of change agents, of opinion leaders and other groups within the potential 'adopter population'. Rather less commonly, individual studies stressed the importance of technical characteristics of new products and processes, compatibility and acceptability of innovations within an existing environment or system, and the role of political lobbies and government policies. [. . .]

From this work on diffusion it seems reasonable to conclude that any innovation which diffuses through half of a potential adopter population, or affects more than two-thirds of the relevant output of a good or service in less than 10 years has enjoyed a rapid rate of diffusion. Most examples of this kind are in fact essentially substitution processes, such as the replacement of metal washing-up bowls with plastic ones or the substitution of integrated circuits for discrete components in various electronic end-products. Substitution processes of consumer durables by households are a type of short replacement cycle which may permit very high adoption rates for a machine or product with some novel features.

More typically, it takes 10 to 30 years for the majority of potential adopters to invest or purchase, or for more than two-thirds of the relevant output to be affected. With completely new products or systems a slow diffusion rate of more than 30 years for the majority of firms or households (whichever is the relevant category) to adopt is quite usual. Examples are the telephone, the tractor, the passenger car and the numerically-controlled machine tool.

The variety of adoption processes involved and the absence even now of satisfactory statistics prevent us from conveying more than an impressionistic notion of the diffusion of microelectronic technology. Table 1 presents such an impressionistic picture and attempts to illustrate the varying diffusion rates for different aspects of the technology, ranging from rapid adoption of new generations of components by the electronics industry itself (on the left-hand side of Table 1) all the way across to very slow adoption rates where very large numbers of potential users are involved, often with minimal skills and knowledge in the areas of potential application. The adoption of microelectronics in the design of new computers, instruments, communication devices, word processors, robots, machine tools, etc. was a necessary condition for the subsequent application of these products in automation and information systems in later decades.

Table 1 Diffusion of microelectronic technology through the economy

Rate of Diffusion / Depth of impact	Rapid (from 1960) High	Medium (from 1965) High	Medium (from 1965) Medium	Slow (from 1970) High	Slow (from 1970) Medium	Slow (from 1970) Low
Product design and redesign	Electronic capital goods. Military and space equipment. Some electronic consumer goods.	Machine tools. Vehicles. Electronic consumer goods. Instruments. Some toys.	Other consumer durables. Engines and motors. Other machinery	Some bio-medical products	Other toys	
Process automation	Some electronic products	Machining and assembly (large batch) especially in vehicles, consumer durables and machinery. Printing and publishing	Continuous flow processes already partly automated: – chemicals – metals – petroleum – gas – electricity	Clothing. Textiles. Food. Machinery and assembly (small batch)	Building materials. Furniture. Mining and quarries	Agriculture. Hotels and restaurants. Construction. Personal services
Information systems and data processing	Specific government, business and professional systems involving heavy data storage and processing in large organisations	Financial services. Communication systems. Office systems and equipment without total electronic systems. Design	Transport. Wholesale distribution. Public administration. Large retailers	Retail distribution. All-electronic office systems. Electronic funds transfer.	Domestic households. Professional services	Agriculture. Hotels and restaurants. Construction. Personal services

The discussion which follows on the main factors affecting the rate of diffusion will clarify the reasons for locating particular applications of microelectronics in various parts of Table 1 and the reader may wish to refer back to it as the discussion proceeds. Almost all social forecasting is subject to wide margins of error and there is plenty of room for argument over the table. Its purpose is illustrative rather than comprehensive, and indicative rather than precise. We turn now to consider first the main factors affecting innovation diffusion processes in general, and then their specific application to the case of microelectronics.

Profitability for potential adopters and producers of new products
This was the main result from Mansfield's work and is of course in line with the theoretical predictions of most schools of economic thought. It must however be qualified by the conclusions of research on early adopters showing conclusively that in the early stages of a diffusion process great uncertainty about actual and future profitability is the rule rather than the exception. Moreover, Metcalfe[5] has shown that even when high profitability has been convincingly demonstrated, large numbers of 'laggards' in the potential adopter popula-

tion may continue to behave 'irrationally' by refusing to invest in innovation. He found in the case of the size-box in the Lancashire cotton industry, an innovation with a pay-back period of less than a year, an extremely high rate of return, a low investment outlay and easy availability, that many entrepreneurs still failed to adopt after many years. This result, which is in line with many of the findings of sociological research, may be particularly important in the British context. It emphasises the point that purely economic analysis cannot be divorced from the social and cultural context, including the educational and other social characteristics of entrepreneurs and decision-makers.

Scale of investment

Again this was one of Mansfield's principal findings and it is hardly surprising that cheaper small-scale innovations should generally (other things being equal) be adopted more rapidly than those requiring a large scale of investment. This point is particularly important in relation to systems innovations where the scale of investment relates not just to the decisions of one potential adopter but to a complementary pattern of investment involving several different and independent decision-makers. Again the qualification must be made that socio-cultural factors and market structure may be influences which outweigh the simple issue of scale of investment outlay. Five or six firms in a technically progressive oligopoly may all adopt a new but very expensive process within the space of two or three years, whereas Metcalfe's study, which has already been cited, shows that a very low investment outlay in a competitive small firm industry does not necessarily lead to a high rate of adoption.

But slowness to adopt is not necessarily an indication of technical or managerial backwardness. Economic 'rationality' may dictate a decision not to scrap an old machine long after a technically superior rival product is on the market. However, completely 'rational' investment decision-making is exceptional (some would say impossible) and in complicated balance in the minds of decision-makers, Mansfield is almost certainly right in believing that a larger investment outlay is usually inversely related to the rate of adoption. Oxygen steel-making is a good example illustrating the point that heavy investment outlays may lead to a comparatively slow diffusion process.

Technical characteristics of a new product or process

Although not emphasised so strongly in some economics research, this is clearly a factor of outstanding importance. Much innovation in machinery has been concerned with the achievement of greater precision and reliability, higher speeds, reduced maintenance and other superior performance characteristics. Although the pursuit of higher standards is not usually in contradiction with the goal of increased profitability, it may sometimes appear to be so, at least in the short run. Users of a new product or process may be ready to pay a higher price for greatly improved performance if increased long-term profitability and convenience seem likely to offset reduced short-term profitability. Again the social and cultural context is important, since there is now much evidence to suggest that this attitude may be more prevalent among purchasers of capital goods in Japan and Germany than in the UK or USA. The military and space market in all countries is of course a special case, where technical characteristics are exceptionally important in the diffusion of innovation.

Environmental acceptability

This is a very wide concept. It embraces both the 'ecological' notion of the environment and the sociological notion of the work environment or domestic environment in which the innovation may be used. Hence it includes 'safety' both in relation to the eco-system (atmospheric pollution, etc.) and in relation to industrial and consumer safety. This factor has become particularly important in the case of drug and chemical innovations, but it is increasingly relevant to almost all major innovations. Clearly it relates to the cultural and political factors investigated by sociologists, as well as to more clearly identifiable and measurable technical characteristics affecting environmental hazards.

Change agents and decision-makers

This concept is wider than that of market structure commonly considered in economic analysis and embraces both suppliers and users of innovations. The number of firms and the nature of competition are one of the important factors but so too are the media, the education system and the policies of governments. Monopolistic market structures on the supply side may delay the rate of diffusion through high price strategies as well as through conservatism in relation to potential new markets. The compulsory licensing of polyethylene to several other US chemical companies in addition to Du Pont helped to accelerate the adoption of this material in many new applications both through price competition and more intense market development. But even in the case of a high degree of monopoly and low profitability, government policies may impose a high rate of diffusion, as has been the case in some countries, at least for short periods in the case of nuclear power. Governments may also decisively affect the rate of diffusion by policies in relation to standards, safety, procurement, environmental regulation, patents, and training. When we are considering an entirely new technology, clearly the availability of the necessary skills in potential adopter firms or other institutions may also be a crucial factor.

Diffusion of Microelectronic Technology

Having considered some of the main factors affecting the diffusion of innovations, as they emerge from the empirical and theoretical work over the past 20 years, we now turn to consider how they may affect the specific case of microelectronics.

Profitability

The process of technical change involves a great deal of gradual improvement, often described as 'incremental innovation'. Improvements in the efficiency of the internal combustion engine or the design of ships are examples of such incremental processes. It is relatively seldom that there is a 'technological leap' which permits an order of magnitude reduction in costs or opens up entirely new and profitable markets. Clearly such breakthrough innovations are likely to stimulate much more rapid diffusion than those innovations whose benefits are measured in fractions of one per cent. The major reason for expecting rather rapid diffusion of microelectronic technology is that it reduces the cost of information processing and other processes not by one but by several orders of magnitude, by comparison with earlier generations of valve technology, at the same time that it improves reliability and ease of operation.

This is a fairly powerful combination and of course it does not mean that the price of all electronic products can be reduced by an order of magnitude. The 'chip' is only one part of an electronic product or a process control system. Even in computer installations it must operate in combination with other peripheral equipment, whose costs have been reduced little or not at all. Moreover the full economic advantages of the chip can be realised only in cases where the same design is used on a large scale. For specialised applications, software and design costs can be very considerable. Nevertheless electronic products are virtually the only major group of products whose prices were falling or constant in the 1970s – a period of very great price inflation.

The advantages of using microprocessors in many *established* applications of electronics are so great that the process of substitution has proceeded extremely rapidly. The electronics industry has already grown accustomed to a fairly rapid process of introducing new generations of equipment in response to technical advances in the components industry. The average price of integrated circuits fell from $50 in 1962 to $1 in 1971. At the same time the percentage of electronic components used in integrated circuits (as opposed to discrete components) increased from 10 per cent in 1963 to 94 per cent in 1971.[6] In value terms the proportion was much smaller because the cost of discrete components remained far higher. This is only one example of the extremely rapid rates of diffusion associated with rapid cost reduction of new types of component which have characterised the electronics industry. [. . .]

Scale of investment

The reduction in cost of central processors and random access memories (RAMs) in the 1970s was so great that it brought the availability of computers within the purchasing capacity of a huge new class of potential users, including small firms and professional people. It also made possible the use of microprocessors in many products and processes where the scale of investment had previously been a prohibitive deterrent. Electronic calculators, toys and domestic computers are examples of markets which have been opened up as a result of this change. However, the problem is more complicated in considering many industrial and office automation and information-processing systems. In the case of industrial automation, there may often be heavy ancillary costs associated with computerisation. These include not only software design, which may be unique or confined to a few applications, but also new types of electro-mechanical equipment which may be needed to derive the full benefits from cheap and efficient computing power. Automobile assembly and machine-shop automation are two examples of industrial processes where the scale of investment required may be high because of complementarities and the need for heavy investment in new machinery as well as software.

There are still other areas where the scale of investment may be enormous, because of the system changes involved. One example is the introduction of fully electronic telephone exchanges, another is the introduction of fully automated shopping and payment systems in supermarkets and other large department stores. The change from a 'paper-based' office system to a fully electronic system for office work is feasible technically but involves substantial costs in new equipment such as word processors, as well as communication equipment and computer systems proper. Even more ambitious objectives such as 'Electronic Funds Transfer' require not only heavy investment in financial institutions (some of which is already in train) but also institutional and legal changes as well as complementary investment in the distribution system.

There is thus a succession of applications moving from simple low investment areas to highly complex system changes and heavy investment. Diffusion starts with very rapid rates on the left-hand side of the spectrum illustrated in Table 1, moving across to slower rates with increased investment outlays and greater system complexity, especially where institutional and legal problems are associated with hybrid systems of heavy public and private investment. Clearly in the areas involving heavy public investment procurement and enabling legislation much will depend on government policies and responses.

Technical advantages

The drastic reduction in power consumption is both a major technical and a major economic advantage and the same is true of the fall in materials consumption associated with miniaturisation. The first generation of large computers required as much power as an electric locomotive, filled a large room, and required a specially controlled environment in which to operate. A microprocessor with greater computing power, faster speeds and greater reliability occupies less space than a fingernail, requires the same power as an electric torch bulb, and is extremely robust. The main technical advantages have been well summarised in the Rathenau Report[7] for the Dutch Government on microelectronics:

> The term micro-electronics is only relatively adequate. It expresses only one of the properties of the electronic products resulting from this new technique, though it is also a very important one: their small size. The surface area on which even highly complicated functions are implemented is no larger than a few millimetres square and efforts continue to reduce that size still further. The development of sub-micron technologies await us in the future
>
> A second characteristic of micro-electronics is that the collection of electronic elements (cells) with their interconnections forms a functional unit and is designed, manufactured and applied as a single whole. Thirdly, micro-electronics might also be called large-series or mass-production electronics. Once designed, an electronic product, however complicated, can and will normally be manufactured in large quantities, so that the marginal cost per additional copy, as in the case of a newspaper, will be practically negligible. Fourthly, micro-electronics is a universally applicable form of electronics. It is difficult in advance to set limits to its application, although it is a fact that its possibilities of application will partly depend on requirements in other fields. Finally, micro-electronic techniques require practically no rare or expensive raw materials.

Environmental acceptability

This is a factor which is favourable to microelectronic technology in almost all its applications. In the first place the technology has low requirements in terms of energy and materials and offers considerable savings in both, whenever it is substituted for electro-mechanical systems. Moreover, the refinement of control which it facilitates means that it is ideally suited to energy-saving applications in existing systems, such as fuel control in the internal combustion engine, the electric motor, marine diesels, central heating systems, and so forth. For these reasons, unlike nuclear power, micro-electronics is often regarded as acceptable, if not benign, by what might be loosely described as the 'ecology lobby' or environmentalist movement.

In the second place the technology appears to involve few if any health hazards whether for users, workers in the industry or third parties. The problems of eye strain and back strain, which have concerned some union negotiators in connection with VDUs, appear to be quite manageable in terms of good work procedures, ergonomics and regulation of work conditions. Again, it is the positive advantages of the technology which seem to far outweigh any known hazards. The applications in robotics and in medical electronics make it possible greatly to reduce health hazards in handling dangerous materials, to monitor and diagnose potential health hazards more rapidly and efficiently, and to develop greatly improved and new equipment for the disabled and sick.

It therefore seems probable that the diffusion of the new technology and the extension of the range of applications will not be retarded in the way that some chemical, drug and food processes and products are now delayed by public anxieties and regulations affecting consumer safety and environmental hazards. [. . .]

Turning to the attitudes of potential adopters as opposed to suppliers of the new technology, the sheer range and variety of potential applications means that these are almost as varied as the entire population. Looking again at Table 1 there is a spectrum of potential adopter groups, ranging from those with very high motivation, a technologically sophisticated approach and ample financial resources, to those who are likely to be very slow adopters, because of attitudes, skills, marginal relevance and poverty.

Summarising the major factors affecting diffusion they add up to a very favourable constellation of circumstances. If a comparison is made with other major post-war new technologies, such as nuclear technology or supersonic flight, then it is clear that both of them emerge far less favourably from this type of analysis. Neither of them has been conspicuously successful on the first criterion of profitability either for suppliers or users. Outside the military applications area losses have been more obvious than profits.

References

1. E. M. Rogers, *The Diffusion of Innovations*, New York: Free Press, 1962.
2. E. Mansfield, *Industrial Research and Technological Innovation*, New York: Norton, 1968.
3. G. F. Ray, and L. Nasbeth, *The Diffusion of New Industrial Processes: an International Study*, Cambridge: Cambridge University Press, 1974.
4. For example, P. Stoneman, *Technological Diffusion and the Computer Revolution*, Cambridge: Cambridge University Press, 1976.
5. J. S. Metcalfe, *The Diffusion of Innovation in the Lancashire Textile Industry*, Manchester: School of Economics and Social Studies, no. 2, 1970, pp. 145–62.
6. J. Kraus, *An Economic Study of the US Semiconductor Industry*, Ph.D thesis, New School for Social Research: New York, 1973.
7. Rathenau Advisory Group, *The Social Impact of Microelectronics*, The Hague: Government Publishing Office, 1980.

Section 5

Government

Introduction: National and Local Government Policies for Design and Innovation

David Wield

Section 5 is also divided into two parts. The first part (articles 5.1–5.5) concentrates on national-level technology policy and industrial strategy. The second part contains three short pieces (articles 5.6(a), (b) and (c)) on science parks and other local innovation initiatives.

Technology Policy and Industrial Strategy

As described in the first section of this book, since the nineteenth century there have been increasing pressures on governments to intervene in the planning, coordination, and control of science and technology. These pressures have been accommodated by gradually increasing government involvement. To the older tradition of controlling the worst effects of hazardous scientific and technological advances, governments have increasingly taken responsibilities for promoting research and its exploitation to produce industrial and technological innovation. Traditionally, such government (or state) interventions have been divided into *promotional* and *regulatory* measures.

By promotional is meant those measures taken to sponsor technological projects and foster the adoption of new technologies. Pressures on governments for such intervention include national security and the development of military technology, and concern over the relative performance of national industry on world markets. The General Introduction at the beginning of this book discusses the historical development of promotional measures. By regulatory is meant those measures taken to avoid or control the 'undesirable social and environmental consequences of technology', as Braun puts it in article 5.1. These include measures to control the health hazards of industrial production, such as health and safety at work legislation, and air and water pollution controls, as well as rules and standards controlling the marketing and use of potentially dangerous technologies like chemicals and drugs. Thus the possibilities for state intervention can range from introducing simple safety standards to nationalizing huge companies and restructuring entire industries. Braun's article contains a useful table, which has a comprehensive list of possible forms of government intervention.

The articles in section 5 concentrate on promotional rather than regulatory measures. We will briefly introduce three themes, although there are many others in the vast and often confusing literature on government policies for technology and industry. The first theme is to what extent increased government intervention is important. The second is in which parts of the innovation process intervention may occur. And the third theme is how to select the technologies and/or industrial sectors for government intervention. Over-shadowing these themes is the issue of how far governments can intervene (or withdraw from intervention), given that their powers are not absolute. There is a mistaken tendency to believe that governments can act straightforwardly both to promote and regulate technological activities without reference to the public, *and* indeed to 'roll back' such inter-vention to allow market influences to drive technological and industrial change.

On the first theme, there is fairly broad agreement that governments should intervene to regulate the production and use of a whole range of potentially dangerous products and processes, from disposal of nuclear waste to the sale of medicines. But agreement has not been so general on the need for government promotion of technological development. Cer-tainly Williams, Roy and Walsh (1982) have argued that 'regulation is a matter of administration, whereas sponsorship is a matter of party political controversy'. The need for government involvement with the production of military technologies has been accepted over a long period, although more recently there has been considerable criticism of what proportion of resources should go to military as opposed to civil technology. What has caused much more controversy has been whether there is a need for coordinated plan-ning for technological innovation. In article 5.2, Williams goes so far as to say that the world 'plan' was in 1984 so ideologically unacceptable in Britain that the more neutral phrase 'national, industrial and technological strategy' had to be used by those arguing for integrated approaches to government technology policy. Certainly a recent article by a civil servant in the Policy Planning Unit of the Department of Trade and Industry (Wallard, 1984) uses just this kind of phraseology.

The example of Japan has been well used both by those arguing for and by those arguing against increased intervention. Those arguing against have pointed to the very high private industrial investment in technological innovation, and to the fact that successful Japanese companies are privately owned and unsuccessful ones are not bailed out. Those arguing for have often used as their example the strategies of Japan's Ministry of International Trade and Industry (MITI). MITI has played a substantial role in assisting companies to coordinate their efforts in the early days of a technological innovation.

MITI's procedure is to find out what a strategically important industry requires to become competitive, then to make a detailed survey of industrial producers and users. It sets up a joint industry–MITI group to decide on work required. The work is then divided among the companies and government research laboratories. The idea is to develop pro-totypes that companies then feel committed to produce. Projects exist on new materials, new electronic devices, and bio-technology, and MITI's programme includes the famous fifth-generation computer project with the involvement of eight companies – Fujitsu, Hitachi, NEC, Toshiba, Mitsubishi Electric, Matsuchita, Oki Electric and Sharp – as well as government research laboratories and the new Institute of New Generation Computing.

What emerges from a study of recent Japanese industrial growth is a strategic product cycle of around fifteen years. This has five phases, according to Bownas (1983). First, a sec-tor is pinpointed as strategic and a new technology is developed with MITI support. Then the domestic market is exploited under protection. This seems to take about five years. Thirdly, the export market is invaded, while domestic markets continue to expand. Then, after about ten years, production begins to be transferred abroad to low-wage or low-skill economies such as Wales and South Korea. Finally, new, technologically more advanced products are produced in Japan.

Although Japanese policies of government intervention have been particularly successful in the recent past, other governments' attempts (for example, in France, Germany and Britain) to promote science and technology have some common attributes and characteristics (Wallard, 1984). They point not so much to a polarization between intervention and non-intervention, but rather to how much intervention is appropriate under the specific conditions prevailing in each country at any particular time.

This relates to our second theme: in which parts of the innovation process should government be involved. Braun divides the innovation process into four phases. *The zeroth phase* – namely, the general 'ambience' within which innovation takes place; *the first phase* – the emergence of an idea, in the main from basic research laboratories, where public funding is often essential; *the second phase* – the development phase which is still within the research and development part of the innovation cycle and thus rather risky, and where public funds have traditionally also been very significant; *the third phase* – the implementation phase, normally the most expensive according to Braun. This phase is the most recent to get substantial government funding, since it was the early parts of innovation, often termed research and development (R & D), that were seen to be most risky, but essential for national economic growth.

In article 5.2, Williams argues that it was not until the 1960s that the view that there was a clear linear relationship between R & D and economic growth was discredited. Until then governments put money into fundamental research in the expectation that eventually growth would occur. Increasingly, the tendency has been for governments to call for more applied research and development, and a closer relationship between R & D and industrial strategy. Braun's article (5.1) includes a detailed description of recent government support in Britain for microelectronics – the Microelectronics Application Project (MAP) – which he presents as an integrated strategy.

Clarke's article (5.5) introduces British energy policies from the mid-1970s, with particular reference to renewable energy technologies like wind, wave and solar. He argues that in this case government shied away from funding renewable energy technology research at the later stages of the innovation process, thereby significantly damaging its chances of competing with other energy technologies like nuclear power and fossil-fuel-based systems, themselves massively funded through state intervention. Both Braun and Williams appear to argue for government support at all stages of innovation. Braun sees his list of policy measures (Tables 2 and 3 in his article) being used in any phase of the innovation process. He visualizes the innovation process as a kind of electrical network in which all the switches have to be appropriately set before the innovation can proceed, and reasons that policy actions can help turn on switches at any phase.

Lorenz's article (5.4), echoing points made in the General Introduction, advocates a more integrated approach to government support for *design,* rather similar to the calls for an integrated approach to *innovation,* through the joining of technological and industrial policies. Lorenz calls for government support for design, linking its support for education, research and industry, so that one branch of government does not take away what another has decided is crucial if design quality is to improve within British industry.

The third theme is the selection of technologies and industrial sectors for government intervention. Here the work of Ieuan Maddock, former Chief Scientist at the Department of Industry, is important (Maddock, 1975). It shows a very great imbalance between government support of R & D in different parts of the economy (see Tables 1 and 2 of Williams' article). He observes the very high concentration of British expenditure on defence R & D. Martin Walker's short article (5.3), summarizing a NEDO report (NEDO, 1983), shows how large a proportion of Britain's R & D expenditure is taken by defence projects – indeed by 1984 it accounted for over half of *all* government R & D expenditure. Walker echoes Cottrell's point, in section 3, that there is probably an upper limit for total R & D expenditure as a proportion of national wealth. Maddock questions whether Britain can afford more R & D, and thus questions the balance between military and civil expenditure; he also shows the high relative state expenditure on nuclear energy, and on basic research like high-energy physics and radio astronomy.

Williams argues that current expenditure on R & D bears a closer relationship to the past than the future. He says that its roots lie in Britain's former status as a leading world power, with major interests in aviation and the new nuclear industries. By comparison, civil manufacturing industry has had much less state support for innovation than Britain's main competitors. He concludes that such support in Britain in the 1980s 'cannot be a matter of political values'. Williams quotes Sir Peter Carey, a former senior civil servant in the Department of Industry, as saying that government cannot pick winners, but ought to be able to identify technologies that are going to be important in the future, rather as MITI in Japan has done. But Carey doesn't go as far as to agree to what he calls 'interference' with industrial companies.

As far back as the 1940s similar views were being put forward. The Federation of British Industry (forerunner to the CBI) was clear that private industry could accept a certain level of state support, but certainly no element of state direction of industry. This tension between acceptance of government support, but antagonism to government control has been a constant theme in Britain since then. It is certainly a different tension to that between MITI and the Japanese industries that have been involved in the so-called economic miracle.

As shown in the second part of this section, on science parks, it is not that successive governments in Britain have not intervened in attempts to improve national industry, but as Williams concludes: the overall balance of Britain's technological effort is meanwhile almost certainly wrong.

Science Parks and Industrial Innovation*

A large number of science parks have been established in Britain in the last few years. In addition to the three established in the early 1970s (Cambridge, Heriot-Watt Research Park in Edinburgh and Birchwood Science Park in Warrington), another eight had begun by late 1983. In early 1984, groups from universities, polytechnics, local authorities and development corporations, government agencies, banks, venture capital and other private firms were planning or discussing over thirty more schemes.

The first use of the term 'science park' was probably to describe Stanford University's Industrial Park, south of San Francisco, which grew rapidly after the Second World War. It now has more then one hundred high-technology firms located on university-owned land, and more than a thousand others close by, in what is now called 'Silicon Valley'. Professor Terman from Stanford encouraged his graduate students, like Hewlett and Packard, to set up businesses close to the university. They were joined by William Shockley, co-inventor of the transistor, who founded Silicon Valley's semiconductor industry. Terman believed that universities could be more than places of learning. 'They are major economic influences in the nation's industrial life', he said in 1950, 'affecting the location of industry, population growth and the character of communities. Universities are a natural resource' (Kehoe, 1983). In article 6.2 in section 6, Peter Hall discusses reasons for the growth of Stanford and similar science parks in the United States. Bullock (1983) from Barclays Bank

---

* This part is drawn from work by Elliott, Massey and Wield (1984).

summarizes clearly what can be learned from the United States in encouraging the growth of small high-technology firms through higher risk loans and venture-capital.

Although the US experience is often referred to by advocates of science parks in Britain, Moreton's article (5.6(a)) shows that Harold Wilson was the instigator of Britain's first parks. In the second half of the 1960s, the government wrote to all universities suggesting that they do more about innovative technology; it was from that initiative that the Heriot-Watt and Cambridge Science Parks began.

The growth in interest in science parks raises a number of interesting issues on the question of industrial innovation. In a sense it ought to be possible to assess the possibilities of their success using a knowledge of the innovation process. We will introduce here three issues that we believe are important in any such assessment. First, improvement of the relationship between academics and industry so that increased transfer of technology takes place from academia to industry. Secondly, the importance of science parks in locally based attempts to increase local economic development. And thirdly, whether the division between R & D undertaken in science parks and production will reinforce not only a locational split, but also a potentially significant split in the innovation process.

Underlying the enthusiasm for science parks is the belief that their establishment will help to solve a key bottleneck in the innovation process – that is, the transfer of technical ideas and expertise from academia to commercial application in industry. British scientists and technologists are undoubtedly good at coming up with novel, potentially profitable ideas. But often new ideas and inventions are not developed successfully or exploited commercially. Science parks are seen as one means of resolving Britain's poor industrial performance by assisting the transfer of ideas between academic institutions and industrial enterprises. The location of science parks near major academic centres is seen as a way of improving this situation by ensuring close links with researchers. In addition, particular attention is paid to the physical environment and working conditions – in the belief that this will attract well-qualified personnel and stimulate creativity.

A key practical question is whether the location of science parks near to universities is likely to lead, or has led, to improvements in the technology transfer process. The indications are that this may not always be so – contact between academia and the science parks has often been minimal (Elliott, Massey and Wield, 1984). The success of the parks in fact seems more likely to depend on the 'image' they attain in the eyes of customers, investors and competitors, because of their location in pleasant environments near to centres of technological/scientific excellence and in areas with an 'innovative climate' such as Cambridge (Bullock, 1985).

This raises a second question, regarding the science park's role in local economic development. Science parks are one of a number of local and regional initiatives aimed at increasing economic growth and reducing local unemployment. One approach is to set up local industrial development offices. This has been done in the main in the poorer parts of Britain, which are able to offer government grants to industrial investors. The biggest such regional operation by far is that of the Scottish Development Agency, set up by the Scottish Office and described in Faux's article (5.6(b)). It has energetically integrated offers of subsidies, cheap or free land, and planning permission to 'capture' footloose companies wanting to locate in Britain. Faux's article describes large foreign investments in semiconductor firms in Scotland by US and Japanese companies, which has led to parts of Scotland being renamed 'Silicon Glen'. From this phenomenon, one could easily draw the conclusion that these regional initiatives dominate investments being made. But it is the south and southeast of England that still have the largest concentrations of semiconductor plants (47 per cent), although Scotland and Wales do have a significant proportion (19 per cent and 11 per cent respectively) (Cooke, Morgan and Jackson, 1984).

Another more recent alternative pioneered by some large local authorities is to set up a Local Enterprise Board. The idea here is to get away from 'capturing' industry in one area at the expense of decline in another. Some Local Enterprise Boards are attempting a more direct approach by investing in selected industries, encouraging innovation in local academic institutions, as well as organizing local communities to articulate what kinds of new product they want, or what products they want to continue to be made in their local area. Mike Cooley's article (5.6(c)) enthusiastically describes the London example, the Greater London Enterprise Board (GLEB): more details may be found in Elliott (1986).

Many local councils now include science parks as a component part of their development policy. Underlying their support is the belief that science parks will help strengthen the local economy, in particular by creating employment. This depends on science parks being replicable from one place to another. Yet there are already a number of reasons for believing that the 'sci-tech' image is not replicable in all parts of the country, For example, it can be extremely hard to create this image in the middle of a semi-derelict conurbation. Aston's park in central Birmingham and Heriot-Watt in Edinburgh are situated well outside the locations favoured for 'sun belt' development in Britain – those areas north and west of London, stretching from Cambridge via Milton Keynes and Reading to Bristol. If they manage to succeed with an image of clean, high environmental quality, high skill and high technology, then an important breakthrough will have been made in terms of the perceived locational advantages of different parts of the country. If, on the other hand, they are not able to succeed in that way, then there is a real danger that science parks can only further exacerbate the geography of social difference within the United Kingdom. Peter Hall's article (6.2) reproduces the view that 'new technology' industries will not grow in what he calls 'yesterday's regions', meaning the already much poorer, old industrial areas like South Wales and north-east England.

The third issue which arises is whether, in their efforts to establish links between industrial and academic research, science parks are not merely promoting a further division between R & D and production. Such a division is of course not new. There are already many examples of large firms that divide R & D and production between different plants. But for science parks this is their whole rationale, indeed there is evidence of production not being allowed into their premises. In terms of the eventual competitiveness of British industry, it may be just as important that there is contact within the firm between R & D and production as that there is contact between industrial R & D and academic research. As several articles in this book testify, the whole chain has to be complete (see Roberts' article (3.5), for example). Examples exist where firms have not moved to science parks precisely because they could not take production along with research and development. The Government has given its backing to science parks. But at the same time one Minister has said that the physical separation of manufacturing industry and its support services, particularly R & D, can lead to inefficiency. The possibilities of reinforcing both a social and a locational split between the research end of British industry and shopfloor production is one with major implications.

References

Bownas, G. (1983) Blueprint for growth that keeps Japanese industry in the lead, *The Times*, 29 July.

Bullock, Matthew (1983) *Academic Enterprise, Industrial Innovation, and the Development of High Technology Financing in the United States*, London: Brand.

Cooke, P., Morgan, K. and Jackson, P. (1984) New technology and regional development in austerity Britain: the case of the semiconductor industry, *Regional Studies*, vol. 18, pp. 277–89.

Elliott, D. A. (1986) Government: renewable energy; the GLC, T362 *Design and Innovation* (Block 4), Milton Keynes: Open University Press.

Elliott, D. A., Massey, D. and Wield, D. (1984) Science parks and industrial innovation in Britain, Technology Policy Group Research Report, *mimeo,* Open University.

Kehoe, Louise (1983) Silicon valley grew out of Stanford campus, *Financial Times,* 21 January.

Maddock, I. (1975) The Seventh Royal Society lecture, Science technology and industry, 12 February, *Proceedings Royal Society,* London: Royal Society, pp. 295–326.

NEDO (1983) *Innovation in the UK,* London: National Economic Development Office.

Segal, Quince and Partners (1985) *The Cambridge Phenomenon,* Cambridge: Segal, Quince.

Wallard, A. (1984) The problematic relationship between research and policy, in M. Gibbons, P. Gummett and B. M. Udgaonkar (eds.), *Science and Technology Policy in the 1980s and Beyond,* Harlow: Longman, pp. 191–209.

Wells, P. (1984) Science parks, *Marxism Today,* July, pp. 2–4.

Williams, R., Roy, R. and Walsh, V. M. (1982) Government and Technology (T361 *Control of Technology,* Units 3–4), 2nd ed, Milton Keynes: Open University Press.

NATIONAL POLICIES

5.1

Government Policies for Technology

Ernst Braun

Introduction

The current interest in technology policy arises out of a paradoxical twin concern: on the one hand the wish by each nation not to be left behind in the race for technology-based prosperity and on the other hand multi-faceted fears about undesirable social and environmental consequences of technology. The belief that better technology means higher economic efficiency and international competitive advantage is a strong driving force behind attempts by most governments to support, foster and accelerate technological innovation. The same recognition also forces governments into a range of policies for infrastructural support of technology in fields as diverse as education and training, telecommunications, fundamental research, patent laws and technical standards. The fears about technology, on the other hand, force governments into a range of regulatory activities, and sometimes also into political battles for their procurement and support efforts.

There is no doubt that in a market economy the major force affecting technology is the market. Nevertheless, public bodies feel compelled to intervene in a regulatory mode in order to avoid damage to health and the environment which unbridled technology might cause, and in a supportive mode to enhance and accelerate the perceived advantages of technology in fields as diverse as administration, defence, public transport and communications. [. . .]

Measures Available to Government

The totality of measures which governments can and do take to influence technology is infinitely varied. However, the infinite variety yields readily to classification and some

E. Braun, *Wayward Technology*, London: Frances Pinter, 1984; extracts from Chapter 5.
Professor Braun was Director of the Technology Policy Unit at Aston University, and now works at the Austrian Academy of Sciences.

insight can be obtained in this way.[1] Government policies can be classified according to three criteria: (i) aim of policy; (ii) type of measures; (iii) target of measures.

Apart from very specific policy objectives there is now a recognised range of technological infrastructural activities which governments accept as their legitimate responsibilities. These include education at all levels, transport and communications, technical standards and regulations about the use of technology, fundamental research, and a legal framework regulating monopoly, competition, professional activities and many more.

In recent years two major convictions have gained considerable strength. On the one hand, it is widely believed that economic growth can only be regained and sustained with the aid of a high rate of technological innovation and on the other hand it is

Table 1   Types of policy measures aimed at stimulating technological innovation

Type of measure	Examples
Financial	Grants, loans, subsidies, financial sharing arrangements, loans and gifts of equipment, provision of free services, provision of buildings
Taxation	Company, personal, indirect and payroll taxation; tax allowances, tax deductible expenditure
Legal and regulatory	Patents, environmental regulations, health regulations, inspectorates, protection of designs, arbitration services, monopoly regulations, planning permissions for buildings and enterprises
Educational	General education, universities, technical education, apprenticeship schemes, continuing and further education
Procurement	Defence purchases, central government purchases and contracts, local government purchases, R & D contracts, prototype purchases
Information	Information networks and centres, libraries, radio and television, freedom of information, advisory services, statistical services, government publications, data bases, museums, exhibitions, liaison services
Public enterprise	Innovation by publicly owned industries, setting up of new industries, pioneering use of new techniques by public corporations, correction of imbalances by public enterprise, participation in private enterprise, investment by public corporations
Public services	Investment and innovation in health services, public building, civil engineering and construction, transport, telecommunications, consumer protection
Political	'Atmosphere', honours system, intervention vs. non-intervention, regional policies, labour policies
Scientific and technical	Technical standards, government research laboratories, testing stations, support for research associations, learned societies, professional associations, research grants
Commercial	Trade agreements, tariffs, currency regulations

Source: After Braun, E., 'Government policies for the stimulation of technological innovation', Working Paper WP–80–10, IIASA, Laxenburg, January 1980.

thought that the required rate of successful innovation cannot be achieved without government support. The corollary of this twin conviction is that any industrial country with an insufficient rate of innovation is bound to suffer relative economic decline and thus government aid to innovation is necessary for a country to hold its own in international competition. We might indeed speak of a new mercantilism, whereby governments no longer attempt to increase their country's share of gold, the old symbol of wealth, but rather their share of the new symbol of affluence: advanced technology.

Thus a specific policy objective of most governments of industrial nations is the stimulation of technological innovation and this objective shall serve as our first example of a technology policy. The first question we must ask is what government can do to achieve this aim. We may distinguish at least ten different types of measures which can be taken towards this general aim. Table 1 lists the types of measures and gives some examples of possible activities.

The different types of measure can be targeted in different ways. If the general objective is the stimulation of innovation as in our example, then we may distinguish four different target areas for policy measures: (i) general ambience, i.e. the economic, social and political atmosphere in which innovation takes place; (ii) industry in general, i.e. policy measures to strengthen industrial performance and industrial investment or measures to strengthen some specific industrial sectors; (iii) innovation in general, i.e. measures aimed at innovative activity rather than at specific innovations; (iv) specific innovations, i.e. support for a particular new technology. Table 2 gives a few examples which show how different types of policy measures can be targeted.

Policy measures can act in any phase of the innovation process. The innovation process may be viewed as a kind of electrical network in which all the switches have to be appropriately set before the innovation can proceed from one phase to the next. The setting of switches, symbolic of the actions required by human actors, visualises the points at which policies can aid the required actions.[2]

The zeroth phase of innovation, in which consideration is given to the nature of the firm and the world it lives in, is strongly influenced by the ambience of free or restricted trade, availability of credits, availability of research facilities, information networks, monopolies and other tangible facts of political and economic life. But intangibles of ambience also play a role – personal honours and social hierarchies, the aura of success, public opinion.

The first phase of the innovation process – the emergence of an idea for innovation consisting of the confluence of a new technical possibility with a market need – requires easily accessible information systems to provide both market and technical intelligence. Many, though by no means all, ideas for innovation emerge from research laboratories and here support from public funds can be of the essence. For the research stage of an innovation can only be financed either from internal or from public funds; normal loan finance, however risk oriented, will shy away from the incalculable risks of research. The support of pure and applied research has therefore traditionally been a cornerstone of public support policies for technological innovation. This support can be in the form of direct or indirect research subsidies, in the form of publicly financed laboratories or publicly aided research associations. Sometimes the best support is in the form of liaison and information, for innovative ideas generally spring from the fusion of facts and ideas from previously separated areas.

To get from the first phase of an innovation to the second – the development phase – requires aid which is somewhat similar to that of the first phase, for more research and development work needs to be done and the product may still not be tangible enough for commercial venture financiers to be very interested. The innovator must have good knowledge of available materials, techniques, specialist suppliers, designers and, last but not

Table 2 Examples of policy measures and their targets

Type of Measure	Target			
	Ambience	*Industry*	*General innovation*	*Specific innovation*
Financial	Ease and cost of credit	Investment in regional factory building	Making venture capital readily available	Supporting specific R & D programmes
Taxation	Supporting entrepreneurial spirit	Making investment allowances	Allowances for innovative investments and R & D expenditure	
Legal and regulatory	Patent laws, monopoly regulations	Factory legislation	Health and safety regulations	
Educational	General educational provision; support for higher education	Technical training schemes		Training schemes in specific new areas
Procurement	Level and type of public expenditure	'Buy at home' policies	Procurement specifications	R & D contracts and orders for specific new equipment
Information	Libraries, broadcasting, government statistics	Technical information services	Liaison services	Information programmes on specific new technologies
Public enterprise	Strength of public sector	Active regional policies	Participation in new ventures, innovative policy in state industry	Public enterprise in new technology; specific innovations in public enterprises
Public services	Transport and communications			Development and use of specific innovations
Political	Access to information; public opinion			
Scientific and technical	Technical standards		R & D availability from public sources	Specific R & D support
Commercial		Tariffs, trade missions		International co-operation in new ventures

Source: E. Braun, 'Government policies for the stimulation of technological innovation', Working Paper, WP–80–10, IIASA, Laxenburg, January 1980.

least, advisers on patents and licences. This means that the innovator needs to be a member of an efficient information network. As at this stage much effort and money are consumed, financial assistance can be crucial.

In the past most government innovation support measures stopped with the end of the second phase, when a finished prototype emerged. Marketing and starting up production of even the most revolutionary products were regarded as purely commercial activities, for which public support was both unnecesssary and repugnant. More recently, thinking has changed a little and some support programmes extend right up to the marketing stage. As the third innovation phase, the implementation stage, is the most expensive and also exposes the greatest weaknesses in the scientific/technical first-time entrepreneur, such extensions of support must be regarded as eminently sensible. Policy measures should not only be designed to stimulate innovation, they should also aim to remove difficulties at specific points of the innovation process. Hence an understanding of the process – a theory of innovation – is an integral part of innovation policy. [. . .]

So far we have considered only government support for technological innovation, but other purposes of technology policy are equally possible. Government may wish to assure maximum safety at work, or the cleanest possible air, or the least consumption of scarce raw materials or any other conceivable objective of technology policy. In each case different sets of possible measures with different target areas can be identified. Let us pick just one more example and assume that a government wishes to ensure minimum consumption of scarce raw materials. The easiest way to achieve this is, of course, to reduce all technical and production activities to a bare minimum. This, however, would be incompatible with other government objectives, such as the maximum possible standard of living for the population, greatest possible influence in world affairs, free trade, and what we have called the new mercantilism. The compatibility or incompatibility of different objectives is one of the many difficult problems of policy making. [. . .]

It is apparent that the consumption of raw materials depends on total manufacturing activity, total consumption, design and life-time of products, degree of recycling of used materials, and efficiency of extraction of materials from their raw source. Policy measures can attack all these aspects. For the sake of brevity the types of policy measures which can be used for the achievement of minimum raw materials consumption and their target areas are listed in Table 3.

The Microelectronics Application Project

[. . .] The basic strategy of all support schemes attempts to stimulate the availability of general advice and information; special consultancy advice; grants for research and development; and aid for product launch or the introduction of new production machinery, as applicable. We shall descibe only the Microelectronics Application Project (MAP) in some detail.

In mid-1978 the British government became extremely concerned about the lack of awareness among British managers of the opportunities and challenges offered by microelectronics. A survey conducted at the time found that only about half of the firms in manufacturing industry were aware of the existence of microelectronics and that only 5 per cent of firms were actually doing something about it. The Department of Industry then initiated several new support programmes for microelectronics under the Microprocessor Application Project (MAP) and one important aspect was 'to alert managements to the industrial scope and potential of microprocessors and to assist in retraining staff'.[3] Under this scheme numerous conferences were organised, supported by DoI funds and publicity material. Training courses in colleges and universities also received support under the scheme. It is thought that the awareness and training programmes, combined with substantial media coverage and the publication of numerous books, did raise the awareness of microelectronics to a pretty high level over a period of about two years.

Other government departments are making their contributions to the training and edu-

Table 3   A few possible ideas for measures to implement a materials conservation policy

Type of measure	Target area				
	Ambience	*Industry*	*Commerce*	*Foreign trade*	*Consumers*
Financial	Cheap long-term loans, expensive short-term money	Support for materials saving design; support for recycling; research grants for materials conservation	Incentives for collection of used materials	Tariffs on selected materials	Incentives for recycling; support for repairs to make goods last longer; long-term loans for consumer durables
Taxation	Lowering of relative cost of labour vs. capital	High tax on new materials; lower rates on recycled materials; similarly, high tax on new machinery, low on reconditioned	Tax incentives on recycling operations and material collecting services		Decreasing tax rates for older durables, including cars
Legal and regulatory		Control of waste materials; controls on writing off of machinery		Import restrictions on selected materials	
Educational	'Old is beautiful' education	Materials engineering courses			
Procurement		Minimum arms purchases; specifications written to save materials			
Information	'Waste not, want not' campaign	Recycling and materials saving information centres			Propaganda on value of materials; 'old and working well is as good as new'
Public enterprise		Research on recycling, on materials substitution and materials on saving design	Public trading companies for materials		Nationwide repair services; material collection services
Political	Peace in our time – for war is the greatest waste of materials			Re-negotiation of trade agreements	
Sceintific and technical		Research institutes for materials conservation; support services for materials conservation and substitution			

cation programmes. The Department of Education, in conjunction with the DoI, is supplying all schools with small computers (though difficulties seem to arise from a shortage of cash to buy software and books). The Department of Employment, through its Manpower Services Commission is supporting many training courses for young people and retraining courses for unemployed older workers. The latest development in this activity are 6 to 12 months courses for young unemployed people, run in conjunction with so-called Technology Centres. These are advisory centres on the use of microcomputers and are administered by the National Computing Centre. The Department of Industry is providing initial financial support for the centres, but within two years they will have to earn enough income from consultancy and training services to become self-financed.

Traditionally, the manufacture of electronic components in Britain has been supported in great measure by the Ministry of Defence (MoD). The MoD has paid and still pays for much advanced research and development work and also is a customer of some substance for electronic components and devices. Plans are rumoured for a huge investment of £50 million into the development of very high speed integrated circuits; a path the US military have also been treading.

The MoD is by no means alone in its support for the electronic component industry. Even in early 1978 DoI support for the industry was running at £20 million per annum – 10 per cent of this for semiconductors. In 1978 a scheme called Microelectronics Industry Support Programme (MISP) was announced.[4] This scheme was supplementary to MAP and pledged £70 million for a five-year programme aimed at helping companies to establish the manufacture of both special purpose and standard microelectronic components. MISP also aimed to support potential industry and perhaps radically new alternative processes which might prove important for the industry.

The picture would be very incomplete indeed without mention of INMOS. This is a company founded in 1978 and financed to the tune of £50 million by the National Enterprise Board (now British Technology Group or BTG).* The purpose of the company is to manufacture VLSI (very large scale integration) chips only, so as to become specialists in the high technology end of this high technology industry. The first products announced by the company were a 16K static RAM and a 64K dynamic RAM. These are probably to be followed by very advanced microprocessors, perhaps single chip microcomputers.

Although BTG money essentially comes from the Treasury, the enterprise was set up by three entrepreneurs and two of these are American. In fact the company has manufacturing and design facilities in both the United States and Britain and, arguably, this dual base and with it access to American know-how, is a necessary condition to make eventual success possible.

Many observers believe – and some official reports support this belief – that for most countries it is more important to apply microelectronics than to manufacture integrated circuits and other microelectronic components. Outstanding among the applications is, of course, the manufacture of electronic equipment of all kinds: telephone exchanges and other telecommunications equipment, computers large and small, instruments of all kinds, process control equipment, electronic office machinery, electronic games and all the rest of the known and still to be invented electronic gadgetry. Many of the items listed above are, in a loose sense, free-standing and self-contained items of electronics. No less important are applications of microelectronics in conjunction with mechanical machinery. To list but a few: computer numerically controlled machine tools, robots, computer controls for automobile engines, electronic weighing machines and computer controlled warehouses.

[* INMOS was acquired by Thorn-EMI in 1984.]

It is argued that the addition of modern electronics to many traditional machines can breathe new life into them and thus create new markets on the one hand and ensure competitiveness on the other. These considerations are reflected in several policy measures. Thus the previously mentioned Microprocessor Application Project (MAP) has, in addition to its information function, two strands in support of microelectronics applications. One consists of help with feasibility studies for new products, normally carried out by consultants registered by the DoI for this purpose. By late 1980 nearly 800 such studies had been completed and 75 per cent of their projects were being further pursued. The second consists of financial help to firms developing products using microelectronics. By late 1980, 345 such developments had been supported with grants of up to 25 per cent of qualifying costs or an alternative cost-sharing arrangement. Currently [in 1983] a 'special offer' of 33⅓ per cent grants is available.

More recently, both the British Technology Group and the Department of Industry have re-stated their faith in microelectronics application and have expressed their willingness to back their faith with money. BTG is giving high priority to firms wishing to develop and market innovations using, or related to, microelectronics and information and the DoI is planning to spend £80 million over the next four years on the support of information technology, under their Product and Process Development Scheme.

The Product and Process Development Scheme straddles the very narrow boundary between the development and diffusion of new technology; that is between the development and implementation phases of technological innovation. While the DoI will encourage the applications of certain new manufacturing technologies, e.g. Computer Aided Design (CAD), Computer Aided Manufacture (CAM), and robots, it will also assist firms wishing to develop these technologies. In fact the interaction between supplier and user of advanced techniques is a vital one and many a project has failed because of inadequacies in this relationship. The Product and Process Development Scheme recognises these facts in assisting user trials through financing pre-production orders for new equipment.

Most recently robotics has been recognised as a priority area for government assistance and this spans the whole wide range of activities from fundamental research to development of robotic technology to the support of users of robotics. It must be stressed here that government is by no means the main agent of change. Robot manufacturers have perfected their products to a viable stage and have developed an adequate infrastructure of support for their potential clients. An association of industrialists and researchers, the British Robot Association, also plays an important promotional role.[5] Similarly, British industrialists have clubbed together to form the United Kingdom Information Technology Organisation and one of their specific aims is to influence government policy on information technology by presenting an industrial viewpoint.

Whether as a result of this lobby or otherwise, the government has become very active in the promotion of information technology, including the designation of 1982 as Information Technology Year. One of the ministers in the DoI has responsibility for this area and the IT Year is coordinated by a committee with industrial membership and an industrial chairman. A great deal of money and publicity were lavished on IT Year, which coincided with far-reaching decisions on the introduction of cable television in many British cities, thus laying the foundations for a vast flood of information, or what passes for it, into every British home and office. Among the many activities of IT Year, DoI has created an electronic show office, where all the latest advanced office equipment is demonstrated and advice given to potential users.

No attempt has been made here to give a fully comprehensive description of all government activity in the field of IT and microelectronics, but even from what has been said the reader will have gathered three strong impressions:

1. the British government is trying very hard and spending a lot of money to help British industry enter the microelectronic age both by using and producing the new equipment;
2. while some uses of microelectronics are relatively straightforward and do not require government intervention, other areas are so fraught with difficulty and risk that even a non-interventionist government feels obliged to intervene;
3. the possibilities for uses of microelectronics – whether in products or processes or as ready-made equipment – are wide and varied, while the possibilities of entry into the manufacture of integrated circuits are very limited indeed.

If the old liberal ideal, prescribing to the state the role of night-watchman, guarding the citizens unobtrusively against evil but not interfering with their legitimate activities, were still to hold, then it would appear that the night-watchmen have become engaged in quiet warfare among their own ranks. For modern governments vie with each other in the support of their industries and international trading competition has partly been superseded by international competition in support policies. Microelectronics and information technology support programmes are a prime example of this trend and very few governments are willing to be seen without such a programme.

References

1. E. Braun, Government policies for the stimulation of technological innovation, Working Paper WP–80–10, Laxenburg: International Institute of Applied Systems Analysis, 1980; E. Braun, Government policies for stimulating technological innovation, in H. Maier and J. Robinson (eds.), *Innovation Policy and Company Strategy*, Laxenburg; International Institute for Applied Systems Analysis, 1982, pp. 51–66; R. Rothwell and W. Zegveld, *Industrial Innovation and Public Policy*, London: Frances Pinter, 1981.
2. E. Braun, Constellations for manufacturing innovation, *Omega*, vol. 9 (1981), pp. 247–53.
3. Department of Industry, *Microprocessor Application Project*, London, 1978.
4. Department of Industry, *Microelectronics Industry Support Programme*, London, 1978. See also a series of brochures currently published by the Department of Industry on various innovation support schemes.
5. J. Fleck, *The Diffusion of Robots in British Manufacturing Industry*, final report to the Leverhulme Trust Fund, 1983; J. Fleck, *The Introduction of Industrial Robots*, London: Frances Pinter, forthcoming.

5.2

British Technology Policy

Roger Williams

Francis Bacon observed that in the declining age of a state, mechanical arts and merchandise flourish. Unfortunately, Britain's case is now almost the obverse of this. The underlying causes of Britain's present circumstances are evidently complex, but a fruitful place at which to begin is with the country's technology policy and its associated research and development orientation.

It is especially alarming politically that the long run decline in manufacturing industry which Britain has experienced, and tolerated, throughout the century has recently accelerated. The country's share of world trade in manufactured goods fell precipitately in the 1960s (15 to 9 per cent), but worse still, there is evidence that imports have on balance become steadily more technologically advanced than exports. [. . .]

Among the factors central to the performance of any country's manufacturing industry is the capacity for rapid, successful and sustained technological innovation,[1] hence concern with what government can do to enhance this, that is with national 'technology policy'.[2] British 'technology policy' is in fact dispersed rather than concentrated and more the indirect aspect of other policies than a direct and self-contained policy in itself, though no criticism is to be inferred simply from this. On the contrary, the fundamental interest of technology to the policy-maker naturally lies in its contribution to, and implications for, more traditional policy areas.

Until the mid-1960s, when the matter was thought about at all, it was tacitly assumed that provided national R & D expenditure as a percentage of GNP was kept high, then all would be well. That there is no necessary correlation between R & D expenditure and growth was realized only later. It is also salutary to remind oneself that in 1964 the UK provided itself with a Ministry of *Technology*, only for the responsible politicians and civil servants to spend much of the ensuing six years discovering that technology was a tool rather than a policy, and that their real focus was properly *industry*. [. . .] But this, as it happens, was only one of two misplaced emphases in that 1964 initiative, and probably not the more important: more telling is the fact that the comparable agency through which, very largely, the Japanese industrial miracle was achieved and maintained was the Ministry of *International* Trade and Industry. In other words, whereas British thinking tended

R. Williams, British technology policy, *Government and Opposition*, vol. 19, no. 1, Winter, 1984, extracts from pp. 30–47.
Roger Williams is joint Professor of Government/Science and Technology Policy at Manchester University.

to be inward looking, the Japanese seem always to have had their sights on international markets even when they could not immediately attack those markets.

Concern with technology is now ubiquitous in British central government, and for departments like Defence, Trade and Industry, and Energy it is central. The public dimension of British 'technology policy' is also much influenced by the nationalized industries and public corporations, and by the Research Council (SERC). The remainder of this paper addresses some of the more important considerations which result. [. . .]

The Pattern of R & D

It is essential to obtain some grasp of the shape of the British R & D effort, and this is the object of Table 1. There are many obstacles to a sound interpretation of the figures. First, R & D expenditures are only part of technology policy – the latter in particular comprises design and production engineering phases as well, not to mention product development, the standards process and purchasing policies. It is also not very satisfactory to have to lump research and development expenditures together as is usually done, since develop-

Table 1   Public support for R & D in the UK: 1981/82 estimates except as otherwise indicated

Body	£m	
DoI	230[a]	
MoD	1680[b]	
DoE	43	1982/3
DoTransport	20	1982/3
Research Councils etc.[c]		
UGC	400	1982/83 estimate
SERC	200[d]	
ARC	37.5	
MAFF	37.4	1980/1
DAFS	16.4	
NERC	42	1980/1 provisional outturn
MRC	73	1980/1 provisional outturn
British Telecom	123.4	1980/1
Energy:		
DEn	48.8[e]	
AEA	205.4	
EC	79.0	
NCB	51.0	
BGC	63.7	
SSEB	2.6	
BNFL	11.4[f]	
BTG	72.7	1980/1

The above list is necessarily incomplete. Many bodies, e.g., British Rail, publish no specific R & D figures.

Notes: [a] About half in industry (increasing) or research associations: most of the rest intramural or European Space Agency.
[b] £260m R, £1420m D; about two-thirds in industry; most of the rest intramural; £1923m for 1982/3 (statement on the Defence Estimates 1982, Cmnd 8529).
[c] 1982/3 Science Vote which covers Research Councils – £480m.
[d] 24% engineering, 32% science, 40% 'big' science; 1981/82 £216.8 net.
[e] Includes £3.0m nuclear, £21.7m offshore, £4.6m coal, £15.0m renewables, £0.8m conservation.
[f] Written off only.

Sources: DoI, MoD, SERC – HL 89–1, 1982/83, pp. 21–3; MRC, NERC – HC 89–11, p. 546; DEn – PQ, 28/10/82: DoTransport, DoE, UGC – Cmnd 8957, p. 8; BTG – HC 89-II, p. 217; All others – annual reports.

ment tends to be much more expensive than research – an average of three times more expensive for instance in the manufacturing sector, and in some cases well over nine times. National R & D figures are also in themselves far from perfect – they are out of date, miss important expenditure especially in small firms, and classification of the relevant expenditures varies considerably from body to body. Technology policy is also, to repeat, effected by mechanisms involving either no direct R & D expenditure, such as tax assistance, or very little such as licensing. Nevertheless, R & D statistics are a useful, and certainly a traditional, point of departure.

In 1978 the £3510m which the UK spent on R & D was about 2.1 per cent of GDP. Of this government funded 48 per cent and industry 43 per cent. Some two thirds was performed in industry and just over a fifth in government establishments. The provisional 1981/2 figures show that around 55 per cent of central government R & D goes on defence, £1726m out of a total of £3316m. [. . .] Sir Ieuan Maddock's diagram (Figure 1), now a

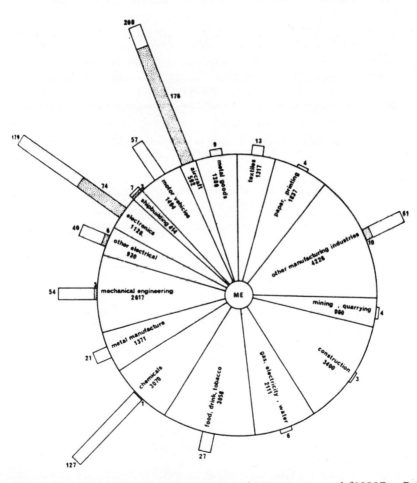

Figure 1   UK net output and R & D (1972, provisional.) Net output total £28897m; R & D total (£820m shaded areas indicate government contribution); nuclear £50m. *Sources:* net output-business monitoring 1000 census of production provisional results 1972; R & D – Trade & Industry, 5 September 1974; nuclear–C.S.O. R & D expenditure.
Sources: Sir Ieuan Maddock, FRS: The Seventh Royal Society Lecture, 'Science, Technology and Industry', 12 February 1975, *Proceedings of the Royal Society.*

decade old but still relevant, provides a dramatic illustration of the imbalance between government R & D expenditure in particular sectors of the economy and the importance of those sectors in the economy as measured by net ouput. Consideration of statistics of this sort has led to observations of the following kind:

(1) There is an *a priori* cause for concern in the UK's concentration on defence R & D, especially in the light of international comparison, where of Britain's main competitors only the US spends proportionately more and West Germany spends much, and Japan very much, less. The most recent word on this issue is again Sir Ieuan Maddock's. In a report[3] to NEDO in 1983, largely underlining his earlier seminal paper, he points out that since most firms are reluctant to move staff from military to civil projects, and since the MoD does the bulk of its work with firms or divisions of firms which deal almost exclusively with it, the technology gap between the military and civil sectors in Britain is continuing to grow. In effect a corollary of this concentration on defence R & D is that spending by the UK government on general industrial R & D, that is civil excluding space and aviation, compares very unfavourably with that by other members of the EEC (Table 2). It is sometimes maintained that UK defence R & D looms large not because it really is large, or at least not large in relation to the UK's defence obligations, but rather because UK civil R & D is so low. Whatever the truth in this, it is open to question that Britain can or will afford more R & D in total in present circumstances. This again is ultimately a political problem, and one which must become a still more pressing one unless the country's decline is otherwise arrested. Entertaining more than marginal expectations of spin-off from the defence sector is, on historical and comparative evidence, unwise, though defence technology is in fact in process of reorganization under a new controller at the MoD who wants to sub-contract more work to industry, with spin-off very much in mind. [. . .]

Table 2   A comparison of industry related R & D by the governments of four European member countries in 1980 (Estimates)

	£ million (% 1980 GDP × 100)			
	United Kingdom	*France*	*West Germany*	*Italy*
Defence	1343.5 (60.1)	1157.4 (41.3)	410.2 (1.6)	21.0 (1.3)
Space	52.3 (2.3)	196.5 (7.0)	173.5 (4.9)	43.7 (2.9)
Civil aviation	69.5 (3.1)	71.8 (2.6)	65.2 (1.8)	5 4 (0.3)
All other industries	85.0 (3.8)	224.1 (8.0)	338.1 (9.5)	130.4 (7.8)

Source: House of Lords Select Committee on Science and Technology, *Engineering Research and Development*, 2nd Report, HL 89–1 1982–3.

Note: Although the table aims to compare like with like there are a number of difficulties in any comparison of this sort, in particular the accuracy of the rate of exchange in the R & D context, and the fact that actual outturns may differ from the above estimates.

(2) Similarly, there are grounds for unease in Britain's heavy emphasis on publicly-funded energy R & D, and particularly as regards the place within this given to nuclear R & D – over half of a £500m total in 1981/82.

(3) There has been complaint about the priority given to basic research and the production of science graduates, as against development and the production of engineers, and above all about the role of the Research Councils in perpetuating this: SERC's expenditures on big science, notably high energy physics and radio astronomy, have been a specific source of criticism in this area. These criticisms are also of course closely related to the relative status of the scientist and engineer in Britain.

(4) The very existence of powerful government research establishments, and of a parallel intramural capability in certain of the nationalised industries, has been questioned on the grounds that such organizations lack commercial discipline and are too much influenced by their perception of the state of technology and their own organizational needs. [. . .]

(5) It has been suggested that, internationally, the recession has led to long term industrial R & D being cut back, only governments and the largest firms being able to sustain the associated costs and risks. But superimposed on this in the British case are other more longstanding problems; in particular the slower growth of GDP *per capita* here than amongst our competitors and the fact that the ratio of civil R & D to GDP has been growing more slowly in Britain than amongst most of those competitors.

(6) The strength of the British service sector is not remotely reflected in R & D expenditures. [. . .]

Overall it needs to be remembered that the current profile of UK R & D activity is the product of an evolution whose roots lie in Britain's former perception of itself as a leading world power with, in particular, major interests in aviation and a determination not to lose out in the new nuclear industries. The country's present pass makes it an especial cause for concern that public support of general, civil, manufacturing industry has always been a poor relation. Yet given that Britain's main competitors support their parallel industries by one means or another, such support in Britain in the 1980s cannot be a matter only of political values. At the very least it has now become an essential of commercial defence. [. . .]

Technology Clustering

A development in which much hope came to reside in the early 1980s is the phenomenon of technology 'clustering'. In the UK as in the USA, new technology-based industries tend, it appears, to cluster geographically, for instance in the 'western corridor' between London and Swindon. The general reasons for this phenomenon are well enough known – government financial inducements; availability of trained (and retrained) manpower (and womanpower); links to universities and other centres of higher education and research; quality (and costs) of the work environment; ease of international access; hard selling by public bodies (like Locate in Scotland); and the nucleus effect of existing high technology companies – but the relative importance and interaction of these various factors is less well understood. The problems faced by local authorities in providing, quickly, the requisite infra-structure, are normally just the sort they welcome. Ensuring continuing comparative advantage for a region, internationally, and also intranationally, is a deeper problem, as is accommodating to the fact that many companies may be subsidiaries of multinationals, with the consequence that the centres of company decision-making lie elsewhere – the 'branch plant economy' syndrome epitomized by the Canadian case. Achieving nationally a positive balance between the rate of creation of new 'new technology' jobs and the loss of longstanding 'old technology' ones is a key goal one which Britain has certainly not yet accomplished.

The 'Route 128 phenomenon', to give it one of its American names, is undoubtedly of major importance. Thus unemployment in the Boston area in the 1940s was apparently twice the national average yet today a third of Massachusetts' employment is in the high technology sector. There have been several attempts to understand the dynamics of the process, and it is clear that it takes time to establish it.[4]

Inherent in the 'silicon valley' or 'Route 128' technology clustering phenomenon is the proximity of one or more higher education establishments, and here two important UK

developments have been teaching company schemes and the emergence of science parks –
several science parks already exist and almost half of Britain's universities have now drawn
up plans for one. These are looked to to assist in overcoming the barriers between the
industrial and university worlds, but as yet judgment on their true significance is not possible. They could so easily become gimmicks of limited relevance, and it is in any case
doubtful whether Britain can support as many as there are now likely to be. [. . .]

With the creation in the City of groups such as Prutech and Cogent, the UK seems to be
moving as regards venture capital investment in high technology in a direction long taken
in the USA. The role performed in the past by NRDC, though the Prime Minister's 'Science
Seminar' in September 1983 made clear this is to be changed, would still seem to be a key
one for BTG, of which of course NRDC is now part. It remains an uncomfortable fact that
NRDC's record would look very different were it not for a couple of items, and it also cannot be assumed that because research work is undertaken in Britain, follow-on development and manufacturing investment will automatically be located here as well. In the same
connection, government also need to consider whether the tax system is giving as much
support as it might to R & D and innovation by small entrepreneurs: technology, unlike
science, cannot be its own reward.

References

1. K. Pavitt, (ed.) (1980) *Technical Innovation and British Economic Performance*, London: Macmillan.
2. C. Carter, (ed.) (1981) *Industrial Policy and Innovation*, London: Heinemann.
3. NEDO (Electronics EDC) (1983) *Civil Exploitation of Defence Technology*, London: National
   Economic Development Office.
4. M. Bullock (1983) *Academic Enterprise, Industrial Innovation and the Development of High
   Technology Financing in the United States*, London: Brand.

5.3

Defence Spending: Britain's Self-inflicted Wound

Martin Walker

The Government has been warned, by one of its own think-tanks, that the defence industry's growing monopoly of Britain's scarce research resources is slowly but surely throttling the rest of the economy.

The warning was delivered last autumn, in a formal paper submitted to the Chancellor of the Exchequer and the other Ministers, industrialists and trade union leaders who sit on the National Economic Development Council.

According to the minutes of the meeting, which was called to discuss innovation in British industry and how to promote it, the role of defence in distorting the rest of Britain's research effort was largely ignored.

The warning was drafted by Mr Geoffrey Chandler, himself a wartime intelligence man, who subsequently became a top oil executive with Shell. Given the etiquette of public affairs, it was couched in blunt terms:

> Given that the UK's defence programme is maintained, fewer national resources are available elsewhere . . . The absorption of a larger number of qualified scientists and engineers by the defence industry requires higher overall numbers of such personnel rather than lower numbers in civil innovation. If our defence commitments cannot be reconciled with these requirements then in time they will themselves become unsustainable as a result of continuing relative economic decline.

An uncanny echo of that warning was delivered this month by Sir Ieuan Maddock, who recently retired as chief scientist to the Department of Industry, in a report commissioned by the electronics industry. Traditionally, the high cost of the defence research and equipment budget has been defended on the grounds that it has a beneficial spin-off effect into civilian industry. Sir Ieuan's report, *Civil Exploitation of Defence Technology*, points out that the reverse is happening:

> What was striking was the distance between the attitudes of the civil and defence-oriented companies even when they existed within the same group. There already exists a large culture gap and it is getting even wider. (NEDO, 1983)

M. Walker, Britannia's self-inflicted wound, *Guardian*, 25 April 1982.
Martin Walker is a *Guardian* correspondent.

He argued that during the Second World War, Britain responded well to the technological challenges because it had a large and healthy civilian industrial base, or platform. 'In the past several decades the level of this civil platform has been subsiding, leaving the defence peaks standing even higher relative to the national electronic engineering plateau', he concluded.

Mr Geoffrey Chandler and Sir Ieuan Maddock are classic figures of the British Establishment, and what is curious about their warnings is how little echo there has been in the academic world or in public life. And given the level of public concern about national economic performance this is remarkable.

There are signs that the issue is reaching a wider public. At the Peace Studies Department of Bradford University, Mr Malcolm Chalmers published a paper on the relationship in Britain between defence spending and economic decline. He noted that in the 1950s, there was widespread public concern about the high cost of British rearmament brought about by the Korean War. The press of the day warned that Britain's high and rising defence budget would leave fewer investment resources that might let us compete with countries like West Germany and Japan.

'The heaviest cost of defence to the modern British economy is not the 10 per cent that it takes of the national income. It is the 60 per cent that defence pre-empts of all the resources this country can assemble to invest in technological progress.' grumbled *The Economist* of 8 September 1956.

'It is remarkable how the debate just died away, never to re-surface. But the correlation between low economic growth and high defence spending was maintained.' Mr Chalmers said. The pattern seems to apply internationally. The two advanced industrial countries with the lowest growth rates, Britain and the US, are the two countries whch spend over 30 per cent of their total R & D budget in the defence sector. The two countries with the highest growth rates, Japan and West Germany, spend less than 7 per cent of their R & D budgets on defence.

In the US, since President Reagan took office, the trade-off between defence and civilian R & D has become sharp. In his first two years, government defence R & D almost doubled to $31 billion, while civilian R & D fell from $17 to $14 billion a year.

Last month, a Japanese study was published in Britain which linked Japan's high growth rate directly to the fact that it has the highest proportion of privately-funded R & D of all industrial countries – over 75 per cent. In Britain, less than half of all R & D is privately funded.

The Ministry of Defence spends as much on R & D as it does on the Royal Navy – 13.6 per cent of the defence budget, over £1,800 million. And defence R & D is growing, from 21 per cent of all British R & D in 1968, to 25.4 per cent in 1978.

But this is just the tip of the iceberg of defence-related research in Britain, because an increasing proportion of our private industry is turning towards the defence market. The Ministry of Defence is currently buying over 28 per cent of Britain's electronic production, almost 40 per cent of the shipbuilding industry's production, and almost 50 per cent of the aerospace industry.

Last month, British Aerospace announced in its annual report that a bare 20 per cent of its £2,000 million business was still in civil aviation. BAC's chairman, Sir Austin Pearce, told the Select Committee on defence last year that two-thirds of his company's product was exported, and two-thirds of that was in the military business.

The Electrical Engineers' Association told the committee that in 1979–80, they sold £540 million in capital goods and another £85 million in components to MoD. In the same year, they exported £400 million of defence electronics. In the current financial year, the association looks like selling over £1,000 million to MoD.

Both carrot and stick are at work. The stick is the recession, which is tightening the civilian markets. The carrots are the guaranteed growth and profits of military business, and the growth is startling. In 1974–5 the MoD spent just 31.3 per cent of its budget on equipment. This year, that proportion has risen to 46.4 per cent, an exceedingly juicy market for private industry over more than £5,000 million.

No businessman worth his salt could avoid that kind of market, particularly when the profits in MoD contracts are virtually guaranteed to exceed the profits in the private sector. Under a series of complex equations worked out by the MoD's 43,000-strong procurement executive, firms which sell to it are expected to make a gross 20 per cent profit, and in 1982, were expected to make a net profit of 3.7 per cent. Then, private industry profits were running at 1.2 per cent.

According to evidence before the select committee, profits usually ran 2 per cent higher than that already generous target. And when MoD accountants back-tracked over 594 contracts, they found excess profits had been made in 360 of them. In February 1982, Air Chief Marshal Sir Douglas Lowe, chief of Defence Procurement, disclosed that falling inflation meant that contractors were this year enjoying £75 million of excess profits. In recession Britain, the MoD is a gravy train.

The defence sector is acting as a giant magnet, sucking unto itself not only a growing proportion of private industry, but also the cream of the high-technology firms, and the highly-trained people. Senior scientists are becoming alarmed at the scale of the process.

'As the university grants are cut, it becomes ever more tempting to go after the external contract work, and these days, that means work for the MoD, the Atomic Energy Authority, and the US Air Force.' says Sir Martin Ryle, the last Astronomer Royal, who is worried whether his former pupils will find any alternatives to defence work.

Reference

NEDO (Electronics EDC) (1983) *Civil Exploitation of Defence Technology*, London: National Economic Development Office.

5.4

Design Policy:
A Resurgence for UK Designers

Christopher Lorenz

To be catapulted into fashion after years of obscurity is a decidedly mixed blessing. Adulation flows aplenty, but so does criticism, exaggeration and misinterpretation. As expectations soar, so do the risks of disappointment and demise, whether the subject of the fashion be individuals, organisations, or ideas.

In the case of British industrial design, it is all three. Over the past two years industrial designers and their consultancies have been propelled into the limelight after decades of languishing at the beck-and-call of the nether reaches of British marketing and engineering.

All of a sudden, designers no longer need to look abroad for the bulk of their commissions. Domestic demand is booming, fortunes are being made overnight as their firms rush to go public, and the designer is emerging as an establishment hero, fêted on television and honoured (in Terence Conran's case, with a knighthood).

It is not surprising that the British designer is suddenly the envy of his peers across the Channel and even the Atlantic.

The fashion owes its existence to a decidedly unlikely congruence of attitudes on the part of government, retailing, finance and industry.

Most of the design fraternity would date it back to a Prime Ministerial cocktail party-cum-seminar in January 1982, which sowed the seeds of a battery of subsequent government measures to boost design: a series of industrial seminars across the country on the theme of 'Design for Profit'; reinforcement of well over £10m to fund free design consultancy for small and medium-sized firms; the support of research to expand and improve design education, not only in schools and colleges, but also at the London Business School; and a flurry of public and private ministerial prodding of the hitherto reluctant businessman.

But the roots of design's resurgence go deeper. In retailing, Conran's chain of Habitat shops had stood almost alone as a symbol of the commercial power of design until the remarkable design-led revival of the Burton Group after 1980. Yet by the time Mrs

C. Lorenz, A resurgence at last for UK designers, *Financial Times*, 23 May 1984.
Christopher Lorenz is management editor of the *Financial Times*.

Figure 1   Two Design Council Award-winning designs: Kitchen Devils' professional kitchen knives; another winner for J. C. Bamford Excavators with the 3CX backhoe loader.

Thatcher took the plunge, Hepworths and a number of other retailers were, in a phrase, following suit, and the City was beginning to show an interest.

Since 1982 the retail fashion for design has accelerated from a slow jog to a headlong gallop: Boots (which is using two top consultancies, Fitch and Pentagram) and House of Fraser (with Aidcom) are just two of the most recent big names to join the rush.

The parallel growth of stock market interest, and the timely development of the Unlisted Securities Market, have enabled Fitch and Michael Peters to join the pioneering Aidcom as public companies.

But the most fundamental change in sympathy has been in the marketplace itself. As the

1970s progressed, the nonsense of industry's old complaint that 'British consumers don't want good design' became clear for all to see.

In one market sector after another, from cameras to cars, and consumer electronics to farm machinery, the public showed its preference for well-designed, reliable foreign products, even if they cost more than their British equivalents.

A number of British manufacturers – with BL the most visible – have at last responded to the belated realisation that quality counts, and that price is by no means everything in international trade, whether at home or on export markets.

As well as reliability and good performance, they have also begun to inject into their products some flair, to remarkable commercial effect.

Such paragons are still in the minority, as was testified by the Mellor Report (1983)* on the design of British consumer goods, and as is underlined by the paucity of yesterday's 1984 Design Council Awards in the 'decorative goods' category. But at least the message is now plain for all to see. With the design bandwagon continuing to roll, even the most recalcitrant manufacturer could hardly ignore it.

The trouble is that for all the bustle and noise – indeed, partly because of it – misunderstanding about design is still rife, even among some of its new supporters. This is not so much a question of Mrs Thatcher's now notorious complaint that Ford's 'car of the future', the centrepiece of a much-trumpeted Design Council exhibition on the motor industry which she opened in March, was too advanced for her.

It is more a matter of a lingering belief on the part of many retailers and manufacturers – plus the City – that design is merely a promotional veneer that can be tacked onto whatever lies beneath, and used as a cure-all.

A number of retailers, for example, have revamped their shops in glossy fabrics and colours, while doing next to nothing to improve their merchandise. A bevy of manufacturers have modernised their packaging, but done little to upgrade the product inside.

With the City 'not really understanding design and not wanting to' (in the words of one of the few stockbrokers who has thoroughly researched the commercial role of design), the risks of a sudden bursting of the design bubble are dangerously high.

What happens, say, if Conran runs into difficulties with Heals, his latest acquisition, or if House of Fraser's fortunes fail to take wing? What, when one of the quoted design consultancies inevitably has a lean year or two?

What if one of Britain's few design-minded manufacturers hits rocky times? Will retailers, industry and the City take a thorough look at what went wrong, or instantly complain that 'design doesn't pay after all'?

If the design fashion is not to prove a nine-day wonder (as it has on several occasions in the last 150 years), there is a great need for more breadth and depth, both in the continuing debate about its commercial potential and in the actions of the various parties involved.

First, more attention should be paid to the linkages between the different aspects of design. It is difficult, for example, for product design or retail design to be successful in isolation from each other: just as a brilliantly designed product will not sell well (or at high margins) in a ramshackle discount shop, so a gleaming department store needs quality products if it is to succeed.

It is equally difficult to run a convincing corporate identity programme if the reality of the company's products, buildings, and ways of doing business, fail to live up to the image; there is nothing worse than a glossily-promoted airline which is unreliable, uncomfortable, and has poor service.

Second, there is an urgent need for convincing empirical studies of the economic impact of design.

In general terms, the case for better engineering design was proved beyond doubt in the late 1970s by research at the National Economic Development Council and the Science Policy Research Unit of the University of Sussex.

But industrial design barely rated a mention, and only very recently have a number of other academics – at Manchester Polytechnic, the Open University, and London University – embarked on sector-by-sector studies. These could prove invaluable in 'educating' the City and industry.

Third, the Design Council should ensure that more of its activities span the whole breadth of design, thereby pre-empting the now widespread criticism from industrial designers that it has become overbiased towards engineering.

To the extent that engineering-based companies have been in greatest need of 'design education' in recent years, the Council's concern with that sector has been justified. But the 'softer' end of design also needs promotion, as the Mellor Report* (which the Council commissioned) made only too clear.

Fourth, the design profession itself would do well to place rather more emphasis on the designer's role as part of a multi-functional team within the client organisation, rather than as a lone hero of the architectural world. Few designers overtly cultivate the hero image, but potential clients often see them in that light and consequently fail to integrate them into the corporate hierarchy; this applies both to consultants and to in-house designers, who are often seen as "outsiders".

Fifth, the government should reinforce its attempts to develop a more consistent strategy for its promotion of design, rather than giving with one hand and taking away with the other. While the Department of Trade and Industry, with prime ministerial backing, has been espousing the cause with its various support schemes, there has been a flurry of cutbacks in the education sector, especially among the most valuable college teachers of all, part-timers who are also practising designers.

The paradox has now been made complete within the Department of Education and Science itself by Sir Keith Joseph's proposal that all school children should study a craft, design and technology course up to the age of 16 – at present only a tiny minority do so.

Though his prime purpose is not to train professional designers, but to rectify the technical and visual illiteracy which has plagued British society as a whole for so long, there is a questionable logic in expanding secondary design education (presuming that, at a time of austerity, the funds can be found), while at the same time cutting tertiary level training.

A degree of co-ordination between the DTI and DES – and with such bodies as the Manpower Service Commission – is now emerging following extensive policy consultations conducted by the DTI minister responsible for design, Mr John Butcher.

In May 1984 it was announced that an additional £10m had been earmarked for the Design Advisory Service Funded Consultancy Scheme (to include the £1.3m for clothing and textile firms and an allocation for firms with fewer than 60 employees outlined in the Budget debate).

A total of £50,000 is also to be made available over the next two years to support research into design and primary education, with further sums of £10,000 going to the London Business School to support the Design Management Unit and £30,000 to help launch a register of apparel designers.

A scheme is also being discussed with the Royal Society of Arts whereby 200 of the brightest and best design graduates would be placed with employers in order to gain industrial experience. [. . .]

[* *Report to the Design Council on the Design of British Consumer Goods* (the Mellor Report), Design Council, 1983.]

There are also plans for a special 'strategy group' within the Design Council to advise on the general policy; this would also include outsiders.*

Together, all these various interest groups have the ability to transform a fragile fashion into a lasting and influential element in British industry and society. But they could also help destroy the fashion, and plunge design back in the shadows from which it has only just emerged.

[* The Strategy Group's report, *Policies and Priorities for Design,* was published by the Design Council in October 1984.]

5.5

Renewable Energy Policy

F. J. P. Clarke

Background

Energy is all-pervasive in our economy. Every activity directly or indirectly involves the use of energy, and expenditure on energy accounts for over a tenth of the UK's Gross National Product. This pervasiveness has made energy policy a focus for discussing a wide diversity of political, social, environmental and even moral issues: for example, environmental impact, nuclear weapons, dislike or like of technocracy or centralization, advocacy of alternative life styles, the moral problem of human poverty. Hence the debate about energy policy or about what should be done in energy research is often not just a matter of (or in some cases not even primarily a matter of) the traditional objectives of energy policy that there should be adequate and secure energy supplies, that they should be efficiently used, and that these objectives should both be achieved at the lowest practicable cost to the nation. Instead the starting point is often concerned with political views about some alternative organization of society and of the consequent way in which human and material resources should be deployed in that society.

The starting point for this chapter will be quite conventional, namely the fact that oil and natural gas which have fuelled rapid growth of economic activity over the past 50 years, cannot continue as major fuels indefinitely. Over the next 20 years and beyond, we will increasingly have to turn to other fuels and sources of energy, namely coal, nuclear power and the renewables. Moreover, the energy use systems – that is, processes that turn energy to meeting users' needs, and the stock of buildings that exist – have developed in a period of cheap energy which is now over. The movement to alternative and more expensive fuels must be accompanied by equally important and complementary changes in energy use systems as well.

Energy policy should be concerned with enabling these changes to evolve smoothly. Many of the changes will require research leading to innovation, by which is meant all the steps through development, design, demonstration and marketing. Energy research policy, therefore, to an important extent, will be a policy about applied research and industrial innovation (it would be tedious continuously to use words like 'research leading to innovation'; hence we shall simply use the term 'research' to cover all these aspects). Such a policy should clearly make some allowance for alternative views of the way in which society may

F. J. P. Clarke, Energy, in M. Goldsmith (ed.), *UK Science Policy*, Harlow: Longman, 1984, extracts from Chapter 5.
F. J. P. Clarke is Director of Management Studies of the United Kingdom Atomic Energy Authority.

develop; but short of a political fiat this will be a secondary consideration, and largely one of ensuring that the resulting research programme is robust to an unexpected turn of events.

The role of government and resulting policies

Governments intervene in the economy because otherwise the market may not allocate resources in the national interest. This is especially true of research. The intervention may be indirect, via some framework of guidelines, incentives or constraints; or it may be very direct with detailed plans, quantified objectives and a substantial commitment of government resources to achieving them. This latter approach is, for example, that adopted in France. A planning council defines targets five years or more ahead for the share of the energy market to be met by each of the conventional fuels or alternative energy sources and allowing for energy conservation targets. Government then intervenes actively to achieve these targets. However, in the UK successive governments have generally eschewed such a prescriptive approach in favour of a market-based policy. The essence of such a policy is that the pattern of supply and use must be left largely to the market-place and to industry, with government concentrating its efforts on providing a framework of guidelines, incentives and constraints aimed at encouraging suppliers and consumers to work towards the government's longer-term energy objectives. All this means that UK energy policy tends to involve qualitative and rather broad objectives, such as achieving flexibility to respond to whatever future occurs, or giving consumers the right price 'signals' about the real cost of energy supply, or encouraging efficiency in the state-owned energy industries.

Such a policy does not give a sharp focus for a derivative energy research policy. However, certain policy guidelines for research can be amplified from what has been said so far. They are:

1. research and innovation will be left primarily to the established industries to promote, with government monitoring progress and plans, especially those in the public sector;
2. in order to monitor effectively there must be an analytical framework that includes all the technologies of future national importance, and that takes account of the fact that the future is imponderable and influenced by a diversity of issues, some of which may be non-technical;
3. the analysis may then identify areas which, due to market imperfections, may not be receiving the attention that is merited on a national view. In this case government will need to decide whether these areas are important enough to merit some form of intervention;
4. because the objective is generally innovation, industrial involvement is desirable even where government does intervene;
5. in some areas government may need research to meet its own requirements, for example, in the safety or regulatory fields.

[We will now consider the case of policy for renewable energy sources.] [. . .]

Renewable energy sources

Attitudes and government policies

The basic policy issue on renewable forms of energy, by which is meant sun, wind, biofuels, etc, is whether the UK is allocating enough resources to their development bearing in mind their huge potential. Advocates of the view that insufficient is being allocated point to the £14m plus allocation in 1981–2 against the £200m plus spent on nuclear energy. And, as in the case of energy conservation, it is argued that renewables need an agency to promote their research and application. This, and possibly the creation of a Renewable Energy Division in the Department of Energy, would help to give these energy technologies the institutional voice which is necessary if they are to hold their own among the other institutional voices at the points of decision. Such basic points have been made again and again over the years. Yet in this area, unlike energy conservation, little progress seems to be made. Why is this?

First, the UK's whole approach to renewable energy is different from that in many countries, such as Sweden and France or the US under the Carter regime. In the UK, the aim of the renewable research programmes is to develop the technology in order by the mid-1980s to define their potential contribution and economics. In contrast, it is assumed under some overseas policies that renewable technologies will definitely make a contribution, and targets are set to obtain this. For example, France has set targets of 3, 4–5 and 10–12 million tons of oil equivalent for the years 1985, 1990 and 2000 respectively, and formed in 1978 an agency or commissariat to help achieve them.

The more tentative UK attitude reflects in part its basic market-determined policies which eschew normative intervention by government. But, partly, it must be that UK governments, ministers and officials are influenced in the way they view these renewable forms of energy, by the relative abundance of the UK's indigenous energy resources, in contrast to the situation of many of its neighbours. Substantial reserves of coal, oil and gas when added to the historical development and potential of nuclear power, tend to produce a more relaxed attitude compared with that of neighbours who have to import substantial amounts of energy. Of course, such a perception is partly illusory because, on a world scale, energy is a traded commodity, so that the UK's own reserves represent a financial asset rather than energy, and one that has to be valued at its opportunity cost on the world market rather than a source of 'cheap' or secure energy.

Another important reason for the UK attitude stems from the fact that the central problem of energy is not its availability but the cost to the consumer of meeting his needs for energy. Analysis of the likely cost effectiveness of these energy forms when deployed in the UK seem not as promising as some of the advocates claim.

Programme determination and review

The process of programme determination and development started in the mid-1970s with a review of each technology that also contained recommendations for research work (Energy Papers 9, 16, 21, 23 and NEL paper EAU M25). These reviews were considered by The Advisory Council on Research and Development for Fuel and Power (ACORD), which made recommendations about the general shape a programme might take. Following this publicity, the Department of Energy's Energy Technology Support Unit (ETSU) invited proposals for work from universities, industry and other laboratories, and stimulated proposals where this seemed desirable. ETSU does not itself carry out any research work but acts as stimulator and manager of the programmes. Once started, the overall shape of the ongoing programme is considered by a steering committee for each area, comprising external industrial, institutional and academic representatives. The recommendations by each steering committee on overall progress for the future programme are submitted yearly to ACORD; recommendations on individual proposals are made to the Department of Energy and ETSU places contracts for those that are approved.

As a result of these activities, the programme grew from virtually nothing in the mid-1970s to about £15m per year by the early 1980s. The growth rate was determined more by technical limitations on the rate a new research programme can be expanded rather than by fundamental limitations. As a result, many of the technologies were, in the early 1980s, moving from the less expensive R & D stage into the more expensive demonstration stage. At this point government policies caused a check in the rate of their further expansion for two reasons. First, their status meant that some were approaching the point of commercial viability; as a matter of policy, industrial money was, therefore, sought to supplement government money. Of course, in the ensuing negotiations both sides tried to limit their financial contribution and obligations; as always, this tended to cause protracted negotiations. Second, this stage was reached when a period of general government restraint in public spending came into operation. The result was that in 1982 ministers were no longer able to say that the rate of expansion of the programme was not limited by availability of funds.

The most rigorous and comprehensive assessments of the results from this programme have been made by ETSU and published in 1982 (ETSU Papers R13 and R14). They were sponsored by the Department of Energy at the request of ACORD and were based upon independent consultants' reports of the markets and up-to-date reviews of the information arising from the government's programmes and elsewhere. They concluded that on current perceptions about UK fuel prices (and, of course, this is always a key assumption and a key unknown) '*all* of the renewable sources examined show some prospects of being cost effective over the next 50 years or so'.

The results suggest that over the next 50 years or so these sources are likely to make a small but useful contribution to UK energy supplies. Their limitation in terms of contribution lies not in the rate at which machinery can be built, but in fact that these natural sources of energy vary so much in their cost effectiveness with geography and climate. This means that as the contribution to national supplies grows, so the technologies have to be applied in less favourable sites and conditions, and so the associated costs rise too. All these factors are consistent, or so it can be argued, with a £15m programme. Against this, critics point out the comparatively vast amount that has been spent on developments in the nuclear area, and the small contribution to energy supplies that nuclear power now provides, or can be expected to provide, until well past the turn of the century. They argue that had only a small proportion of this nuclear development money been placed behind the renewable sources of energy then many of the shortcomings that have been highlighted by the government programme would have been overcome.

Institutional representation for renewable technologies

ACORD's role has been crucial in all these developments. In the initial stages it strongly supported programmes to obtain information on the contribution and economics that might be obtained. As results began to flow in from the many university, industrial and independent research workers in the field, the council played a key role in highlighting both the strengths and weaknesses of the different technologies. And this in turn began the process of focusing the programme so that money was concentrated increasingly in fewer areas. All this culminated in the much publicized critical review which took place in 1982 based upon the ETSU Paper R13. Much of the press misrepresented ACORD's role at this critical meeting as one of cutting back the research money allocated to these technologies. In fact, what it did was to recommend priorities so that the best possible programme could be fitted into a constrained budget.

Some of the advocates of renewable energy technologies see a problem in the balance of institutional representation on ACORD. Now in fact, over the past five years, ACORD has steadily broadened its representation beyond what is recognized as the established energy institutions, and towards more independent representation from both general industry and academia. Thus, of 17 members in 1983, 10 were from outside the energy industries themselves. Nevertheless, many critics feel, as in the case of energy conservation, that while the strengthening of independent interests is a step in the right direction, this is no substitute for an active institutional champion of the renewables' cause at the point of decision.

What balanced view can be taken of this situation? There is always a chicken-and-egg problem over introducing new technology. Until it is proved people will be reluctant to back it too strongly, and this may inhibit successful uptake; but until it is successful one cannot have the confidence to back it. It is difficult to see what better one can do than ask an independent but knowledgeable group to make a judgement and a recommendation. That group will, of course, need to hear all sides of the argument. ACORD has in the past operated in this way, and the independent members are sufficiently numerous and outspoken, that any natural bias of the established institutional interests that may be present does not dominate.

Having said this, it is perhaps a pity from the viewpoint of a technical innovation that government funding had to be restricted at the point of demonstration, when funding overall normally has to increase quite dramatically. The argument that the technologies are approaching the commercial state and so private industry should take a greater share is valid; but no-one should underestimate the speed of their movement to maturity. Beyond this argument, however, there is a genuine public concern over nuclear power and an accompanying public wish that all the alternatives are seriously explored.

LOCAL AND REGIONAL INITIATIVES

5.6(a)

Science Parks

Anthony Moreton

There is a tendency to believe that science parks are the answer to all Britain's problems. The uncommitted observer might be forgiven for thinking, such is the interest in the subject, that attracting high-technology concerns to sylvan settings will transform the economy.

In past years there has been a rash of announcements about the inauguration of such parks. Universities such as Warwick, Keele, Surrey, Swansea and Southampton are following in the footsteps of Heriot-Watt in Edinburgh and Trinity College, Cambridge.

The Scottish Development Agency has talked about having five or six in Scotland and the Mid-Wales Development Board is looking at the possibility of one in Aberystwyth.

Local authorities and new towns

Local authorities, new towns and urban development corporations such as the Wirral, Warrington and London Docklands, which lack direct ties with universities, are claiming how near they are to these institutions for their schemes. Private developers are becoming less interested in industrial estates. Science parks are the thing.

The country is rushing, helter-skelter, into a new technological world of laser beams, electron beams, computer hardware and software, microfoils, fibre optic technology and diagnostic reagents.

There are even parts of Britain which proudly, if inaccurately, proclaim themselves as the British Silicon Valley. The West of Scotland got very annoyed when Glenrothes and Fife generally adopted the name and the M4 axis towns looked on with amusement.

A bubble which has not been blown up can hardly be said to have burst, but there is a grave danger that too high expectations are being put on science parks.

Experience in America, where the parks started 30 years ago, should be a salutary guide. There has been a high failure rate and the benefits take a long time to materialise. Many small high-technology firms have remained small high-technology firms.

One of the problems discovered in America is that small companies, led by brilliantly

A. Moreton, Science parks, *Financial Times*, 21 January 1983.
Anthony Moreton is regional affairs editor of the *Financial Times*.

innovative graduates, are not always the best at marketing their products. For every Hewlett Packard there are countless others which have sunk.

Despite these reservations the experience of those British parks which have been in existence for a few years is sufficiently encouraging to merit further development.

The Cambridge Science Park on the outskirts of the city, set up in 1970 and officially opened in 1975, now houses 25 companies and provides work for 750 people. The number should be up to 1,000 by the end of the year and the park is aiming for around 2,000.

If the local authority planners had been more receptive to the whole idea a decade ago, it would have housed the European research centre for IBM, a large employer.

The Heriot-Watt Research Park at Riccarton outside Edinburgh has also got off to a good start. It allows only scientific and technological research and eschews mass production. It claims to be the only research park on a campus site in Europe.

Science parks in Britain appear to have originated not so much as a response to the American model, as to the desire of the then Mr Harold Wilson to translate his 'white-hot technology' speech, made while Opposition leader in the early 1960s, into practical politics when he became Prime Minister in 1964.

In the second half of the 1960s the Government wrote to all universities suggesting they do more about innovative technology and it was from that initiative that the first steps were taken in Britain at Heriot-Watt and Trinity College in Cambridge.

The different titles, and slightly different approaches, of these two parks lead to an important question of definition. What is a science park?

A pure science park would be one set up within the confines of the university's grounds, where the companies on the park dealt only in pure research and where there was close involvement on the part of the university and a direct interface between academic staff and the companies involved on the park.

It is probable that no such institution exists in the world to meet these criteria, though one or two in America, such as the Research Triangle Park, associated with the three universities of North Carolina, and the University of Georgia Research Park at Athens, Georgia, come nearest to it.

Heriot-Watt is called a research park, merely a synonym for science park. But there are other names. A technology park would be an area where there was a high proportion of applied research, perhaps involving a university; a business park could have a proportion of commercial activities; and an industrial park is often just another name for an industrial or trading estate.

In practice, the title hardly matters. What is of greater concern is the link with the university or other academic institution and the physical layout.

If a science park is anxious to attract in high-technology concerns, then it has to lay out the estate in such a way that working and environmental conditions are maximised. Each park sets its own standards but at Cambridge only 20 to 25 per cent of the space is given over to buildings. The rest is services and landscaped surroundings.

Of more importance is the interaction between university and production unit. Dr John Bradfield, senior bursar of Trinity College, says:

> A science park should allow for the interchange of ideas between firm and university. If a scientist or technologist comes up against a problem then he should be able to turn to someone in the university for help.
>
> Many of the companies on science parks are operating at the frontiers of technology and identifying what is happening is frequently a very complex thing. This is where we can help.
>
> Sadly, there is too little of this meeting of minds. Much of British industry tends to

have an anti-academic bias and within the universities there is too often a feeling of not wanting to get hands dirty with industry.

Fortunately, there have been enormous changes for the better over the last 20 years. There have been huge changes in attitudes within the universities and almost everyone is willing, indeed anxious, to help.

But I wish that more British concerns would see us in the universities as listening posts for them.

US experience, according to Mr Nick Segal of Job Creation is that the most common type of science park is essentially a high-quality property development in a strategically excellent business location.

The tenants would then largely comprise mobile R & D and high-technology projects of major companies, university research institutions and small but fast-growing advanced technology manufacturing companies.

That this is what has happened in the US is beyond doubt, despite the failures there. California alone has 15 such parks, Colorado five, Maryland six, Massachusetts seven. Some have become very large, the Research Triangle Park in North Carolina probably having over 10,000 working on it.

American experience and its message

American experience on the whole, though, suggests that it is a long time before numbers rise into four figures and so it would be wrong to expect British parks to be offering an immediate or even medium-term solution to the country's employment problems.

Science parks are necessary, according to Dr Bradfield, so that Britain:

> Shall not slip into the peasant economy category.
>
> The sheer amount of effort being put into high-technology work elsewhere is staggering. The Japanese have just produced a programme in conjunction with 45 universities.
>
> The Americans have a $2bn programme linking almost every university and important technology-based company.
>
> We in Britain are, tragically, failing to capitalise on our research. Fortunately, there are signs of change. The penny is beginning to drop. We have shown at Cambridge what can be done on a well-run science park and if we can do it others in Britain ought to be able to achieve comparable results.

5.6(b)

Regional Development: High Technology in Silicon Glen

Ronald Faux

It is easy to sniff a confident air in Scotland these days. Although the first into the economic doldrums, Scotland now promises to escape ahead of other areas which, historically, have been more resilient.

North Sea oil has generated a huge spin-off in services and expertise. Even though much of this has centred on subsidiaries of multinational and foreign companies, some at least will survive long after the oil fields have been emptied.

What may prove even more important in the long run, oil spurred a political dynamic for devolution which may have stopped short of a Scottish Assembly, but probably led directly to heightened autonomy via the Scottish Office and the creation of a powerful restructuring force in the shape of the Scottish Development Agency (the SDA).

English regions left as supplicants for general regional aids may well envy Scotland's direct voice in the Cabinet, which has helped protect the Ravenscraig steelworks and the Scott Lithgow yard on the lower Clyde, as well as Scotland's high-level decision-making on local capital spending and its strong powers to promote industry and development in a coordinated policy of its own.

The SDA may well be the most valuable product of the drive for devolution. In three years, it has proved the main coordinating arm of government in acting as a catalyst for change both in industry and the environment for industry and has scored vital successes in attracting new high technology industries north of the border.

Now it is trying to take those industries beyond the status of mere manufacturing arms of multinational companies. Increasingly, Scotland is building a base that covers the full range of activity from research and development through to the final product, harnessing the strength of universities and traditional professional skills to make Scottish factories less vulnerable as the distant offspring of parent companies whose hearts lie elsewhere.

Dr George Mathewson, the SDA's chief executive, has made a straightforward response to the decline of traditional industries: 'Unless you create an alternative industrial base, you are dead. We have to rebuild our strengths and get jobs in industries we are good at.'

R. Faux, How the SDA nurtured high technology in Silicon Glen, *The Times*, 27 April 1984.
Ronald Faux is *The Times'* Scottish regional reporter.

[Recently, in one month] alone, Scotland netted a £100m investment in semiconductors by National Semiconductor at Greenock and another £30m factory to be built by Shin-Etsu Handotai, the Japanese producer of semiconductor silicon at Livingston.

The latter was a classic case of the SDA at work. Shin-Etsu Handotai had inquired about a site somewhere in Britain as a European base, among others with the SDA, but its application form was percolating through the procedures of the Department of Trade and Industry in London.

When Dr Ian Robertson, director of Locate in Scotland, the SDA arm operated jointly with the Scottish Economic Planning Department, discovered that the Japanese company was 'on the boil', that its board had fully approved the decision to invest in Britain, he caught the next aeroplane to Tokyo to conduct immediate talks. Belying the image of lengthy Oriental negotiation, a complete deal was prepared within a week and signed by the Japanese and the SDA nine days later.

The SDA may not offer any more than the usual government supplied incentives for persuading industrialists to move to places they might not otherwise choose but, like the co-ordinating bodies set up to aid former steel towns, it gains from being able to marshal all the help behind one door.

The SDA also has direct financial links with 725 Scottish companies, although it now eschews full ownership. It owns and administers more than 1,000 industrial units and has formed a technology transfer group to help Scottish companies license or make joint ventures for new products with overseas companies. It invested £8m in large and a growing £4m in small companies [in 1983].

But its heaviest, direct spending goes into improving the environment through land renewal and development projects for derelict industrial areas on the principle that foot-loose companies with the world tugging at their sleeves will not be attracted to shabby locations that smack of failure. In 1983, the cumulative value of such schemes reached £163m and the SDA is backing a rising number of reclamation projects.

The results, though still patchy, are beginning to show. Some 28 American electronics groups have either chosen or short-listed Scotland for new facilities and expansion. The latest figures for inward investment transactions show 45 projects involving £186m have been finalized. And although the United States remains the biggest source of foreign capital, the SDA has now begun to home in on the potential of growing Japanese interest.

The battle to turn round Scotland's industry is still to be won however. The famous silicon glen (stretching west from Fife through the belt of central Scotland), the oil industry construction yards and a rash of new building represent a new prosperous economy.

The other economy of heavy engineering, steel, coalmining and shipbuilding that made Scotland the workshop of the empire still suffers the traumas of contraction and dissolution. The problem of unemployment against a background of change and decay of urban communities remains unresolved.

New technology can provide a new future but cannot absorb the numbers rejected by the old, labour-intensive industries. Even so, the electronics industry has grown to become a bigger employer than shipbuilding, coalmining or steel, which, where they have survived, are increasingly themselves taking on new techniques with smaller more flexible workforces.

The worst industrial news north of the border has probably now broken, leaving Scotland with the benefit, in a sense, of having lost out before industrial decline seemed universal and inevitable. But much of what is left of the old Scottish industry is now slimmed down, rationalized and, in some sectors, re-equipped to compete from a smaller base as part of a broader and better balanced economy.

The main hopes for throwing off Scotland's image as a depressed outlying region once and for all still lie mainly with electronics and computer-based industries. Here there are signs that Scotland has begun to father an accelerating momentum of its own quite different from the traditional product of expensive Whitehall-based regional policies that scatter a random series of grant-aided marginal factories and sub-offices round the periphery of Britain.

A recent comparison of Scotland, the San Francisco Bay area and the south-east of England showed Scotland emerging as a centre for the growth of new technology based particularly on small companies. Scotland led silicon valley in the employment of research and development staff and there were notably closer links with local universities as a source of product innovation.

Other reports suggest Scottish people are more at home with high technology appliances than the rest of the United Kingdom and that Scotland is ahead of the rest of the United Kingdom (outside London) in computer usage.

The electronics industry may not be a newcomer; Ferranti opened its plant in Edinburgh during the last war. But growth in recent years has been phenomenal. More than 200 companies provide 40,000 jobs with an investment of £500m.

The newly announced Shin-Etsu Handotai and National Semiconductor factories will be of special significance in consolidating Scotland's position as the leading semi-conductor producer in Europe. Already Scotland satisfies 21 per cent of European needs and with projects in the pipeline the SDA calculates this could rise to 50 per cent.

Silicon glen has already reached the point at which the snowball effect takes over as one or two leading companies in the same specialist field attract others. Health care products already employ 7,000 people in 60 companies. Although many of the factories are mere outposts of drug companies, again several leading manufacturing names have now become involved in research with Scottish universities and teaching hospitals.

This process of tying manufacturing with local research and services bodes well. Raw assembly work is often vulnerable to recession or changing corporate plans, but once Scottish companies or subsidiaries of foreign companies are tied to crucial stages of product development, they are more secure and have more spin-off effects in the surrounding economy.

This exciting potential transformation of the Scottish economy can provide arguments for both sides of the current debate over regional policy. Certainly, the Scottish Office and the SDA have spent public money and used central financial aids. They might not have succeeded otherwise.

But in Scotland this money may have seeded a revolution rather than merely provided an expensive palliative and if this proves to be the case it will be because local interests have come together to identify the areas where Scotland might excel rather than merely take part. They have in effect formed a business plan for their country and have harnessed all parts of the community, from capital to academic, in a plan to develop new growth points rather than merely plonk jobs in hard-pressed areas.

5.6(c)

Local Enterprise Boards: Tapping London's Skill Resources

Mike Cooley

Faced with the plight of the capital's 353,000 jobless, the Greater London Council set up the Greater London Enterprise Board (GLEB) in November 1982 with some £30m to invest in industrial developments in the area in 1983.

The board is one of the agencies charged with implementing the GLC's industrial strategy, aimed at providing 'socially useful and sustainable jobs'. Other agencies include a Popular Planning Unit, which aims to mobilize trade unionists and community groups to participate actively in planning the future of their city. The Economic Policy Group is developing a London industrial strategy, to provide a framework within which to begin to resolve the economic problems born of restructuring, recession and unmet needs.

Wherever the GLEB intervenes to save or create jobs, it insists in conventional companies, on enterprise planning agreements involving the workforce. It also encourages cooperatives and other innovative forms of democratic industrial structures. So those who work in London's threatened factories and workplaces need to know what other product ranges or services they could work on, if the GLEB's intervention were to give them a say.

Sadly, traditional management power structures have prevented 'the workers' (including design staff, technologists and researchers), from having any significant say in planning the future of their enterprise, or any access to research and development facilities to develop new products.

But Greater London's three universities and seven polytechnics offer a range of scientific and technical resources, material and human, hard to match anywhere else in the world – and with little real contact with the community outside on which they depend, apart from limited contact with one side of industry.

Technology networks are the GLEB's innovative attempt to bridge this gap. In the current year, it is providing £3m to set these up and buildings are now being completed and staff recruited for the first two.

M. Cooley, Tapping London's skill resources, *Times Higher Educational Supplement*, 1 July 1984.
Mike Cooley is Director of the Greater London Enterprise Board's Technology Division.

Two of the networks are geographically based: the north and east network, in conjunction with the North London Polytechnic, and the south-east network, in conjunction with Thames Polytechnic and other local education institutions.

Two of the networks are product based. An energy network is being set up in conjunction with the Polytechnic of Central London and the Polytechnic of the South Bank. A new technology network involves academics from Queen Mary College, Imperial College, City University, the Polytechnic of Central London and researchers at St Thomas's hospital.

Networks will provide a creative interface between the community and those engaged in academic and scientific work. Each network has one or more buildings close to a campus but not on it. Experience shows that industrial workers and community groups find university environments rather alienating (probably most students do likewise but then they haven't as yet got much choice).

They provide an environment in which both can meet on neutral ground and where the projects undertaken tend to be developmental and more activity based than research orientated.

Each network will have a small core staff which can demystify science and technology and appreciate the deep tacit knowledge of industrial workers and those who live in the community. They could be specially recruited, or seconded from a university or a polytechnic.

Each network will be run by a democratically-elected management committee which will represent community groups, the academic institutions, the borough councils, trades councils, cooperative development agencies and other groups and enterprises with a legitimate interest in the running and use of the networks.

In most cases the networks will have some sort of 'shop front' which builds on the tradition of science shops in Holland but links this with real development and manufacturing capabilities.

In some instances workers in declining factories may be involved in setting up small embryo units which when they become self-sustaining can move out as self-sufficient industrial units. Even at this embryonic stage there are strong indications that the networks will meet real requirements in the community.

Workers from sections of the electromechanic industry are now working on motor-controller devices using micro-processors in a system which is likely to result in significant energy savings and provide a means of moving from electro-mechanical equipment to electronics.

The new technology network is working on an expert computer system in conjunction with a major teaching hospital to develop software which will diffuse the most up-to-date knowledge of the consultant back into the community and general practice.

Tenants' groups are working with the energy network on products to reduce energy costs in their buildings. Societies for the disabled and some of their advisers have produced a list of 457 products and services which they require. Some women's groups have suggested training courses based on the recycling of electro-mechanical and electronic equipment so that they could develop the deep diagnostic skills involved in the repair of such equipment. Some of these groups suggest that they would prefer a separate all-women working environment until they develop the competence and confidence to work with male colleagues.

Where a network helps a group to develop a new product range, they could gain some royalty or other payments once the activity becomes profitable. These can be recycled into the network to provide more extended services to the community.

Those who question the ability of industrial workers and community groups to develop new products in this fashion could be reminded of the remarkable plan for socially useful production by the workers at Lucas Aerospace in 1976. Faced with structural unemployment they proposed over 150 socially useful products which in the long term would conserve energy and material and provide socially useful work. In a sad reflection on our institutions, several of the products proposed are now being manufactured in West Germany, Japan and elsewhere, but not in Britain.

The technology networks should provide mechanisms in which the community can identify its own needs and use its practical experience and knowledge to begin to address these. For academics it will provide an opportunity to serve the community which provides for them.

The Centre for Alternative Industrial and Technological Systems (CAITS) originally located at the North-East London Polytechnic and now at the North London Polytechnic, and the unit for the development of appropriate products (UDAP) at Coventry (Lanchester) Polytechnic, both institutions set up by the Lucas workers, indicate that these organizations also provide fascinating students' projects and that students working on them display much higher levels of motivation than on some artificially contrived project – provided simply to meet the project requirement.

The proposal for technology networks in London has already attracted international attention and a conference in May 1983, hosted jointly by the GLEB, the CAITS and the north and east network attracted academics and community workers from Denmark, Holland and the United States and showed clearly that an international requirement exists for facilities of the kind proposed in London. Underlying the whole philosophy of the networks is a belief that the most important asset in solving London's problems is the skill and ingenuity of its people. The networks are one way of unlocking that.

Section 6

Future Directions

THE PATTERN OF INNOVATION

Section 6 is concerned with two broader issues affecting the future direction of technological change. Articles 6.1–6.3 in this first part are concerned with the question 'is there a pattern to the evolution of designed objects and technological innovation and, if so, what are the implications for the innovative strategies of firms and for the growth of the economy?' The second part (articles 6.4–6.8) deals with the question 'is technological innovation necessarily a "good thing" and what can be done to steer it in socially desirable directions?'

Introduction: Design Evolution, Technological Innovation and Economic Growth

Robin Roy

Patterns of technological change may be discerned at varying levels both in historical time and geographical space. In article 6.1, Abernathy and Utterback provide some very interesting ideas about how major technical innovations evolve. Several examples,

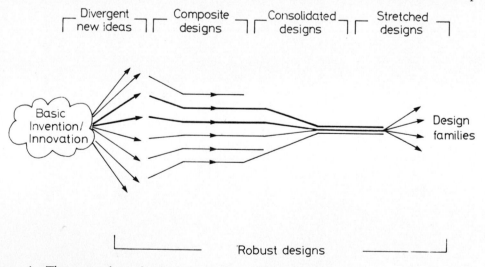

Figure 1   The stages through which dominant, 'robust' designs evolve.
Source: Adapted from Gardiner and Rothwell (1985).

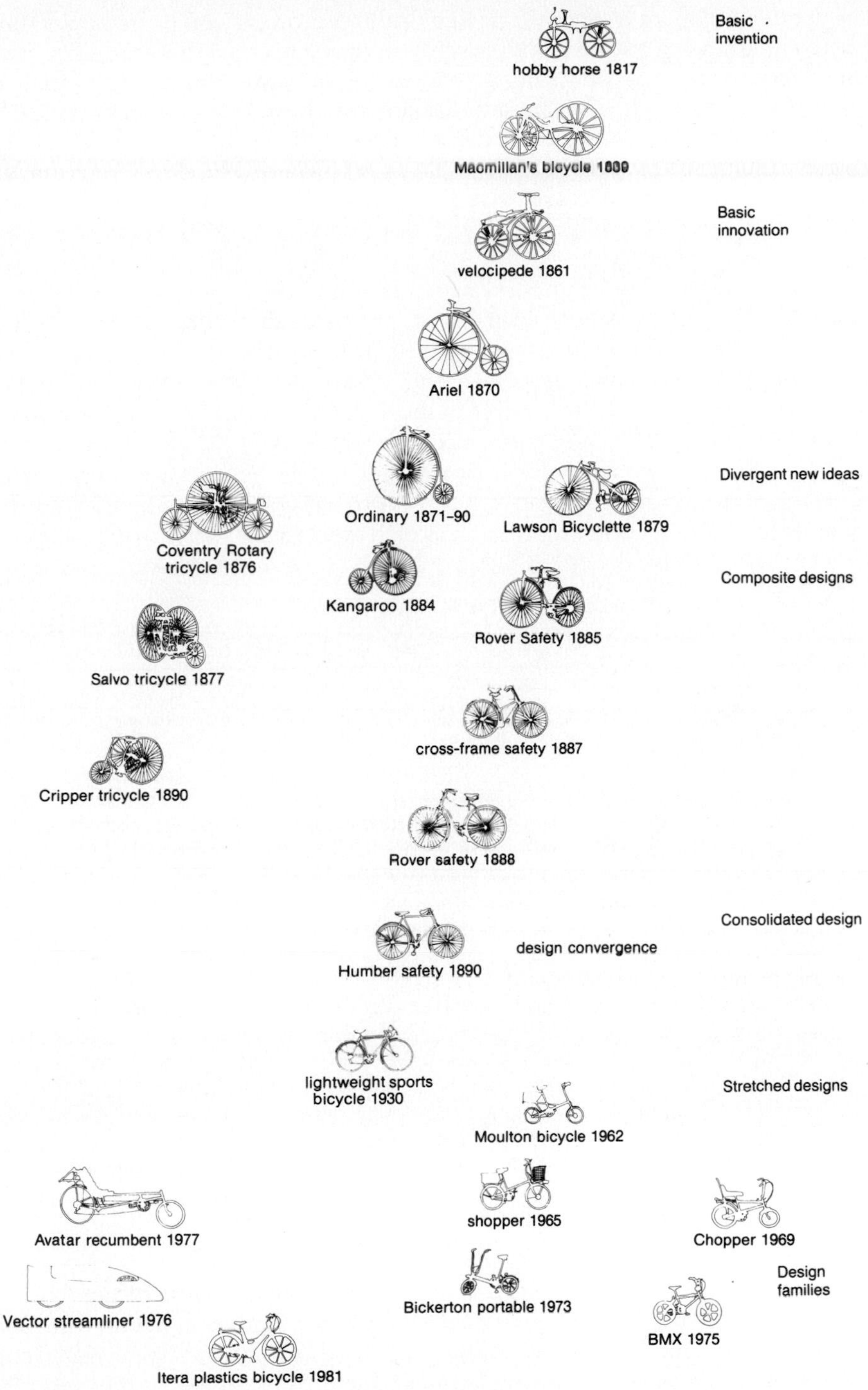

hobby horse 1817

Basic invention

Macmillan's bicycle 1839

velocipede 1861

Basic innovation

Ariel 1870

Coventry Rotary tricycle 1876

Ordinary 1871–90

Lawson Bicyclette 1879

Divergent new ideas

Salvo tricycle 1877

Kangaroo 1884

Rover Safety 1885

Composite designs

Cripper tricycle 1890

cross-frame safety 1887

Rover safety 1888

Humber safety 1890

design convergence

Consolidated design

lightweight sports bicycle 1930

Moulton bicycle 1962

Stretched designs

Avatar recumbent 1977

shopper 1965

Chopper 1969

Vector streamliner 1976

Bickerton portable 1973

Design families

BMX 1975

Itera plastics bicycle 1981

Figure 2   Evolution of the bicycle through divergent, convergent and divergent phases between 1860 and 1980.
Source: Adapted from Roy and Cross (1983).

such as the development of the electric light bulb, motor car and passenger airliner, are used to show that as an industry matures from its early origins, there is a shift from 'radical' product innovation, producing a great variety of competing designs, to 'evolutionary' improvement of a few 'dominant' designs that have emerged from the early variety. In parallel with this shift from radical to evolutionary innovation, Abernathy and Utterback argue that there is a transition from *product* to *process* innovation, as firms expand and begin to manufacture in larger volumes. Eventually both product and process innovation became evolutionary with dominant designs competing on price and quality.

Abernathy and Utterback's model (which is shown in Figure 1 at the beginning of their article) has had a major influence on innovation theory. It is nevertheless a simplification, and more sophisticated models of how products evolve over time have since appeared. In particular, Gardiner and Rothwell (1985) have developed Abernathy and Utterback's model to analyse in more detail the process that leads to and follows from the emergence of a dominant design. Gardiner (1984) has illustrated his and Rothwell's model (shown in Figure 1) by reference to the evolution of engineering products such as motor cars and aircraft, but it is probably best explained by considering a simpler object. The bicycle provides an excellent example. As shown in Figure 2 and described in Box 1, its design evolution closely matches Rothwell and Gardiner's model.

Box 1   Design evolution of the bicycle (based on Roy, 1984)

> 1. *Divergent new ideas.*   The basic invention was the 'hobby horse' of 1817, and the first successful innovation was the 1861 'velocipede'. The quest for improved performance led to the rapid development of components (e.g. wire-spoked wheels, chain drive, tubular steel frames) which were combined into literally hundreds of two, three and four-wheel cycle designs.
> 2. *Composite designs.*   Through trial and error three basic design configurations emerged by 1880 – the fast, but unsafe, Ordinary ('penny-farthing'); the safe, but cumbersome, tricycle; and the rear chain-driven bicycle.
> 3. *Consolidated designs.*   The introduction of the diamond-frame, chain-driven Rover 'Safety' bicycle in 1885 soon displaced all earlier designs, and when equipped with the pneumatic tyre of 1888, resulted in a 'dominant' design which survived virtually unchanged for 60 years.
> 4. *Stretched designs.*   New and improved components, materials and accessories (e.g. hub gears, alloy steels, dynamo lights) allowed manufacturers to produce a variety of machines suited to different uses – general utility, racing, touring, etc.
> 5. *Design families.*   The introduction of the Moulton small-wheeler in 1962 led to a whole family of bicycle types aimed at different markets: the 'Shopper'; the folding bike; the 'Chopper' and BMX fun bikes for children, etc.
>   By the late 1970s radical new forms of pedal-powered vehicle were emerging, including recumbent 'streamliners' and electrically assisted cycles, suggesting a return to the initial 'divergent ideas' stage of evolution.

Explanations of design history, such as Box 1, may appear to be of purely academic interest, but in their booklet *Design and the Economy*, Rothwell, Schott and Gardiner (1985, pp. 27–8) argue that products (like the bicycle), which have evolved through the various phases of composition, consolidation and stretching, are what they term 'robust' designs (see Figure 1). They argue further that robust designs are more likely to succeed commercially, because they are more flexible and adaptable than highly refined 'lean' designs. Rothwell, Schott and Gardiner compare the highly successful 'robust' Ford Cortina family of cars with the less successful 'lean' BL 1100/1300 series to make their point, but it is possible to think of more spectacular examples of sophisticated, but inflexible 'lean' designs (e.g. Concorde, the Advanced Passenger Train; see Potter and Roy, 1986) that failed in comparison with less complex, but more adaptable designs (e.g. the Boeing 747 'Jumbo' airliner and the French Railway's Train à Grande Vitesse (TGV)). The benefits of flexible technologies are discussed further in article 6.5 by Collingridge.

So far we have been concerned with the evolution of individual products and processes. Abernathy and Utterback relate the pattern of product and process innovation to the evolution of whole industries, from an early 'fluid' stage dominated by small innovative firms, through a 'transitional' stage in which larger firms emerge to produce the dominant designs, to a 'specific' stage dominated by large corporations producing standardized products in large volumes. The articles by Hall (6.2) and Ray (6.3) attempt to link the evolution of whole industries and technologies to the growth of the economic system.

Both Hall and Ray start with the idea that industrial economies are subject to 50-year 'Kondratiev cycles' of recession and recovery. They note that, according to Schumpeter and Mensch, the main force behind these 'long cycles' is the appearance, during periods of economic decline, of a 'wave' of major technological innovations whose diffusion causes an upswing in the economic system. Although both authors favour the idea of innovation-induced economic cycles, Ray examines the theory much more critically than Hall. Ray observes that it is very difficult to pin-point the origins of any major innovation, and even more difficult to say precisely when it evolved to the point of major economic importance. So the cause-and-effect relationship between the appearance of major innovations and the recovery of the economy is never certain. It also means that it is difficult to say precisely which innovations of today will be crucial to the next Kondratiev cycle. Hall clearly believes that the next Kondratiev cycle (forecast to peak in the late 1990s) will be based on the cluster of innovations resulting from the application of microprocessors, biotechnology, new materials and energy technologies. Ray believes there are many other contenders for the 'super-class technologies' of the future, although in a later article he argued that the key innovations behind past and future economic cycles are all *energy*-related (Ray, 1983). What is perhaps most important, Freeman (1983) has suggested, is not individual innovations, but the 'clustering' of related innovations into what he calls 'new technology systems' (see article 4.6).

Neither Kondratiev nor Schumpeter properly addressed the question of why the *location* of waves of innovation and economic growth shifts from country to country and region to region. Hall, as a geographer, is particularly concerned with this question, but Ray also discusses it in some detail. Both describe how the centre of innovative activity and economic power shifted from Britain in the eighteenth and early nineteenth centuries to Germany and the United States in the nineteenth and twentieth centuries, and most recently to Japan. The interesting question is where will the innovative wave for the predicted fifth Kondratiev of the 1990s occur? Hall clearly believes that the key innovations of the future will emerge from those nations and regions that exploit the results of the latest scientific and technological research. The main source of these innovations, Hall believes, will be new enterprises linked to universities – on the pattern of the science parks described in section 5 of this book. Ray notes that is not just the *creation* of innovations that is vital to economic recovery, but their *adoption* and *diffusion* into use. He believes that the reason for Britain's economic decline was not any lack of inventiveness, but a failure of its industries and institutions to exploit and adopt the latest technologies. He is hopeful, however, that it may be 'Britain's turn now' as a leader in areas such as finance and insurance, computer software, and scientific research and development.

Finally, both Hall and Ray point out that the headlong international pursuit of innovation cannot occur without social cost. They point to the problems of alienation at work and growing unemployment, especially in regions reliant on more 'mature' industries like coal and steel. The issues that such social problems raise for the control of technological innovation are the subject of the articles in the second part of section 6.

References

Freeman, C. (1983) Design and British economic performance, *mimeo*, lecture given at the Design Council, 23 March, Science Policy Research Unit, University of Sussex.

Freeman, C., Clarke, J. and Soete, L. (1982) *Unemployment and Technical Innovation: a study of long waves and economic development*, London: Frances Pinter.

Gardiner, J. P. (1984) Robust and lean designs with state-of-the-art automotive and aircraft examples, in C. Freeman (ed.), *Innovation, Design and Long Cycles in Economic Development*, London: Design Research Publications, Royal College of Art, pp. 215–248.

Gardiner, J. P. and Rothwell, R. (1985) Tough customers: good designs, *Design Studies*, vol. 6, no. 1, January, pp.7–17.

Potter, S. with Roy, R. (1986) T362, Research and development: British Rail's fast trains, *Design and Innovation* (Block 3), Milton Keynes: Open University Press.

Ray, G. F. (1983) Energy and the long cycles, *Energy Economics*, January, pp. 3–7.

Rothwell, R., Schott, K. and Gardiner, J. P. (1985) *Design and the Economy: the role of design and innovation in the prosperity of industrial companies*, 3rd edn, London: Design Council.

Roy, R. and Cross, N. (1983) *Bicycles: Invention and Innovation*, T263: *Design Processes and Products* (Units 5–7), Milton Keynes: Open University Press.

Roy, R. (1984) Design and innovation in a mature consumer industry, in R. Langdon (ed.), *Design Policy Vol. 2: Design and Industry*, London: Design Council, pp. 91–5.

6.1

Patterns of

Industrial Innovation

William J. Abernathy and James M. Utterback

How does a company's innovation – and its response to innovative ideas – change as the company grows and matures?

Are there circumstances in which a pattern generally associated with successful innovation is in fact more likely to be associated with failure?

Under what circumstances will newly available technology, rather than the market, be the critical stimulus for change?

When is concentration on incremental innovation and productivity gains likely to be of maximum value to a firm? In what situations does this strategy instead cause instability and potential for crisis in an organization?

Intrigued by questions such as these, we have examined how the kinds of innovations attempted by productive units apparently change as these units evolve. Our goal was a model relating patterns of innovation within a unit to that unit's competitive strategy, production capabilities, and organizational characteristics.

This article summarizes our work and presents the basic characteristics of the model to which it has led us. We conclude that a productive unit's capacity for and methods of innovation depend critically on its stage of evolution from a small technology-based enterprise to a major high-volume producer. Many characteristics of innovation and the innovative process correlate with such an historical analysis; and on the basis of our model we can now attempt answers to questions such as those above.

A Spectrum of Innovators

Past studies of innovation imply that any innovating unit sees most of its innovation as new products. But that observation masks an essential difference: what is a product innovation by a small, technology-based unit is often the process equipment adopted by a large unit

W. J. Abernathy and J. M. Utterback, Patterns of industrial innovation, *Technology Review*, vol. 80, no. 7, 1978, extracts from pp. 41–7.
William J. Abernathy was Professor of Business Administration at Harvard Business School; James Utterback is a research director at the Center for Policy Alternatives, Massachusetts Institute of Technology.

to improve its high-volume production of a standard product. We argue that these two units – the small, entrepreneurial organization and the larger unit producing standard products in high volume – are at opposite ends of a spectrum, in a sense forming boundary conditions in the evolution of a unit and in the character of its innovation of product and process technologies.

One distinctive pattern of technological innovation is evident in the case of established, high-volume products such as incandescent light bulbs, paper, steel, standard chemicals, and internal-combustion engines, for example.

The markets for such goods are well defined; the product characteristics are well understood and often standardized; unit profit margins are typically low; production technology is efficient, equipment-intensive, and specialized to a particular product; and competition is primarily on the basis of price. Change is costly in such highly integrated systems because an alteration in any one attribute or process has ramifications for many others.

In this environment innovation is typically incremental in nature, and it has a gradual, cumulative effect on productivity. For example, Samuel Hollander has shown that more than half of the reduction in the cost of producing rayon in plants of E. I. du Pont de Nemours and Co. has been the result of gradual process improvements which could not be identified as formal projects or changes. A similar study by John Enos shows that accumulating, incremental developments in petroleum refining processes resulted in productivity gains which often eclipsed the gain from the original innovation. Incremental innovations, such as the use of larger railroad cars and unit trains, have resulted in dramatic reductions in the cost of moving large quantities of materials by rail.

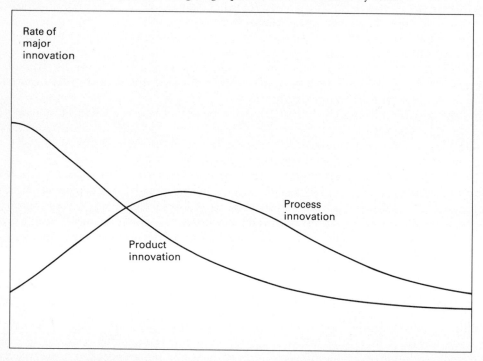

Figure 1   The changing character of innovation, and its changing role in corporate advance. The authors focus on three stages in the evolution of a successful enterprise; its period of flexibility, in which the enterprise seeks to capitalize on its innovations where they offer greatest advantages; its intermediate years, in which major products are used more widely; and its full maturity, when prosperity is assured by leadership in several principal products and technologies.

Table 1

	Fluid Pattern	*Transitional pattern*	*Specific pattern*
Competitive emphasis on:	Functional product performance	Product variation	Cost reduction
Innovation stimulated by:	Information on users' needs and users' technical inputs	Opportunities created by expanding internal technical capability	Pressure to reduce cost and improve quality
Predominant type of innovation:	Frequent major changes in products	Major process changes required by rising volume	Incremental for product and process, with cumulative improvement in productivity and quality
Product line:	Diverse, often including custom designs	Includes at least one product design stable enough to have significant production volume	Mostly undifferentiated standard products
Production processes:	Flexible and inefficient; major changes easily accommodated	Becoming more rigid, with changes occurring in major steps	Efficient, capital-intensive, and rigid; cost of change is high
Equipment:	General-purpose, requiring highly skilled labor	Some subprocesses automated, creating 'islands of automation'	Special-purpose, mostly automatic with labor tasks mainly monitoring and control
Materials:	Inputs are limited to generally-available materials	Specialized materials may be demanded from some suppliers	Specialized materials will be demanded; if not available, vertical integration will be extensive
Plant:	Small-scale, located near user or source of technology	General-purpose with specialized sections	Large-scale, highly specific to particular products
Organizational control is:	Informal and entrepreneurial	Through liaison relationships, project and task groups	Through emphasis on structure, goals, and rules

In all these examples, major systems innovations have been followed by countless minor product and systems improvements, and the latter account for more than half of the total ultimate economic gain due to their much greater number. While cost reduction seems to have been the major incentive for most of these innovations, major advances in performance have also resulted from such small engineering and production adjustments.

Such incremental innovation typically results in an increasingly specialized system in which economies of scale in production and the development of mass markets are extremely important. The productive unit loses its flexibility, becoming increasingly dependent on high-volume production to cover its fixed costs and increasingly vulnerable to changed demand and technical obsolescence.

Major new products do not seem to be consistent with this pattern of incremental change. New products which require reorientation of corporate goals or production facilities tend to originate outside organizations devoted to a 'specific' production system; or, if originated within, to be rejected by them.

A more fluid pattern of product change is associated with the identification of an emerging need or a new way to meet an existing need; it is an entrepreneurial act. Many studies suggest that such new product innovations share common traits. They occur in disproportionate numbers in companies and units located in or near affluent markets with strong science-based universities or other research institutions and entrepreneurially oriented financial institutions. Their competitive advantage over predecessor products is based on superior functional performance rather than lower initial cost, and so these radical innovations tend to offer higher unit profit margins.

When a major product innovation first appears, performance criteria are typically vague and little understood. Because they have a more intimate understanding of performance requirements, users may play a major role in suggesting the ultimate form of innovation as well as the need. For example, Kenneth Knight shows that three-quarters of the computer models which emerged between 1944 and 1950, usually those produced as one or two of a kind, were developed by users.

It is reasonable that the diversity and uncertainty of performance requirements for new products give an advantage in their innovation to small, adaptable organizations with flexible technical approaches and good external communications, and historical evidence supports that hypothesis. For example John Tilton argues that new enterprises led in the application of semiconductor technology, often transferring into practice technology from more established firms and laboratories [see also article 3.7 by Rothwell]. He argues that economies of scale have not been of prime importance because products have changed so rapidly that production technology designed for a particular product is rapidly made obsolete. [. . .]

A translation from radical to evolutionary innovation

These two patterns of innovation may be taken to represent extreme types – in one case involving incremental change to a rigid, efficient production system specifically designed to produce a standardized product, and in the other case involving radical innovation with product characteristics in flux. They are not in fact rigid, independent categories. Several examples will make it clear that organizations currently considered in the 'specific' category – where incremental innovation is now motivated by cost reduction – were at their origin small, 'fluid' units intent on new product innovation.

John Tilton's study of developments in the semiconductor industry from 1950 through 1968 indicates that the rate of major innovation has decreased and that the type of innovation shifted. Eight of the 13 product innovations he considers to have been most important during that period occurred within the first seven years, while the industry was making less than 5 per cent of its total 18-year sales. Two types of enterprise can be identified in this early period of the new industry – established units that came into semiconductors from vested positions in vacuum tube markets, and new entries such as Fairchild Semiconductor, IBM, and Texas Instruments. The established units responded to competition from the newcomers by emphasizing process innovations. Meanwhile, the latter sought entry and strength through product innovation. The three very successful new entrants just listed were responsible for half of the major product innovations and only one of the nine process innovations which Dr Tilton identified in that 18-year period, while three principal established units (divisions of General Electric, Philco, and RCA) made only one-quarter of the product innovations but three of the nine major process innovations in the same period.

In this case process innovation did not prove to be an effective competitive stance; by 1966 the three established units together held only 18 per cent of the market while the three new units held 42 per cent. Since 1968, however, the basis of competition in the industry has changed; as costs and productivity have become more important, the rate of major product innovation has decreased, and effective process innovation has become an important factor in competitive success. For example, by 1973 Texas Instruments which had been a flexible, new entrant in the industry two decades earlier and had contributed no major process innovations prior to 1968, was planning a single machine that would produce 4 per cent of world requirements for its integrated-circuit unit.

Like the transistor in the electronics industry, the DC–3 stands out as a major change in the aircraft and airlines industries. Almarin Phillips has shown that the DC–3 was in fact a cumulation of prior innovations. It was not the largest, or fastest, or longest-range aircraft; it was the most economical large, fast plane able to fly long distances. All the features which made this design so completely successful had been introduced and proven in prior aircraft. And the DC–3 was essentially the first commercial product of an entering firm (the DC–1 and DC–2 were produced by Douglas only in small numbers).

Just as the transistor put the electronics industry on a new plateau, so the DC–3 changed the character of innovation in the aircraft industry for the next 15 years. No major innovations were introduced into commercial aircraft design from 1936 until new jet-powered aircraft appeared in the 1950s. Instead, there were simply many refinements to the DC–3 concept – stretching the design and adding appointments; and during the period of these incremental changes airline operating cost per passenger-mile dropped an additional 50 per cent.

The electric light bulb also has a history of a long series of evolutionary improvements which started with a few major innovations and ended in a highly standardized commodity-like product. By 1909 the initial tungsten filament and vacuum bulb innovations were in place; from then until 1955 there came a series of incremental changes – better metal alloys for the filament, the use of 'getters' to assist in exhausting the bulb, coiling the filaments, 'frosting' the glass, and many more. In the same period the price of a 60-watt bulb decreased (even with no inflation adjustment) from $1.60 to 20 cents each, the lumens output increased by 175 per cent, the direct labor content was reduced more than an order of magnitude, from 3 to 0.18 minutes per bulb, and the production process evolved from a flexible job-shop configuration, involving more than 11 separate operations and a heavy reliance on the skills of manual labor, to a single machine attended by a few workers.

Product and process evolved in a similar fashion in the automobile industry. During a four-year period before Henry Ford produced the renowned Model T, his company developed, produced, and sold five different engines, ranging from two to six cylinders. These were made in a factory that was flexibly organized much as a job shop, relying on trade craftsmen working with general-purpose machine tools not nearly so advanced as the best then available. Each engine tested a new concept. Out of this experience came a dominant design – the Model T; and within 15 years 2 million engines of this single basic design were being produced each year (about 15 million all told) in a facility then recognized as the most efficient and highly integrated in the world. During that 15-year period there were incremental – but no fundamental – innovations in the Ford product. [. . .]

This shift from radical to evolutionary product innovation is a common thread in these examples. It is related to the development of a dominant product design, and it is accompanied by heightened price competition and increased emphasis on process innovation. Small-scale units that are flexible and highly reliant on manual labor and craft skills utilizing general-purpose equipment develop into units that rely on automated, equipment-intensive, high-volume processes. We conclude that changes in innovative pattern, production process, and scale and kind of production capacity all occur together in a consistent, predictable way [see Box 1]. [. . .]

Box 1 Dominant designs

The milestone in all the examples of transition in the accompanying article is a dominant new product synthesized from individual technological innovations introduced independently in prior products. This dominant design has the effect of enforcing standardization so that production economies can be sought. Then effective competition begins to take place on the basis of cost as well as of product performance.

Similar product design milestones can be identified in other product lines: sealed refrigeration units for home refrigerators and freezers, effective can-sealing technology in the food canning industry and the standardized diesel locomotive in the locomotive and railroad industry. In each case the milestone signals a significant transformation, affecting the type of innovation which follows it, the source of information, and the size, scope, and use of formal research and development.

Dominant designs are likely to display one or more of the following qualities:

– Technologies which lift fundamental technical constraints limiting the prior art while not imposing stringent new constraints.
– Designs which enhance the value of potential innovations in other elements of a product or process.
– Products which assure expansion into new markets.

Managing technological innovation

If it is true that the nature and goals of an industrial unit's innovations change as that unit matures from pioneering to large-scale producer, what does this imply for the management of technology?

We believe that some significant managerial concepts emerge from our analysis – or model, if you will – of the characteristics of innovation as production processes and primary competitive issues differ. As a unit moves toward large-scale production, the goals of its innovations change from ill-defined and uncertain targets to well-articulated design objectives. In the early stages there is a proliferation of product performance requirements and design criteria which frequently cannot be stated quantitatively, and their relative importance or ranking may be quite unstable. It is precisely under such conditions, where performance requirements are ambiguous, that users are most likely to produce an innovation and where manufacturers are least likely to do so. One way of viewing regulatory constraints such as those governing auto emissions or safety is that they add new performance dimensions to be resolved by the engineer – and so may lead to more innovative design improvements. They are also likely to open market opportunities for innovative change of the kind characteristic of fluid enterprises in areas such as instrumentation, components, process equipment, and so on.

The stimulus for innovation changes as a unit matures. In the initial fluid stage, market needs are ill-defined and can be stated only with broad uncertainty; and the relevant technologies are as yet little explored. So there are two sources of ambiguity about the relevance of any particular program of research and development – target uncertainty and technical uncertainty. Confronted with both types of uncertainty, the decision-maker has little incentive for major investments in formal research and development.

As the enterprise develops, however, uncertainty about markets and appropriate targets is reduced, and larger research and development investments are justified. At some point before the increasing specialization of the unit makes the cost of implementing technological innovations prohibitively high and before increasing cost competition erodes profits with which to fund large indirect expenses, the benefits of research and development efforts would reach a maximum. Technological opportunities for improvements and additions to existing product lines will then be clear, and a strong commitment to research and development will be characteristic of productive units in the middle stages of development. Such firms will be seen as 'science-based' because they invest heavily in formal research and engineering departments, with emphasis on process innovation and product differentiation through functional improvements. [. . .]

A small, fluid entrepreneurial unit requires general-purpose process equipment which is typically purchased. As it develops, such a unit is expected to originate some process-equipment innovations for its own use; and when it is fully matured its entire processes are likely to be designed as integrated systems specific to particular products. Since the mature firm is now fully specialized, all its major process innovations are likely to originate outside the unit. [. . .]

The organization's methods of coordination and control change with the increasing standardization of its products and production processes. As task uncertainty confronts a productive unit early in its development, the unit must emphasize its capacity to process information by investing in vertical and lateral information systems and in liaison and project groups. Later, these may be extended to the creation of formal planning groups, organizational manifestations of movement from a product-oriented to a transitional state; controls for regulating process functions and management controls such as job procedures, job descriptions, and systems analyses are also extended to become a more pervasive feature of the production network.

As a productive unit achieves standardized products and confronts only incremental change, one would expect it to deal with complexity by reducing the need for information processing. The level at which technological change takes place helps to determine the extent to which organizational dislocations take place. Each of these hypotheses helps to explain the firm's impetus to divide into homogeneous productive units as its products and process technology evolve.

The hypothesized changes in control and coordination imply that the structure of the organization will also change as it matures, becoming more formal and having a greater number of levels of authority. The evidence is strong that such structural change is a characteristic of many enterprises and of units within them.

References

Abernathy, William J. (1978) *The Productivity Dilemma: Roadblock to Innovation in the Automobile Industry;* Baltimore: Johns Hopkins University Press.

Bright, James R. (1958) *Automation and Management;* Boston: Division of Research, Graduate School of Business Administration, Harvard University.

Clarke, R. (1968) Innovation in liquid propelled rocket engines, Doctoral Dissertation, Stanford Graduate School of Business.

Enos, J. (1967) *Petroleum Progress and Profits;* Cambridge: MIT Press.

Hogan, W. T. (1971) *Economic History of the Iron and Steel Industry in the United States,* Vol. 3; Lexington: Mass.: D. C. Heath Books, p. 1011.

Hollander, S. (1965) *The Sources of Increased Efficiency;* Cambridge: MIT Press.

Jenkins, Reese V. (1975) *Images and Enterprise: Technology and the American Photographic Industry, 1839 to 1925;* Baltimore: Johns Hopkins University Press.

Knight, Kenneth E. (1963) A study of technological innovation: the evolution of digital computers, Doctoral Dissertation, Carnegie Institute of Technology, Pittsburgh.

Little, Arthur D., Inc. *Patterns and Problems of Technical Innovation in American Industry: Report to the National Science Foundation*, PB 181573, US Department of Commerce, Office of Technical Services, Washington D.C.: US Government Printing Office, September.

Miller, R. E., and D. Sawers, (1970) *The Technical Development of Modern Aviation*, New York: Praeger.

Stigler, G. J. (1968) *The Organization of Industry*, Homeward, Illinois: Richard D. Irwine.

Tilton, John E. (1971) *International Diffusion of Technology: The Case of Semiconductors*, Washington D.C.: Brookings Institution.

von Hippel, E. (1978) Users as innovators, *Technology Review*, January, pp. 30–34.

Further Reading

For readers who wish to explore this subject in greater detail, the authors recommend:

Abernathy, W. J. and Townsend, P. L. (1975) Technology, Productivity and Process Change, *Technological Forecasting and Social Change*, vol. 7, no. 4, August, pp. 379–96.

Abernathy, W. J. and Wayne, K. (1974) Limits of the learning curve, *Harvard Business Review*, vol. 52, no. 5, September–October, pp. 109–19.

Utterback, James M. (1974) Innovation in industry and the diffusion of technology, *Science*, vol. 183, February, 1974, pp. 620–6.

Utterback, James M. and Abernathy, W. J. (1975) A dynamic model of process and product innovation, *Omega*, vol. 3, no. 6, pp. 639–56.

6.2

The Geography of the Fifth Kondratieff Cycle

Peter Hall

A spectre is haunting Europe – and the United States also. It is the spectre not of communism, but of a Moscow professor who died over 50 years ago in one of Stalin's jails. In 1925 the economist, Nikolai Kondratieff,* published his paper with its extraordinary discovery: that capitalist economies everywhere, from the Industrial Revolution onward, followed a regular long cycle. About every half century, it appeared, they went full circle from bust to boom and back to bust again.

Kondratieff merely presented the facts – of commodity prices, wages, production of industrial basic goods, and trade. He did not explain them. But in 1939 Joseph Schumpeter, who had discovered Kondratieff's work in Germany before he moved to Harvard, made it one of the main planks of his monumental work on business cycles. Then other economists began to attack the whole idea and it fell into disfavour. But now it is getting a lot of attention again – and it seems poised for a radical reappraisal. For it seems to tell us a great deal about the current sad state of western economies – and about the likely way out.

As interpreted by Schumpeter, each Kondratieff long wave represented a new industrial revolution, based on a new group of technologies. The so-called industrial revolution was simply the first of these: it was based on Abraham Darby's discovery of smelting iron ore with coal and on the mechanisation of the Lancashire cotton industry, and it ran from the 1780s to 1842. The second Kondratieff was the age of steam, railways and Bessemer steel: it ran from 1842 to 1897. The third, the Kondratieff of electricity and chemicals and cars, ran from 1898; it was barely complete when Schumpeter was finishing his great book [Figure 1].

The point about the Kondratieff waves was that each ran a regular course. A long climb up from a depression was followed by a fairly sharp slide into another one – though the process was complicated by shorter waves. Thus the great depression at the end of the Napoleonic wars – the time of Peterloo; the great depression of the 1870s and 1880s, culminating in Bloody Sunday in Trafalgar Square; and the world depression of the 1930s.

---

* This name is transliterated from the Cyrillic alphabet and can also be spelt Kondratiev, as in article 6.3.

P. Hall, The geography of the fifth Kondratieff cycle, *New Society*, 26 March 1981, pp. 535–7.
Peter Hall is Professor of Geography at Reading University.

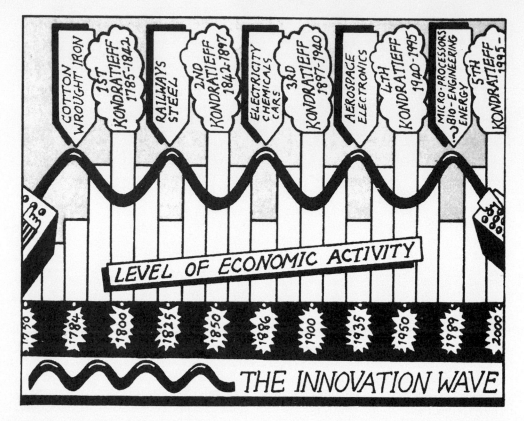

Figure 1   50 to 60-year-long cycles in economic activity identified by Kondratieff and linked to innovation waves by Schumpeter and Mensch.

Neither Kondratieff nor Schumpeter embarked on speculation about the future. But if they had, then the operation of the cycle would have predicted a steady expansion of the world economy from about 1945 onward – and then a descent into depression from the 1970s. It is small wonder that some people are starting to take Kondratieff seriously again.

There is another odd feature about the Kondratieff waves; that each expansion phase tends to culminate in major wars. Thus the Napoleonic wars, the American civil war and the wars in central Europe in the 1860s, the First World War and Vietnam. Only the Second World War was out of character. Perhaps that was because it was started by a madman: or perhaps it counts as a continuation, after a truce, of the First World War?

Schumpeter's analysis provides a full explanation of what Kondratieff had mentioned only in a pregnant sentence. During the recession period of each long wave, there is an exceptional cluster of new inventions – which are, however, applied only at the start of the next upswing. Schumpeter distinguished between invention and its application – which he entitled *innovation*. Innovation, he argued, was the basic function of the capitalist and the basic dynamic of capitalist production. As each innovation wave worked its way through, the market for a particular range of goods became saturated, profits fell and unemployment rose. The wars that arose at this point could be interpreted as struggles for markets.

All this appears like pure grist to the Marxist mill. (Ernest Mandel's major tome on late capitalism, for instance, hinges upon it.) But the problem is that it explains perfectly why, each time, the falling rate of profit is followed by a burst of innovation that compensates for it – and so postpones, perhaps indefinitely, the crisis of capitalism. Kondratieff, in fact, explains precisely why Marx's original thesis was wrong. No wonder Stalin locked him up.

There is another especially good reason for taking Kondratieff seriously again. In 1979, a translation appeared in America of a remarkable German book originally published in 1975. Gerhard Mensch's *Stalemate in Technology: innovations overcome the depression* is a complete vindication and explanation of the Kondratieff process. But more than that: it is also, in many ways, the most original contribution to macro-economic theory since Keynes.

With a terrifying display of Teutonic scholarship in the Schumpeter tradition, Mensch demonstrates that throughout the last 200 years the pace of technological innovation has systematically varied. This is because – as Kondratieff hinted and Schumpeter specified – the rate at which inventions become innovations is subject to what Mensch calls the *wagon-train effect*. Inventions may and do occur in big bunches within a short time, causing the rise of new processes, new products and new industries.

Mensch traces this process in detail for hundreds of innovations. He shows that there is great statistical regularity in the way this happens. Innovations peaked in certain years, which he calls the radical years of history: 1764, 1825, 1886 and 1935. Each came in the middle of a depression and was directly responsible for the succeeding Kondratieff upswing, which began between eleven and seventeen years later [Figure 1].

But precisely because the process is so regular Mensch is able to predict it. He does this not by a simple averaging from the past, but by a complex technique that is based on the observed rate of innovation in the recent past. According to him, by the mid-1970s some half of all innovations for the next Kondratieff had already reached the commercial feasibility stage.

The next radical year of history, he announces with some confidence, will be 1989 – but the decade of maximum innovation should start in 1984.

The 1989 peak year

It is important, of course, that 1989 will be the year of peak innovation. The effect on the world economy is unlikely to be visible for a decade or more after that. The depression, on the Kondratieff–Schumpeter–Mensch theory, is likely to last well through the 1980s.

But there is one element that the theory does not very specifically account for. That is the fact that evidently the locus of innovation shifts. In the first Kondratieff of coal and iron, it was unambiguously Britain. In the second of steam and steel, Britain was starting to share pride of place with Germany and the United States. In the third of chemicals and cars, the lead had passed to the United States – though Germany still played an important role. In the fourth Kondratieff – the post-1945 cycle based on aircraft, weaponry and computers as well as a range of new service industries – the centre of innovation was undoubtedly the United States, but with Japan playing a progressively greater part.

The question everyone must ask is where is the locus of the fifth Kondratieff cluster of innovations? For, as each major nation becomes the innovation leader, so does it become world economic leader.

There is general agreement among experts about the nature of emerging high-technology industries on which the next long wave will be based. One, of course, is the microprocessor and the whole range of machines that will incorporate it. A second is genetic engineering. And a third is concerned with new ways of winning raw materials, above all those concerned with energy.

It is not merely the technical experts who are in agreement here. So, in certain parts of the world, are the speculative investors. On 14 October 1980, American stockbrokers released to the public for the first time the shares of Genentech, a small San Francisco firm specialising in genetic engineering. This is the manipulation of basic genetic structures to produce new forms of biological behaviour, including bacteria that would interact with their environment to produce new drugs, new forms of nutrition and alternative energy sources.

Starting at $35, within one minute the Genentech share price rose to $80. It closed at just over $71 the same evening. The firm's two founders became multimillionaires in 60 seconds. Even the partner in the broking firm, who had managed the underwriting of the shares, could only describe it as feverish speculation. For Genentech had absolutely no history of earnings at all – and there was no other company in the field to which it could be compared.

But perhaps the most significant facts about Genentech are its location and its origin. For the two are closely connected. The firm is in San Francisco, because it resulted from inventions at the local campus of the University of California – with which Genentech are now locked in a legal row about patents. Another major centre of this kind of research is Harvard–MIT – though Harvard have said that they do not want commercial involvement in this work. This provides an exact parallel to the history of the microprocessor industry, which spun off from fundamental research at MIT and at the campus of Stanford University in California.

Research on the Silicon Valley microprocessor phenomenon, in fact, has shown that the university link there was not at all fortuitous; it was planned. Silicon Valley is currently the fastest-growing industrial area in the United States if not the world. Over the past decade it has added 40,000 new jobs per year. It is next door to Stanford, and it is the result of the genius of Frederick Terman, a Stanford professor.

In the 1920s, Terman began to encourage his graduates in radio engineering to stay in the area and start companies there. Two of the first, William Hewlett and David Packard, started theirs in a garage near the campus, in 1939; it is now one of the world's largest electronics firms. But the bulk of the Terman spin-offs were born in the 1950s and 1960s – by which time, Terman had persuaded his university to develop a special industrial park for them [see Introduction to Section 5].

This gives the clue to the new industry. To a greater degree than ever before, the innovations of the fifth Kondratieff will come directly from fundamental research advances – so a university link is essential. But also important is the right kind of environment – for the new industries will employ scarce and very valuable researchers, who must be tempted by what the economists like to call *psychic income*. In comparison, traditional factors of industrial location are almost irrelevant.

All this has been partly true, too, of the industries of the fourth Kondratieff: the aerospace and computing complexes of America's Sunbelt. But they were born out of the defence expenditures of the cold war, and the location in large part reflects that fact. The growth of northern Florida, of Houston, of Phoenix in Arizona, has followed the presence of Cape Kennedy, the Houston Space Centre, and the great research and testing grounds of the Arizona desert. The evidence so far suggests that this kind of link will play a smaller part in the future than in the past. Government money is still likely to play a major role. But it will go to the universities, and in particular to those among them that specialise in the new research areas.

The point here is that university systems, even in a country as dynamic as the United States, have a great deal of built-in inertia. Giant top-rank universities like Harvard and MIT, Berkeley and Stanford, are not likely to lose their status in a hurry. Few other institutions, in the present bleak financial climate, seem destined to join them: the best bets would seem to be state universities in states with buoyant tax revenues, like Virginia, Texas or Colorado. So, in addition to Greater Boston and the San Francisco Bay Area, the strong possibilities for growth are areas like the research triangle of North Carolina, next to the University of North Carolina and Duke University; Austin, the home of the University of Texas; and the Boulder/Colorado Springs belt of Colorado. Significantly, these were three of the areas picked by experts in the *Wall Street Journal* last year for fastest growth in the 1980s.

Hamburger heaven

Not all the innovative growth is likely to take place in such areas – for not all of it will be of the high-technology type. In the fourth Kondratieff, the great new Sunbelt industries included not only companies like Texas Instruments and Hewlett Packard, but also firms like Holiday Inns and Ramada Inns, McDonalds Hamburgers and Kentucky Fried Chicken, Disneyland and Six Flags over Texas – companies that grew not on the state-of-the-art technology so much as organisational innovation. Especially in the burgeoning recreation industry, that is likely to continue to be the case. But again, as in the last 30 years, the innovations are likely to arrive in the American regions that are generally innovatory – that is, the Sunbelt of the south and west.

Not, in other words, the declining American industrial heartland of the north east and mid-west. There, the evidence seems clear that the failure of the traditional industries – the industries of the second and third Kondratieffs, such as the steel of Pittsburgh and the automobiles of Detroit – was a failure to continue innovating in response to challenges from abroad. Pittsburgh should have switched to specialised high-quality steels. But it stayed producing bulk steel in plants that were inefficient compared with its overseas competitors. Similarly, Detroit failed to grasp fast enough the challenge of the switch to small cars.

The great Victorian economist, Alfred Marshall, said it all – as so often he did – nearly a century ago. The spirit of enterprise, of innovation, is something that is in the air of a place, and the air may go stale. Altering the metaphor, the Glasgow economic historian, Sidney Checkland, described the economy of that city in terms of the Upas tree of the Pacific which is so spread out as to kill all new vegetation in its shade. So complacently did the Glaswegians build ships, that they neither saw the need to build them in new ways nor to develop new industries to take the ships' place.

So there is a major lesson for Britain here. Perhaps an even bigger one than for the United States. Both are now economies in trouble. But the troubled part of the British economy is far larger, in relation to the whole, than the American one. And the good part is conversely much smaller. For the vast American Sunbelt – from Washington through Atlanta to Miami, Houston, Phoenix, Los Angeles and San Francisco – substitute London's new towns plus burgeoning cities like Bristol. It makes a disturbing comparison.

Yet the irony is this. Now, as before, the fundamental inventions of the fifth Kondratieff are still being made in Britain. British scientists are still featuring (in much greater numbers than the country's reduced circumstances would warrant) in the Nobel prize lists. Some of them are in the very areas – such as genetic engineering – that are almost certain to spawn the new industries. The problem is that someone else, as before, may turn their genius into commercially exploitable innovation.

What, then, to do? The first and essential point is to realise that many of the older industries, the products of the previous Kondratieff waves, cannot – and should not – be saved. The job should be to do what the Japanese Ministry of International Trade and Industry have been doing so brilliantly this last quarter century: systematically identify the growth industries of the next wave, pump the necessary governmental research and development money into them, and wait for the private sector to come in and exploit the results. That, however, will need a new kind of civil servant: the kind who can spot future commercial winners, rather than backing Concorde.

Goodbye Port Talbot

But the spatial implications must be clear. The new industries are not going to be born in Port Talbot or in Consett. They might just be born again in Glasgow and Manchester and Birmingham, or any other British city with a prestigious university – provided such a research centre were linked directly with an enterprise zone of the right kind, one that systematically encouraged spin-off industry of the Silicon Valley variety. But most likely they will originate in the south.

This is not mere speculation. The important work of John Goddard and his colleagues at the University of Newcastle-upon-Tyne shows that the rate of innovation can be measured – by using such indices as the Queen's Award for Exports, or the innovation records of the Science Policy Research Unit at the University of Sussex. And the result comes out loud and clear: the innovating firms are overwhelmingly in the south-east. The Newcastle team hazard an explanation. Unsurprisingly, it is that this region has the right climate of innovation.

But, as so often in studies of this kind, the authors end with a perverse conclusion: that policy should try to disperse and spread this climate to the other regions of the country. This wins conventional political applause. But there is no evidence at all that it might work.

Rather, what Britain needs now is someone to say what President Carter's departing team on urban policy told him – what he promptly disowned but what Reagan may yet buy: that tomorrow's industries are not going to be born in yesterday's regions, and that the aim of government should be to start planning for a massive move of people from the old areas and cities to the new. Britain's future, if it has one, is in that broad belt that runs from Oxford and Winchester through the Thames Valley and Milton Keynes to Cambridge.

Of course, this is going to be maximally unpopular. The Labour legions in the north will hate it. So will the Tory squirearchy in the shires. But if Britain is to have any industrial prospects in the fifth Kondratieff – which will also be the time when the oil runs dry – that surely is the way it is going to be.

6.3

Innovation in the Long Cycle

George Ray

The sharp break during 1974–5 in the long period of postwar prosperity and the slow recovery from the deep 1975 recession with all its allied phenomena – particularly rapid inflation and high unemployment – have drawn attention, once again, to the much debated long waves in economic activity. They have come back into the forefront of interest – and with them the controversy which has surrounded them ever since their conception.

The concept of long cycles has become associated with Kondratiev,* who analysed the development of long-term trends in selected economic indicators. In 1925 he published his finding, that there exist half-century long cycles, together with an attempt to explain them theoretically. His study, originally in Russian, was translated into German in 1926 and into English in 1935 (Kondratiev, 1935).

Kondratiev's cycles were based on his study of a large number of indicators, most of them concerning the movements of prices and interest rates. Presumably he used them as proxies for any other measure of economic activity; indicators of real output, as it is understood in our time, were not available to him in the 1920s.

Our interest lies, not directly in the theory of the Kondratiev cycle, but in its relevance to innovation. Scientific and technological advance, innovation and investment implicitly play an important part in all the studies concerning the long cycles, and more explicitly in those of Schumpeter (1939). Innovation is indeed a cornerstone of Schumpeterian business cycle theory, according to which its economic impact is immense.

Schumpeter's thesis, in its most simplified form, stated that the upturn in the first Kondratiev cycle (1790–1813) was *largely* due to the dissemination of steam power, the second (1844–74) to the railway boom, and the third (1895–1914/16) to the joint effects of the motor car and electricity. These all fitted Kuznets' requirements of an all-pervasive influence on all, or many sectors of the economy (Kuznets, 1940). In order to develop the argument, let us accept their thesis (knowing only too well that it has not met with general approval).

* This name is transliterated from the Cyrillic alphabet and can also be spelt Kondratieff, as in article 6.2.

G. Ray, Innovation in the long cycle, *Lloyds Bank Review*, no. 135, January 1980, extracts from pp. 14–28. George F. Ray is a Senior Research Fellow at the National Institute of Economic and Social Research.

Innovation cycles?

An interesting debate can be followed in the recent German economic literature about the innovation-generated nature of the Kondratiev cycle. It started with the publication of various works by Mensch (1978 and 1979) who suggested the need for a 'new push of basic innovations' to lift the world economy up from its present unsatisfactory state, characterized by high and rising structural unemployment. He mapped the last 200 years of economic/technical history and found a clustering of basic innovations* in four periods: around the years 1770, 1825, 1885 and 1935 – and nothing much since.

A comparison of this clustering with Kondratiev's long waves is of obvious interest. Table 1 presents a simple attempt at this. It indicates peaks and troughs in Mensch's series of basic innovations and in Kondratiev's original data.

Table 1   Peaks and troughs

	Mensch's basic innovations	Kondratiev's (original) cycles†
Trough years	1795	1790
	1845	1844
	1905	1895
	1955	
Peak years	1770	
	1825	1814
	1885	1874
	1935	1916
Number of years:		
trough to trough	50,60,50	54,51
peak to peak	55,60,50	60,42
trough to peak	30,40,30	24,30,21
peak to trough	25,20,20	30,21

† Kondratiev used 25 series, of which ten concerned the French economy, eight the British, four the USA, one (coal) the German, and two (pig iron and coal production) the world economy.
Source: Mensch (1979); Kondratiev (1935).

Long cycles do seem to appear, albeit with no great regularity and not simultaneously in both areas. Without going into the details of any theoretical discussion of whether we are forced with some statistical coincidence, or whether deeper causes resulted in the long cyclical movements, let us consider the plausibility of any connection between Mensch's innovation 'clusters' and Kondratiev's cycles. The lags between Mensch's innovation peaks and Kondratiev's are approximately 40 years (more precisely: 44, 49, 41 and 32 years respectively).

---

* Mensch defined basic innovations as 'technological and social innovations that create completely new social benefits, new lines of service or industrial products in the public sector and in private business, for which there exists a need, and for which the manufacture and distribution necessitate the creation of new markets, many new jobs, and profitable investment possibilities.'

Kondratiev's three troughs followed the innovation-poor periods with an even more uniform lag of about 50 years. Given the difficulties of measurement, this apparent regularity provides food for further thought since, if the high 'technological content' of each of the long-wave theories – most explicitly Schumpeter's – is considered, this surely must be *the* most important macro-economic aspect of innovation.

This kind of approach can, of course, be commented on (indeed, objected to) on several grounds, only one of which – believed to be decisive – will be detailed here, and this concerns the very complex nature of the innovation process.

The Schumpeterian identification of Kondratiev's cycles in steam power, railways, electric power and the motor car may serve as a starting point. Freeman (1979) has already pointed to some snags, for example, to the very different development in time of the automobile industry as an employer in America, Europe and Japan. He has also pointed to the need for 'basic science coupled to technical exploitation', followed by 'imaginative leaps' – all preceding the Kondratiev upswing. Indeed, Schumpeter – as well as Dupriez, another analyst of long cycles – emphasized the view that whilst there *is* a relationship between innovation and economic development, it is a very complex one. One invention or innovation is followed by another and the long chain eventually produces new products or processes which are again further developed and/or replaced by others. If the 'new' product or process is important enough, it generates activity in many allied areas and cascades through the whole fabric of economic and social life.

Whether we take steam power, the railways or any really major new phenomenon (each of them the result of many innovations), they have led, in their time, to very marked changes in either industry or the lives of ordinary people. They have created new industries, helped to industrialize new regions, made people move around, and so forth.

Historical evidence and questions arising

The country that implements any of these truly major, epoch-shaping innovations on a large scale, speedily disseminating it and creating the conditions – within its natural endowment – favourable to the cascading of investment stemming from it, can be reasonably supposed to do better and advance faster than another country which is later or slower in this process. In what follows, an attempt will be made to test this innovation-centred hypothesis.

The 'industrial revolution', for a variety of reasons, started in Britain. In the first half of the nineteenth century she dominated the, then still narrow, world economy and during the period 1820–50 she produced two thirds of the world's coal, one half of its iron, more than one half of steel, one half of (commercially produced) cotton cloth, and 40 per cent of all hardware. Britain had no competition: the United States was too young, France had been set back several decades by the Napoleonic wars, neither Germany nor Italy were even geographical entities; the Austrian Empire of many nations was a century behind, and the smaller countries in Europe were typically agricultural communities, with the possible exceptions of Belgium and Switzerland.

Even before the end of the eighteenth century, Dr Johnson wrote that 'the age is running mad with innovation' and 'every Master Manufacturer hath a new Invention of his own and is daily improving on those of others'. This classical period somewhat preceded and largely overlapped the first Kondratiev upswing (1790–1813), with the result that 'the British (by the 1820s) were the masters of the Continent in fire, reflecting the advantages of physics and chemistry; the Napoleonic wars were a hothouse of developments in metallurgy and machinery for Britain, but not for the Continent'.

Indeed, historical statistics demonstrate British supremacy and show coal and iron output incomparably higher than in any other country, including those with much higher populations. Britain, the innovator, was riding high on the Kondratiev wave of steam power, with more steam engines than the whole of the rest of the world put together. She needed them mostly – but by no means exclusively – for operating her cotton industry, which, already leading the world at the end of the eighteenth century, continued to raise its output very quickly and retained over-all leadership to the end of the nineteenth century.

But what was the 'basic' scientific advance, invention or innovation that eventually sparked off the first (steam-power induced) Kondratiev upswing around 1790? Many would attribute it to Watt – but then Watt did *not* invent the steam engine, it had been introduced long before by Newcomen (1712), or indeed in a simpler form by Savery (1698); Watt's further developments (separate condenser in 1769 and the rotative engine in 1781) were undoubtedly very significant improvements. In fact, none of these great men invented the steam engine; their merit as innovators is beyond question – but whose was the 'basic' innovation? The first scientist who understood the power of steam and the uses of cylinder and piston was Hero of Alexandria, in the first century AD.

The second Kondratiev cycle was identified by Schumpeter as the railway boom. It again started in Britain and, in 1844, at the beginning of the upturn of the cycle, was already well under way there, whilst in a more primitive phase elsewhere. According to Kondratiev it lasted until 1874. During that period, until about 1860, British industrial production rose rapidly, faster than elsewhere, but towards the end of the upswing – roughly, in the 1860s and 1870s – its growth rate started to lag behind that of Germany, Italy, and some smaller countries such as Sweden, Switzerland and Belgium. The railway system had matured – the innovation having spent its initial driving force – earlier in pioneer Britain than in the countries which were 'followers'. However, at the end of this upswing Britain was still in the lead in many respects, unsurpassed in coal, iron (and the newcomer, steel) and cotton cloth production, but it was becoming clear that in other areas her leadership was being seriously challenged.

We can now ask the same question as before: what – and whose – was the basic invention? Kondratiev's second upswing started in 1844 – Stephenson's first locomotive came into operation in 1814; the Stockton–Darlington railway, the first to be built for public transport and not for coal haulage alone, opened in 1825. But Trevithick had built a locomotive to haul coal in Wales as early as 1804. Who was the 'basic innovator' – or does the laurel go back to the steam engine?

Continuing the pursuit of British economic history: reports from various international exhibitions (e.g. Paris in 1855, etc.) had already mentioned sharpening competition when the Commission of Enquiry into the Economic Depression stated boldly in 1886: 'We are no longer alone. More active rivals with better equipment are springing up and leaving us behind.' The continental industries had been forging ahead at twice the rate of the British ones from about 1870–75. Soon after the turn of the century they were catching up with or overtaking Britain in coal, iron, and especially steel, as well as the new driving force, electricity. The USA had reached maturity and become another powerful competitor. The cotton industry, in which Britain still retained her lead, had lost its paramount importance to steel; in the years just preceding the Great War, German steel output was more than twice that of Britain (and a good part of it was sold to Britain).

Although none of the scholars involved in the long cycles attributed any upswing to steel, it is clear that it was the basis, not only of the war efforts, but also of the third Kondratiev upswing (1895–1914) in which, according to Schumpeter, electricity and the automobile played the major parts. There were many factors contributing to the German advance in steelmaking, among which, however, the early adoption and rapid diffusion of two important innovations – the open hearth of Siemens (1866) and the Thomas process (1878) – by Germany were of outstanding significance.

This was no longer the age of cotton. Structural changes from the first to the last quarters of the nineteenth century, and during the first decades of the present one, were spectacular. The metalworking industries expanded vigorously, branching off into many spheres, and they were joined by the even faster-growing chemical industry – an area in which Germany took the lead, again as the consequence of innovative activity.

The third Kondratiev cycle was attributed by Schumpeter to electricity and the motor car. Again, Germany was more active than Britain in taking up electricity. Yet, in order to assess the British position in electricity early this century, we had better turn to the other newcomer, America. Even in 1907 when Britain was still a great industrial power, and much more comparable with the USA in terms of total industrial output than at any later time, her electricity production was only one fifth of that in the USA, and she was well behind in the adoption of electricity in industry. Among the many consequences was the delayed introduction in Britain of power tools – a seemingly small innovation which, however, greatly raised productivity in construction, shipbuilding, engineering and many other applications.

Electricity was a truly major new development, fitting exactly the requirement that the innovation triggering off the start of a long cycle must be an all-pervasive one, widely affecting various sectors of the economy. Looking at other areas, however, we find a similar relative lagging in Britain around this period of the Great War: in telephones, office machinery, the motor ship, and so forth.

The great new industry, electrical engineering, which sprang up with the spread of electricity, was definitely lagging behind in Britain as compared with the German and American companies: Siemens, Westinghouse and General Electric. The case of the new transport medium, the automobile, was similar. Although a long way behind the USA, Britain was not really behind the continental European countries in the use or production of automobiles, but as early as 1910 the largest car maker in Britain was Henry Ford, producing more cars than the next two largest firms combined. Thus, British entrepreneurs lagged in this major industry too, and this is likely to have had a considerable impact on the general development of a number of branches of the engineering industry in view of the requirements of automobile production in terms of machine tools and new machines – as well as some other newer and older industrial sectors such as rubber and instruments, or even textiles and timber (for upholstery and seats).

Turning again to the originators of the upswing in the third cycle, Lanchester and Austin designed their first cars in 1895 but they had been preceded by Daimler (1887) as well as by others in Europe (France produced 500 cars in 1893). The idea of the internal combustion engine, however, originated much farther back in time, in the late seventeenth century. With electricity we are in similar trouble: the 'basic' invention may have been that of Faraday in 1831, but he could not, of course, have presented his theories in that year without the outstanding achievements of scientists like Benjamin Franklin (1749), Galvani (1791), Volta (1800), Ampère (1822) and others. It was a long way from Faraday's work to the large-scale electricity industry that 'created new social benefits, new markets and new jobs'. [. . .]

Neither Kondratiev nor Schumpeter lived long enough to see the peak of the recent upswing in the 1960s; it was probably attributable to a combination of major innovations in the chemical industries (catalytic cracking, synthetic materials and fibres, antibiotics), in aircraft (jet engine, helicopters), in the electrical/electronic industry (TV, computers), and perhaps even in the peaceful application of nuclear power.

These questions are not answerable on objective criteria. The above will make the point that:

– a long chain of previous inventions helps the 'basic' inventor (whoever he is);

– innovation may follow the relevant basic invention by several decades (Schumpeter and/or Mensch mention mostly innovations);

– from the point of view of its impact on the economy, it is not the basic innovation but its *diffusion* across industry or the economy, and the *speed* of this diffusion, that matters. Only the widely-based rapid diffusion of some major innovations can be assumed to play any part in triggering off the Kondratiev – or any other – long-term upswing;

– with the benefit of hindsight, it may be possible to assess the fundamental importance of some major innovations, but it is difficult to classify any relatively new development in this super-class without the historical perspective. After all, Watt's simple aim was a better engine, Daimler's was a new vehicle he could sell, whilst Faraday and Rutherford were types of scientific genius. Probably none of them was thinking of his achievement as the start of a new era, nor were their immediate contemporaries thinking it. The great new opening was usually recognized somewhat later by others – such as Henry Ford in the case of the automobile.

To sum up; it appears legitimate to say that:

– the outstanding importance of innovation, and its great impact on economic development, finds support in the works of major students of long cycles *á la Kondratiev*, although the causal relationship, or even the identification of those outstanding major innovations which have sparked off a major upswing in the past, is very complex and intricate, as well as difficult to establish;

– the innovation process is, in the main, a continuous one and, although major scientific discoveries and inventions may or may not come in 'jumps', their implementation and large-scale diffusion is usually a gradual process;

– it is difficult, and probably impossible, to take a firm view of whether any one or several of the innovations already known are likely to become *the* driving force of the next upswing (or indeed whether there will be such an upswing in the near future), and, if so, which one(s)? [. . .]

The next round?

There are many who believe that the next great innovation, following in significance the motors of earlier Kondratiev cycles and comparable to them in width and depth of impact on the economy, will be the microprocessor [see for example article 4.6 by Freeman]. The importance of micro-electronics can be seen already in many areas and it is not surprising that the 'microprocessor revolution' has begun to merit serious discussion. It has been emphasized that the microprocessor is a chameleon and that it takes on the character of whatever program has been fed into it; it can direct a guided missile, operate a coffee dispenser, regulate the use of petrol in a car or control an industrial process. If properly programmed it can be used almost anywhere, in communications, in metal machining, and in widely varying applications, from libraries' bibliographies to medical diagnosis. It is conceivable that it could be a candidate to lead a technological upheaval, giving the necessary push for a swing up out of Mensch's 'technological stalemate'.

But are we really in a stalemate position? It cannot be denied that economic performance in the Western world has not been outstanding in the past few years and that the immediate outlook is not cheerful either. There are many factors contributing to this state of affairs and it is impossible to allocate the blame to any single one. It is similarly difficult to say where we stand now in respect of the long cycles. We can do no more than speculate.

Mensch's innovation peaks followed each other with a lag of 50–60 years; the most recent one was in 1935 – the next, on this basis, should follow some time after 1985. Kondratiev's cycles required about 25 years from trough to peak; if we consider 1975 as the trough, the peak will only be reached by 2000, but in the meantime should come the upswing. On past experience, if the indications in Table 1 are accepted, this is too soon after the innovation peak in 1985, since earlier there used to be 40 years between the peaks of the two series – but then the time lag between the innovation peak (1935) and the economic trough (1975) was also shorter than the 50 years observed earlier. This kind of speculation cannot lead us very far, especially when all the ambiguities and uncertainties of the estimates are borne in mind, as well as the doubts surrounding the existence of the long cycle at all.

Let us therefore turn to another approach, which seems perhaps more plausible, if only slightly. Whilst it is difficult to do no more than guess about the future and attempt to choose the winner in a race which is likely to have several runners-up, and possibly many other 'also rans', it is not that difficult to point to areas where considerable developments are likely to occur in the not too remote future, as well as to others where the need for innovatory change is pressing.

Such pressing needs are obviously present in the developing areas of the world. Their agriculture needs upgrading and desert areas need to be made fertile by means of suitable technologies and equipment, if mass undernourishment is to be reduced and prevented. The industrialization of these countries – which has, benefits aside, produced the ugliest face of this process in the overcrowded, inhuman squalor in some districts of Calcutta, Rio and elsewhere – will require productive equipment suited to local conditions.

In the last hundred years, major and minor innovations have helped to improve efficiency markedly in the use of primary energy, but in the present predicament much more will have to be done in this area – and possibly in the very near future. Nuclear energy on a really large scale is still a possible promise for the future, in a form which puts public opinion at rest in all allied areas (including waste disposal) and utilizes more than the present 1–2 per cent of the energy in nuclear fuel. We are still a long way from the large-scale introduction of any really new form of energy, whether it be solar, wind, wave, or nuclear fusion. Problems very similar to those in the energy field may arise, though probably somewhat later, in the supply of some of those raw materials which are at present indispensable to the operation of industry – as we know it today. Environmental protection may hit industry, but will also create demand for new products and processes.

Services, the 'tertiary' sector, have been the most rapidly expanding major part of the economy in all developed countries, East or West. In the 10–15 years to 1976, about five million persons a year transferred to services from other sectors in the industrial countries. In an active area there is always more scope for innovation than elsewhere. Services are no exception to this rule and the great changes in data processing and communications are likely to make themselves felt strongly in many service industries. This may concern especially the largest employer among them: government and other public services.

This is precisely the sort of important area where microprocessors, in computers and otherwise, are likely to have a major part to play. One of the large US corporations said in their annual report, characterizing the 'inundation of the country with paperwork' as the outcome of encroaching bureaucracy, that more than 10,000 federal forms are in existence; to complete and return them costs the private sector some $20 billions a year – the government spends another $20 billions processing them. In Europe, the situation is not much better: according to Shell, a plant extension in the Netherlands 10–20 years ago would have taken some 10 man-days of effort to produce the necessary permit – now they have to deal with 12 official agencies and it takes 5 man-years. Surely this area is ripe for some radical innovation.

Will the last two decades of this century go down in future history as the nuclear era, or that of the microprocessor, or something else, such as a breakthrough in agriculture, biochemistry or allied fields? Although no more than guesses can be expressed in this respect, there does seem to be enough in the pipeline for thinking that the outlook is perhaps somewhat better than a stalemate.

Two Particular Problems

No doubt, whatever the main directions of future development, many problems will arise. Two of them, believed to be of particular significance, are mentioned here.

The first concerns the social and psychological impact on the labour force of modern industrial work. This is characterized by the gradual disappearance of the craftsman, 'assembly line' type of boring work, and – in general terms – the lack of job satisfaction and the resulting alienation of the worker from his work, performed in ever larger and more impersonal surroundings. The idea of 'small is beautiful' may have its attractions, but there are great difficulties in its implementation. This problem probably requires an innovatory approach by other disciplines but the economic impact of any solution will have important consequences, as will the lack of it.

The second major problem area is employment. Industrial innovations may create new jobs, but they also work in the opposite way by reducing labour requirements. Both have been at work in the past and it is not known where the balance will lie in the future. The great fear in the 1950s and early 1960s – that mechanization and automation would make millions unemployed – did not materialize. What will be the effect, in such terms, of the electronic revolution? The few examples already known and the estimates available indicate very big savings in labour requirements. The economy will probably feel this effect only gradually. If the service industries cease to absorb labour at their present high rate, the advanced world may face the problem of structural or technological unemployment on a very large scale indeed, unless some organizational or institutional innovation prevents its emergence.

Reference

Freeman, C. (1979) The Kondratiev long waves, technical change and unemployment, *Proceedings of the OECD experts meeting on Structural Determinants of Employment and Unemployment* (March 1977), Paris: OECD. [See also Freeman, C. *et al.* (1982) *Unemployment and Technical Innovation*, London: Frances Pinter.]

Kondratiev, N. D. (1935) The long waves in economic life, *Review of Economic Statistics*, November. (Reproduced in *Lloyd's Bank Review*, no. 129, July 1978.)

Kuznets, S. (1940) Review of 'Business cycles' of Schumpeter, *American Economic Review*, vol. 30, June.

Mensch, G. (1978) 1984: A new push of basic innovations, *Research Policy*, vol. 7, pp. 108–22.

Mensch, G. (1979) *Stalemate in Technology*, Cambridge: Ballinger.

Schumpeter, J. A. (1939) *Business Cycles: A Theoretical, Historical and Statistical Analysis of the Capitalist Process*, New York and London: McGraw Hill.

CONTROLLING TECHNOLOGY

Introduction: The Control and Direction of Technological Change

David Wield

This concluding subsection is about a most complex and often confusing topic – that of the control of technology. It is a topic that generates highly polemic and often apparently unconnected contributions. But it has also engaged the attention of many thoughtful scientists and technologists who are worried about the uses made of their work.

One such scholar was Dennis Gabor, who was responsible for significant basic research on laser technology. But he was also a scientist who wrote and communicated on a wide range of topics, advocating the control of technological change. In article 6.4, he asserts that technological innovation has become a 'compulsive' and accelerating activity that will result in a 'catastrophe', if it is not restricted. Gabor does not attempt to advocate the wholesale stopping of the use of inventions, but he suggests that contemporary forms of control are too rudimentary to be more than a crude set of restrictions. Persuasive as this argument may appear, Gabor does not attempt to detail a logical approach to control of technology, although he acknowledges that control involves in practice the control of some groups by other groups in society.

The articles chosen in this subsection present different, sometimes highly polarized views on two issues concerning control: how to build up an understanding of the future impact of technologies; and whether more flexible and 'alternative' technologies can be developed.

In article 6.5, Collingridge sees one problem of control as that of avoiding harmful future impacts of technologies. Setting himself the question of how to improve the control of technology, he suggests that one approach is to try to predict potentially harmful effects at an early stage of technological development and diffusion. But he believes that understanding of interactions between technology and society is so poor that harmful social consequences of a fully developed technology cannot be predicted confidently enough to justify anything other than very arbitrary controls. And even if, by chance, a technology does perform as expected, like Rogers in article 4.5, Collingridge thinks it will always generate other unanticipated and unwelcome social consequences, making its control by this time expensive and slow.

Collingridge believes that the control of technology involves a whole new approach to decision theory. In the book from which this extract is taken, he shows that due to vested interests and other factors, many technologies (like nuclear technologies and addition of lead to petrol) very rapidly become inflexible to change and extremely difficult to stop. He develops a theory of decision-making, such that decisions taken about technology are always reversible or flexible, so that, if they are later found to be wrong, they can be changed. His conclusion is that controlling technology cannot be a question of forecasting its social consequences; rather, it must be a question of trying to retain the ability to modify a technology, 'even when it is fully developed and diffused', so that harmful consequences 'can be eliminated or ameliorated'.

Clearly that in itself is a very large task. But Collingridge does not address an even more complex problem raised by Gabor – who is controlling technological change; and how to control the controllers?

Many scientists disagree fundamentally with the view that technology can or should be controlled. Their vision of a scientific and technologically determined future can sometimes appear more akin to science fiction. And sometimes the most 'science fictional' view of the future is the one accepted for massive funding. An example of this was the successful Apollo project to land a man on the moon before 1970. A more recent vision of a future determined principally by technical advance, rather than social, economic or political factors, was a contribution to a United States Congressional hearing on the future by Sir Clive Sinclair, a well-known inventor/entrepreneur. As shown in a case-study by Boyle (1986), he has already had some influence on the future through his development of personal computers and other electronic products. In his contribution, he talked of machines with more intelligence than humans, and a 'man'-made world in space provided by intelligent robots.

> Quite soon, in only 10 or 20 years perhaps, we will be able to assemble a machine as complex as the human brain, and if we can we will. It may then take us a long time to render it intelligent by loading in the right software or by altering the architecture, but that too will happen.
>
> I think it certain that in decades, not centuries, machines of silicon will arise first to rival and then surpass their human progenitors. Once they surpass us they will be capable of their own design. In a real sense they will be reproductive. Silicon will have ended carbon's long monopoly. And ours too, I suppose, for we will no longer be able to deem ourselves the finest intelligence in the known universe . . .
>
> Further ahead, by a combination of the great wealth this new age will bring and the technology it will provide, we can really begin to use space to our advantage. The construction of a vast, man-created world in space, home to thousands or millions of people, will be within our power and, should we so choose, we may begin in earnest the search for worlds beyond our solar system and the colonisation of the galaxy. (Sinclair, 1984)

Sinclair would certainly not agree with Gabor or even Collingridge's views on control. 'In principle,' he says, 'it could be stopped. There will be those that try, but it will happen nonetheless. The lid of Pandora's box is starting to open.'

Opposed though Sinclair and Gabor's views are on the need for control, they have at least one thing in common – their views are technologically deterministic. Technological determinism is 'the idea that social development is determined by the technological changes a society invents or develops . . . Societies are classified according to the stage of technological development they have reached: from the stone, iron and steam ages of the past up to the atomic or computer age of the present' (Roy, 1976, p. 281).

Back to earth, the article by John Davis, a mechanical engineer, takes a non-technologically-determined approach with his advocacy of 'appropriate technology'. His article uses a series of examples in an attempt to demonstrate not only that appropriate technology (AT) requires just as ingenious engineers as more complex technologies, but also that AT

> does not aim to replace everything big with everything small . . . It simply seeks to find the best balance of big, medium and small, to avoid unnecessary complexity with a mixture of high and low technologies, and to be as responsible as possible by eliminating unnecessary waste.

His examples of feasible small-scale technologies include not only well-known examples of AT like wind pumps, but the successful small-scale sugar production units in India, expansion in the numbers of mini-steel mills in the United States, and the spread of small bread-baking and beer-brewing units in Britain.

He also gives an example of a building construction system developed for the Third World using unfired bricks encapsulated in a thin but strong rendering of glass-fibre reinforced cement. This technology has been taken up in Devon, with one-fifth of the energy cost and one-half of the financial cost of a conventional brick wall.

His advocacy of practical and flexible technologies gives concrete expression to Collingridge's more theoretical exposition of modifiable technologies, whilst taking a clear position on the need for a new approach to technology. 'Engineering' must serve the people and the needs of the time if it is to be appropriate technology.

The final two articles take up the call for 'alternative technologies'. Perhaps the best-publicized campaign for the production of alternative products is that of the Lucas Aerospace workers' plan, to which Mike Cooley's article (6.8) refers. He describes the campaign at Lucas to avoid the redundancy of skilled workers. The Lucas plan proposed 150 socially useful products which Lucas could design and manufacture, including scarce equipment for medical uses such as kidney dialysis machines (Wainwright and Elliott, 1982). He also eloquently and sympathetically describes the attempts to develop computer-aided machine-tool systems, which enhance the skills of machine operators. He insists that operators as end-users should have a say in the design of the machines they will use.

We also include an article (6.7) which is an excerpt from a trade union pamphlet arguing for the conversion of military manufacture to alternative, socially useful products. This is an attempt by the Transport and General Workers Union to demonstrate that jobs can be protected, while still supporting disarmament.

Those advocating these alternative approaches to design and innovation would argue that they are all feasible, although they would certainly admit that some are more worked out than others. But most would also argue that they are working against a very unbalanced form of control of technological resources, including unbalanced control of financial and technical knowledge (see, for example, Clarke, in article 5.5, where he contrasts the CEGB's championing of centralized electricity supply systems with the missed commercial opportunities in wind and wave systems).

Certainly, most of those who believe that the direction of technical change is not pre-determined, or determined by a 'straightforward' selection of the 'best' possibilities in a 'simple' choice, have no illusions that alternative routes will easily emerge. And there are few models of successful alternatives for those who do wish to become more flexible. Changes in direction will not be easy and will certainly involve changes in society.

A report jointly written by engineers and social scientists, trade unionists and manufacturing managers for the Council for Science and Society (1981) put it as follows:

> The historical development of steam engines, spinning and weaving machinery, machine tools, electrical power, computers, factory organisation and the rest – has led to a quite astonishing increase in productivity and the material wealth. We have no experience of any alternative way in which technology could have been developed. We therefore are easily led to believe that the historical path was the only one which could have been pursued.
>
> In particular we may be led to believe that the development of technology must inevitably lead to the elimination of skill in the jobs which it directly affects. The conclusion which the Report suggests is a different one. The new technologies of micro electronics and computers and communications can be used to reinforce and extend the historical process of subordinating men and women to machines, and of eliminating their initiative and control in their work. They can also be used to reverse this process, to develop a technology which is subordinate to human skill and cooperates with it. Which of these outcomes will ensue depends upon the struggles of those concerned. What is important is to believe in the possibility of a second outcome, and to believe that the full use of human abilities is a higher and more productive goal than the perfection of machines.

One link between the very different opinions in the articles in this subsection is a belief in the possibility of another outcome.

References

Boyle, G. (1986) Marketing: Sinclair microcomputers, T362 *Design and Innovation* (Block 2), Milton Keynes: Open University Press.

Council for Science and Society (1981) *New Technology: Society, Employment and Skill*, CSS.

Roy, R. (1976) Myths about technological change, *New Scientist*, 6 May, pp. 281–2.

Sinclair, C. (1984) Coming soon – a robot slave for everyone? *Guardian*, 24 August.

Wainwright, H. and Elliott, D. A. (1982) *The Lucas Plan: a new trade unionism in the making?* Alison and Busby.

6.4

Compulsive Innovation

Dennis Gabor

What is now called innovation still has an element in it of the instinct that drove primitive man to produce such wonderful inventions as the bow and arrow, or to devise such complicated social arrangements as totemism. Mechanical inventions and social innovations have remained indispensable but uneasy partners from prehistory to modern times. They were devised by two different kinds of human minds, and both were suppressed for long periods by the third type of man, who cared neither for technology, nor for social progress, but only for power. Regrettably, history records mostly the deeds and misdeeds of this third type of man.

We have now come to a point in the evolution of civilization when such history must have a stop. After a long and mostly tragic epoch of darkness, some three hundred years ago the empirical technology of the craftsman was joined by the systematic knowledge of nature and the method of reasoning from facts, not from fancies, which we call science. The confidence of techniques and theoretical science created applied science, which gradually became synonymous with modern technology. As we all know, this has now reached a stage at which it can destroy all civilization, at least temporarily, or create a new and happier world – if only Man is fit for such a world. The prospects for 'Post-historic Man', as painted by Roderick Seidenberg (1950), are dim. When looking into the far future, one is torn by doubt whether Man, this fighting animal, can ever settle down to be happy. One doubts also whether such a state, in which most of our instinctive and historic values are negated and in the end even human intelligence becomes unnecessary, is worth achieving. Many of our greatest minds, such as Albert Einstein, have come to the conclusion that *Homo sapiens* is approaching his end.

But the instinct that has brought us from the naked ape in the forest to modern man tells us that we must not give up. Man has fought nature and his own kind for perhaps a hundred thousand years; now he will have to fight his own nature. Ultimately this must be the aim of any far-sighted innovator of our times and in the years to come.

At the present there is a terrifying imbalance in innovations. After many centuries in which innovation was almost imperceptible, and after a few in which all technical progress was identified with human progress, we have now reached a stage in which innovation has

D. Gabor, *Innovations: Scientific, Technological and Social*, Oxford: Oxford University Press, 1970; extracts from Introduction.

The late Dennis Gabor was the inventor of holography and Professor of Applied Electron Physics at Imperial College, London.

become *compulsive* – but only technological innovation. A large vested interest has been created, even apart from the military–industrial complex, embodied in the *avant-garde* industries and research organizations, which believes that it must 'innovate or die'. Man's landing on the Moon is the epitome of this development; a splendid triumph of applied science and of a brilliant cooperative organization, at a time when most thinking Americans are filled with grave doubts about the sanity of their society.

The time of inventions in the ordinary sense is not yet over. The scientific–technological complex that we have created will produce more, almost automatically, by its own inertia, which is very different from conservatism. We can with certainty expect from it cheaper and more abundant power, an even closer communications network, and revolutionary improvements in information processing. But novelties far more important in their human and social implications can be expected from the biological sciences, which as yet have much smaller establishments. They have already upset our world equilibrium by death control; now they can upset it even further by creating enough food for many more billions of people, until the world becomes intolerably overcrowded. They will also almost certainly provide us with powerful methods of mind control, which can be used for good or evil.

Shall we be able to control the Controllers? There is no simple answer to this. The time of simple solutions is long behind us, though we are still suffering from the simplistic slogans of the nineteenth century, such as 'Free Enterprise', or 'Common Ownership of the Means of Production'. We have long ago entered the era of compromises, of the piecemeal reconcilement of human nature and of vested interests with desirable aims, national and international. Unfortunately, the struggle with these contrary forces has recently visibly shortened the 'lead times', and we can see the governments planning for months ahead, rather than for ten to twenty or more years.

If this goes on much longer, we are certain to run into a catastrophe, even without an all-out nuclear war. With the present trend, the world population will double by the year 2000 – and AD 2000 ought not to be the end of the world. Unfortunately all our drive and optimism are bound up with continuous growth; *'growth addiction'* is the unwritten and unconfessed religion of our times. In industry and also for nations, growth has become synonymous with hope. Undoubtedly, quantitative growth will have to go on for many more years, but unless we prepare for a turning-point well before the end of the century, it may by then be too late.

History must stop, the insane quantitative growth must stop; but innovation must not stop – it must take an entirely new direction. Instead of working blindly towards things bigger and better, it must work towards improving the quality of life rather than increasing its quantity. Innovation must work towards a new harmony, a new equilibrium; otherwise it will only lead to an explosion. [. . .]

Quantity changes into quality

It would be highly misleading to consider the change from the individual inventor in his garret, who took his little knowledge out of old textbooks of physics and chemistry, to the modern super-team, with all grades of men, from the mathematician to the mechanic, and the compression of time from a generation or two to a few years, merely as a change in quantity. A change on such a scale means *a change in quality,* a change in the intrinsic nature, aims, and consequences of the process of invention and innovation.

There are three factors in this; the change of the time-scale, the change in the magnitude and of the social consequences of the innovation, and the change in scope or aim. Let us take them in turn.

The natural time-unit in social life is one generation. Apart from rare exceptions, people cling throughout life to the values they have made their own in childhood or as young men. Medieval man expected his grandchildren and his great-grandchildren to live as he did, with the same values. Technological change was almost imperceptible. [. . .] A radical change came about only in the nineteenth century, at the time when the mechanical loom threw thousands of weavers into abject poverty, but it affected only a small fraction of the population; for the great majority it was still a stable world. It grew slowly. Even in the last century, the population in the industrial countries doubled in about a hundred years. Only in our time did the doubling time come down to about one generation, and essential changes in the mode of living to a fraction of a generation.

Second, the change in magnitude. Gunpowder, the Maxim gun, even the bomber plane with high explosive could kill only a fraction of a population; the hydrogen bomb, and perhaps also some devilish viruses bred in biological warfare establishments, could kill as good as the whole of it. The mechanical loom made a few thousand weavers jobless. Modern mechanization, rationalization, and automation could *reduce* the labour force in the United States by about a million a year, instead of having it take up the annual surplus of births over deaths. In the United Kingdom the 'saving' could be about the same, though the labour force is only about one-third that of the USA, because of the considerable overmanning in British industry. Of course this will not happen, and cannot be allowed to happen so long as we do not know what to do with the millions of unemployed. It demonstrates clearly the imbalance between technological and social innovation, and the unhealthiness of a state in which most of the best brains still try to improve technology when the bottleneck has shifted long ago to the mismatch between technology and society.[. . .]

When Alexander Graham Bell invented the telephone, he could rightly feel that he had satisfied an archetypal desire of mankind. But now the telephone is with us, to the extent that we could not do without it, and men (and even more women) spend a sizeable fraction of their time on the telephone. What is there left to invent? The Picturephone, to see the speaker, not only to hear him? Yes, certainly, but then we shall have reached saturation.

Rather worse then saturation is what is likely to happen to the equally archetypal desire of man to travel fast and to fly. The motor car has crowded some cities to the extent that traffic jams are everyday experiences, and it would pay to walk – were it not for the difficulties of crossing the traffic. In the air near-misses are equally common, and collisions can only just be avoided by the most ingenious devices that men can contrive. But air traffic is now doubling every seven years, and if it goes on increasing like this it may be impossible, even with the most ingenious computer-controlled devices and with the biggest airbuses carrying 500 to 1000 passengers, to prevent either traffic jams in the airports or frequent collisions in the air. [. . .]

Traffic jams and deaths on the road or in the air, to which we could add the problems of pollution of the air, the rivers, the lakes, and the sea, take us to what is probably the most important difference between the inventions of the past and of the present day. *The most important and urgent problems of the technology of today are no longer the satisfactions of primary needs or of archetypal wishes, but the reparation of the evils and damages wrought by the technology of yesterday.*

Elsewhere I have expressed this by saying that we cannot stop inventing because we are riding a tiger. Fossil fuels are threatened by exhaustion; so we must have nuclear power. Death control has upset the balance of population, so we must have the pill. Mechanization, rationalization, and automation have upset the balance of employment; what is it we must have? For the time being we have nothing better than Parkinson's Law and restrictive practices.

Reference

Seidenberg, R. (1950) *Post Historic Man*, Chapel Hill; University of North Carolina Press.

6.5

The Dilemma of Control

David Collingridge

There is almost universal agreement that the greatest success stories in the annals of technology are the American Manhattan Project which, in three years of research and development of an intensity never equalled, produced the world's first atomic bomb, and the American landings of men on the moon. These achievements will always stand as monuments to our ability to bend recalcitrant nature to our purposes. They reveal the depth of our understanding of the material world and our power to grasp the workings of the most complex physical systems. These successes engendered great optimism about the future benefits obtainable from technology. If the power of the atom could be understood and harnessed, and if men could walk upon the moon, what obstacles could there be to technological progress? The horizon seemed limitless – all that was required was the organization, the skill, the dedication, the tenacity, and the willingness to invest in success so characteristic of these great triumphs, and there could surely be no barrier to technology satisfying almost any human purpose. All that could stand in the way of curing disease, prolonging healthy life, feeding the hungry, providing an abundance of energy, giving wealth to the poor, was lack of will and organization. [. . .]

Technological progress enables workers to produce far more than previously, to the benefit of all, but only because the work has been robbed of all individuality and reduced to meaningless repetition. The technology of nuclear weapons may have prevented war for a few decades, but at the cost of vastly greater destruction when the war finally comes. Food production has increased through the use of chemicals, but at the cost of the future collapse of agriculture due to damage to the soil and to its supporting ecosystem. Modern transport is so cheap that many can enjoy their leisure in beautiful parts of the world, but at the cost of reducing the local population to serfdom and forcing their surroundings into the deadly uniformity of hotel fronts.

Thus, technology often performs in the way originally intended, but also proves to have unanticipated social consequences which are not welcome. What I wish to ask is how can we make decisions about technology more effectively; how can we get the technology we want without also having to bear the costs of such unexpected social consequences, and how can we avoid technologies which we do not want to have? To put it another way:the

D. Collingridge, *The Social Control of Technology*, Milton Keynes: Open University Press, 1980; extracts from Chapter 1.
David Collingridge is a lecturer at the Technology Policy Unit, Aston University.

problem is how technology can be controlled in a better way than at present. There is a central problem concerning the control of technology which we may now focus on, which I shall refer to as the *dilemma of control*.

Two things are necessary for the avoidance of the harmful social consequences of technology; it must be known that a technology has, or will have, harmful effects, and it must be possible to change the technology in some way to avoid the effects. In the early days of a technology's development it is usually very easy to change the technology. Its rate of development and diffusion can be reduced, or stimulated, it can be hedged around with all kinds of control, and it may be possible to ban the technology altogether. But such is the poverty of our understanding of the interaction of technology and society that the social consequences of the fully developed technology cannot be predicted during its infancy, at least not with sufficient confidence to justify the imposition of disruptive controls. The British Royal Commission on the Motor Car of 1908 saw the most serious problem of this infant technology to be the dust thrown up from untarred roads. With hindsight we smile, but only with hindsight. Dust was a recognized problem at that time, and so one which could be tackled. The much more serious social consequences of the motor car with which we are now all too familiar could not then have been predicted with any certainty. Controls were soon placed on the problem of dust, but controls to avoid the later unwanted social consequences were impossible because these consequences could not be foreseen with sufficient confidence.

Our position as regards the new technology of microelectronics mirrors that of the Royal Commission in 1908. This technology is in its infancy, and it is now possible to place all kinds of controls and restrictions on its development, even to the point of deciding to do without it altogether. But this is a freedom which we cannot exploit because the social effects of the fully developed technology cannot be predicted with enough confidence to justify applying controls now. Concern has been expressed about the unemployment which may result from the uncontrolled development and diffusion of microelectronics, but our understanding of this effect is extremely limited. The future development of the technology cannot be foreseen in any detail, nor can its rate of diffusion. Even if these were known it would be impossible to predict the number of workers displaced by the new technology in various sectors. This depends upon a whole bundle of unknown factors; the demand for the displaced labour from other sectors of the economy, the number of jobs created by the economic savings from the new technology, the number of jobs involved in making, developing and servicing microelectronics and so on. It is hardly surprising to find that forecasts of the effects on employment of the uncontrolled development of microelectronics cover a huge range.[1,2]

It is clear that our understanding of how this infant technology will affect employment when it is fully developed is so scanty that it cannot justify the imposition of controls on its development and diffusion. It will not do to suggest playing safe and imposing such controls to avoid the unemployment which *may* result from the new technology. This is to forego all the benefits from microelectronics for the avoidance of the unquantifiable possibility that it will cause serious unemployment. Such caution would effectively eliminate technological change.

The motor car in 1908 and microelectronics now are typical of technologies in their infancy. They may be controlled easily in all sorts of ways, but control can only be arbitrary. Our understanding of the interactions between technology and society is so poor that the harmful social consequences of the fully developed technology cannot be predicted with sufficient confidence to justify the imposition of controls. This is the first horn of the dilemma of control.

The second horn is that by the time a technology is sufficiently well developed and diffused for its unwanted social consequences to become apparent, it is no longer easily controlled. Control may still be possible, but it has become very difficult, expensive and slow. What happens is that society and the rest of its technology gradually adjust to the new technology, so that when it is fully developed any major change in the new technology requires changes in many other technologies and social and economic institutions, making its control very disruptive and expensive.

The motor car may again be taken to illustrate the point. As this technology has gradually diffused adjustments have been made by other modes of transport. The provision of buses and passenger trains has adjusted to the existence of more and more private motor cars, and villages have been urbanized and outer suburbs grown as travel has become cheaper and more convenient. The ability of more and more workers to move around easily has led to more and more offices and factories moving from city centres. The importance of crude oil for the production of petrol has also grown, and the chemical industry has adjusted to the existence of the motor car by learning how to exploit the residues from crude oil after petrol extraction. The existence of the motor car also made alternatives to the internal combustion engine very unattractive so that, for example, very little research and development on electric motors for road transport has been done. The list could be extended indefinitely.

Imagine now that severe controls must be placed on the use of motor cars because, for example, crude oil has suddenly become very expensive and scarce, or because some pollutant of motor car exhaust is suddenly found to be many times more toxic than previously thought. Society has adjusted to cheap transport in its siting of houses, factories and offices, so sudden restrictions on transport would inevitably be very expensive. Large parts of the economy would virtually collapse. A significant shift from private cars to buses and trains would be impossible in the short term because their number has adjusted to the existence of a transport system dominated by the private car. It would take many years, and huge investment before buses and trains could carry anything but a tiny fraction of those previously using motor cars. Confident of crude oil supplies, the chemical industry has invested very little research and development into using any other feedstock than crude oil residues, and so many years would be needed before it could adjust to the new situation, making the change extremely expensive. Similarly, little work has been done on substitutes for the petrol-fuelled internal combustion engine, so that a long time would be needed before substitutes could be used on a large scale.

The example is an extreme one, but as such it makes the point: the interaction between technology and society works in such a way as to make the severe control of a major established technology very expensive and necessarily slow. As for the motor car, so for microelectronic technology. By the time this is fully developed and diffused, all sorts of adjustments will have been made which will make its control very costly and slow.

The dilemma of control may now be summarized: attempting to control a technology is difficult, and not rarely impossible, because during its early stages, when it can be controlled, not enough can be known about its harmful social consequences to warrant controlling its development; but by the time these consequences are apparent, control has become costly and slow.

The concern of this book is the efficient control of technology and so its natural starting place is the dilemma of control. The customary response to the dilemma concentrates on the first horn, and amounts to the search for ways of forecasting a technology's social effects before the technology is highly developed and diffused, so that controls can be placed on it to avoid consequences not wanted by the decision makers and to enhance the consequences they desire. On this view, the key problem about the control of microelectronics is the lack of a powerful forecasting device able to give firm information about the employment consequences of this technology's uncontrolled development. Many of the forecasting techniques which have been devised are extremely dubious, but what lies behind the expenditure of effort in this direction is a serious misconception about the quality required of forecasts. To be of any use, a forecast of future unwanted social effects from a technology now in its infancy must command sufficient confidence to justify the imposition of controls now. It is not enough for the forecast merely to warn us to look for bad social consequences of a particular kind in the future, because by the time they are discovered their control may have become very difficult and expensive. The prediction of social effects with such confidence demands a vastly greater appreciation of the interplay between society and technology than is presently possessed. I doubt that our understanding will ever reach such a sophisticated level, but even if this is possible it will only be as the outcome of many years of research. Thus even an optimistic view leaves us with the problem of how to improve the control of technology in the period needed for the development and testing of adequate forecasting methods. A pessimist like myself regards this period as effectively infinite.

The only hope seems to be in tackling the other horn of the dilemma of control. If the harmful effects of a technology can be identified only after it has been developed and has diffused, then ways must be found of ensuring that harm is detected as early as possible, and that the technology remains controllable despite its development and diffusion. On this view, the key problem of controlling microelectronics has nothing to do with forecasting, but is to find ways of coping with any future unemployment which it may generate, and ways of avoiding overdependence on the technology which would make it difficult to control.

This is an enterprise requiring nothing less than a wholly new way of looking at decisions concerning technology. [. . .]

The essence of controlling technology is not in forecasting its social consequences, but in retaining the ability to change a technology, even when it is fully developed and diffused, so that any unwanted social consequences it may prove to have can be eliminated or ameliorated. It is, therefore, of the greatest importance to learn what obstacles exist to the maintenance of this freedom to control technology.

References

1. Sleigh, J. *et al.*, *The Manpower Implications of Microelectronic Technology*, London: HMSO, 1979.
2. International Metalworkers' Federation, *Effects of Modern Technology on Workers*, Geneva: IMF, 1979.

6.6

Appropriate Engineering

John D. Davis

It would be all too easy writing about appropriate technology in an engineering journal to concentrate on hardware and completely miss the point. Appropriate technology certainly has concrete expression in terms of hardware, but its true significance is to be found in the people and the communities which the hardware serves. Technology is not neutral; it is a dominant factor in shaping the kind of world we live in. It carries its own culture with it, and it even influences our system of values. As George McRobie, the Chairman and co-founder with Dr E. F. Schumacher, of the Intermediate Technology Development Group (IDG) said recently: 'The choice of technology is the most important collective decision confronting any people.'

The appropriate-technology movement had its origin in London in 1965 and has since become worldwide. Its emergence signals a growing awareness and concern that all is not well with the traditional course of technological development, which increasingly dominates people rather than being subordinate to them as it becomes larger, more costly, complex and prodigally wasteful of limited resources. The more costly it becomes the less accessible it is to ordinary people, and the more it divides the rich and the poor. The economies of scale are a diminshing return, but the diseconomies increase with size. We are fascinated by complicated devices and lose sight of the beauty of simplicity. And a technology designed for a world of cheap and apparently unlimited natural resources becomes a catastrophe when shortages begin to appear and prices go out of control as demand exceeds supply.

The appropriate technology movement is searching for technologies that meet the needs of people in all their different circumstances. To plagiarise from the subtitle of Dr Shumacher's book *Small is Beautiful*,[1] the search is for 'economic development as though people mattered' instead of striving for big, complex, powerful, costly, sophisticated and highly concentrated technologies, the search is to see, for each particular state of economic development, how small and dispersed, how simple and resource economical, how congenial and humane it can be, using all that is available from the knowledge gained in science and technology and all that is best in human skill and ingenuity.

Engineering is concerned with harnessing the resources of nature for the benefit of man. The appropriate-technology approach is a sound engineering approach plus something

J. D. Davis, Appropriate engineering, *Engineering*, December 1979.
John D. Davis is an engineer with the Intermediate Technology Development Group in London.

extra. It is for the benefit of all mankind and for the responsible stewardship of our physical environment. Appropriate technology does not aim to replace everything modern with
everything primitive. It simply seeks to find the best balance of big, medium and small, to
avoid unnecessary complexity with a mixture of high and low technologies, and to be as
responsible as possible by eliminating unnecessary waste.

Scale

In most economic activities there are economies and diseconomies of scale. They are different in each industry: but always they are limited and in many it would appear the
optimum size of a production unit is remarkably small. Even in a society as advanced as
the United States the average number of people employed in a factory is only 90.

The industries that gain most from large-scale operations are mostly in materials production – coal, oil, chemicals, metals, etc. But even in some of these there are signs of a
move to smaller units. By the standards of 1950, some very small oil refineries have been
built in recent years. (A comparison between the small new Shell refinery in Switzerland
and the huge old Pernis refinery in Holland is a striking example.) Many modern paper-
manufacturing plants have capacities of between 300 and 800 tonnes per day; yet opinion
is growing that units of 50 to 100 tonnes per day make better sense.

Even in the steel industry – though not yet in the British Steel Corporation – a bright
future is foreseen for the mini-mill. The move is well under way in the US where there are
already 60 such mills. In Britain there are a handful in the remnants of the private steel sector.
Bill Sirs, the steelworkers' leader, was right when he said, 'BSC is hypnotised by the Japanese
example of the super-plant. There is a good future for the small plant on a modernised
basis. For only a small outlay you can have an efficient small works geared to small orders.'

Dr F. E. Jones, the Chairman of the Mechanical Engineering Industry NEDC, has published an interesting comparison between British and Japanese industries in a paper to the
Institution of Mechanical Engineers entitled *'The economic ingredients of industrial suc-*

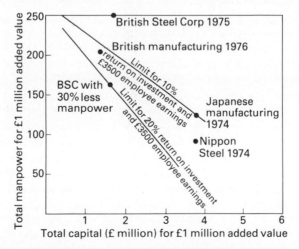

Figure 1   A comparison of the different mixtures of manpower and capital used to create £1 million
of added value in British and Japanese manufacturing industries overall and in the British Steel Corp.
(BSC) and Nippon Steel. Average industrial employee earnings (gross) were about £3500/yr in 1974
in Japan and in 1976 in Britain. Mixes of manpower and capital on a limit line are equally cost-effec-
tive in adding value for the particular rate of return on total capital for that line.

cess'. In it there is a comparison between the British Steel Corporation and Nippon Steel. I have reworked his data to a common base of £1 million of added value and have plotted them as Figure 1.

BSC management is understandably impressed by the Japanese performance, which produces the same added value with one-third the number of employees and double the capital investment. They would be less impressed if the comparison was made with the present overmanning in BSC eliminated. Under those circumstances the return on investment in BSC would be higher than for Nippon steel despite the fact that BSC would still have a much higher manning level and an unchanged capital investment. A high capital/labour ratio suits the Japanese, who enjoy among other things a low cost of money and consequently require only a low rate of return on investment. It is quite inappropriate in the UK where the cost of money is high. Big plants and highly capital-intensive operations may suit the Japanese, but in Britain smaller units with a low-capital/high labour mix are a much better fit.

Perhaps the most dramatic example comparing big and small production units is to be found in Indian sugar mills. There the share of the national market supplied by some 5000 small mills has grown to 20%. The maximum daily output of a small sugar plant may be only 80 tonnes compared with 1250 tonnes for a big mill, which requires 40 times as much capital as a small one. For the same output of sugar, 20 small mills require only half the amount of capital needed by a big one and they employ five times as many workers. In a situation where labour is cheap and money expensive, the unit operating costs of the big and small plants are not very different. Since the small ones enjoy certain trading advantages on purchases and sales they can compete very well.

It is quite proper to protest that what might suit the low-labour-cost environment of India would hardly apply in Britain. But in fact we can find similar and equally significant examples here. For instance, we have recently witnessed the withdrawal of one of the three big bread-baking companies, while small local bakeries increase their market share and multiply. Similarly, while major garment manufacturers are in trouble, it is possible for a low-distribution cost local manufacturer using homeworkers to flourish.

One of the most striking recent developments has been the appearance, for the first time for a century, of a number of very small new breweries. At Blackawton in Devon, Nigel FitzHugh runs a one-man brewery supplying about 1000 gallons per week to 25 pubs and clubs within a 30-mile radius. The plant consists of second-hand equipment costing £6000 installed in an old blacksmith's shed. But it is not only in the depths of rural Devon that this formula works. Patrick Fitzpatrick has set up Godson's Brewery at Gunmakers Lane in the heart of London, where he is supplying 50 pubs and clubs.

A very large part of what we have come to know as economic growth has occurred with the transfer of what are basically domestic technologies into industrial systems. Many favourable factors contributed to that transfer at the time that it was occurring. Now the balance between the economies and diseconomies of scale is changing, a new opportunity is coming to reconstruct the production system for a wide range of finished products – garments, food and drink processing, furniture – in smaller, local units, with some of it returning to its place in our homes.

In any discussion of industrial scale, inevitably car production is mentioned. Over the past 30 years most of the small and some of the big car manufacturing firms have disappeared and most of the few that remain survive only on a massive scale. Although in many ways a modern car is superior to its predecessors, it has not become a cheaper product since World War II as a result of the trend to giant industrial systems. For those who recognise that the cost of motoring is more related to the rate of depreciation than to the initial cost of a car, it is worth noting that it is some of the small-volume production cars like the

Morgan, which is still hand-built at a rate of about 400 per annum, that are the cheapest cars to run.

It is reasonable to ask why it is that the theoretical economies of high-volume production do not always result in a cheaper product. There is no simple answer even for a single product. But there is little doubt that too little attention has been given to the enormously costly and wasteful distribution and marketing operation that is an inevitable consequence of highly centralised, high-volume, capital-intensive manufacture. Furthermore, it is this distribution system that invites a flood of foreign imports.

Simplicity

I believe it was Charles Kettering of General Motors – or was it Henry Ford? – who said: 'Build simplicity into it.' Whichever said it, they both practised it with great ingenuity. However, as one looks around at modern engineering products the wisdom of ingenious simplicity seems to have been forgotten and in its place complexity rules. [. . .]

I was not surprised that the owner of an £11,000 Jaguar was almost apoplectic when he was told that his car would be unserviceable for a week because his electric windows would not work and spare parts were not available. [. . .]

A really outstanding example of ingenious simplicity is the Humphrey pump in which the water to be pumped also acts as the piston of the power source (Figure 2). Several very large examples were installed early in this century. Lately, interest in various sizes has begun to develop and the Intermediate Technology Development Group has supplied designs for several overseas organisations. It is a particularly appropriate device for a Third World country. It is extremely simple to make, there is no piston or cylinder to machine, it requires no lubrication and suffers no wear. Consequently it can easily be made, operated and repaired locally. These same attractive properties are equally valuable here in Britain.

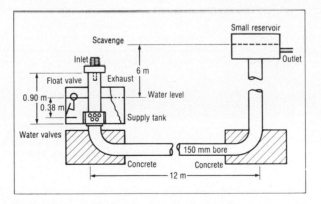

Figure 2 A 225 m/³h, 9m head Humphrey pump developed at Reading University and sent by ITDG to the National Research Centre in Egypt for field trials as an irrigation pump.

Another example of ingenious simplicity has been produced by ITDG in their design of pumping wind machines. The head mechanism of a conventional design is a gearbox assembly. The ITDG design is a link mechanism bearing some similarity to a beam-engine drive linkage. For a 6½ m diameter rotor machine the conventional head weighs 1500 kg. whereas the ITDG design weighs approximately 180 kg. The conventional design requires regular lubrication service, while ITDG has used sealed-for-life ball bearings for the main

Figure 3   Comparison of a traditional hydro-electric governor (a) and a modern electronic replacement (b) for use in small-scale hydropower installations. The electronic unit is more reliable and costs only about one-thirtieth the cost of the mechanical unit.

drive shaft and self-lubricating bushes elsewhere, all of which are over-size for the anticipated loads. Naturally it has been possible to design a light-weight head gear. Overall there is a big saving of metal and cost.

For more than a century very little attention has been paid to the very considerable number of small-scale water-power sources that exist in almost every country. During the Victorian era small water turbines were to a considerable extent scaled-down versions of big hydrosystem turbines, using the same principles of mechanical control. A different design approach has been adopted in the development of modern small-scale (1 to 10 kW) water-turbine units. All attempts to control water flow by mechanical means have been abandoned in favour of an electronic black-box controller (Figure 3). As a result there has emerged a very much simpler machine to build, maintain and operate, which of course is very much cheaper per kilowatt capacity than its Victorian predecessors. The use of modern technological developments – self-lubricating bearings, electronic controllers – when properly applied can contribute to the recapture of simplicity.

Those engineers who are engaged in finding simple solutions to human problems very soon discover that the work is much more fascinating and challenging than the conventional approach, which seems remorselessly to lead to increasing complexity for even the simplest of tasks. It really is not very clever to use a powered auger costing several thousand pounds to dig a hole in which to set a telegraph post. [. . .]

Resource economy

The magnitude of unnecessary waste in our modern economic system is almost too big and too complex to comprehend. Recent energy problems have forced us to look closely at that part of the system, and it is not an exaggeration to say that the cost of our unnecessary waste of energy alone is the equivalent of our total expenditure on education or health or defence – £8000 million per annum. In terms of our depleting reserves of fossil fuels it amounts each year to about 160 million tonnes of coal equivalent. Detailed justification for that claim can be found in Gerald Leach's *A Low Energy Strategy for the United Kingdom*.[2]

Almost half of our total primary energy consumption of six tonnes coal equivalent per capita is used for space heating of buildings which emit radiation into space instead of retaining heat within the building. The appropriate technologies that we need to apply to buildings to conserve energy are primarily the simple and inexpensive techniques of sealing, insulation and heat exchange. Secondarily, we can utilise the waste heat from power generation (combined heat and power) – which is easiest with a small neighbourhood power station – and use directly gas-fired heatpumps, such as are being developed by Glynwed, whenever there is no waste heat available.

Thirdly, particularly for new buildings, we need to apply much more effective means of utilising solar energy. The French Trombe wall is an outstanding design of passive solar heating which even in our climate is known to halve the annual space-heating requirements of houses (Figure 4). The south-facing wall of a house is enclosed in a glass casing and becomes a heatsink for solar radiation. The glass produces a greenhouse effect and air from the house is circulated between the glass and the outer surface of the wall. Houses incorporating this principle are being tested at Bromborough in the Wirral.

When the energy-use system has been reconstructed and unnecessary waste eliminated we shall only require a fraction of the primary energy that we now use. Then alternative energy sources, which are inevitably more expensive than the cheap fossil fuels that we have been used to, will be useful when used in small quantities and with care. Solar photovoltaic cells are still a very expensive method of obtaining electricity, but it is expected this will reduce to about $1.00 per peak kilowatt generated within five years or so. When that target is reached domestic solar electricity will become a very interesting alternative to centrally generated power. As human animals, our individual electricity requirements are small and widely dispersed. Solar cells sited on a house roof sidestep the distribution problem of centrally generated power.

Product durability and what I term the four Rs – repair, reconditioning, re-use and recycling – call for a great deal of attention by designers in the search for resource economy. A recent study of long-life cars with which I have been concerned suggests that the intrinsic durability of existing popular cars is usually good and a combination of initial

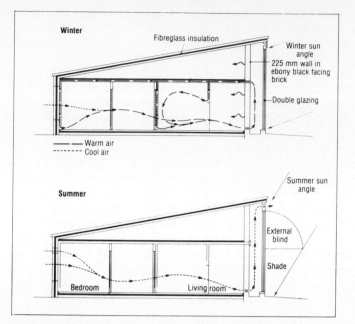

Figure 4   The Trombe wall

Figure 5   Devon County Council's building construction system.

protective treatments and a tight remedial maintenance schedule would be the most economical way of extending the average life of popular cars from 10 to 20 years. A design approach which facilitates maintenance, repair and reconditioning may well prove to be a more effective and cheaper way of reducing the mountains of scrap that modern society produces, than a frontal attack aimed at increasing intrinsic durability of products. The detail panel design of the Ford Fiesta was a move in the right direction.

A greater use of local natural materials alone, or in association with synthetics, is another way in which benefits can be achieved. The Severn Valley Brick Company provides a useful example of the use of local clay as raw material and the waste products from the nearby Philips carbon-black plant as its source of energy. The design thinking for this undertaking came from John Parry who heads the ITDG building materials team.

Another of his concepts, originally developed for the Third World, is a building-construction system which combines unfired bricks encapsulated in a thin but very strong rendering of glass-fibre-reinforced cement (Figure 5). The Devon County Architect has adapted this concept in some experimental buildings which he has constructed. Compared with an equivalent conventional brick wall the energy cost of the new system is about one-fifth and the financial cost is less than one-half. When engineers turn their minds to becoming ingenious a little can be made to go a very long way.

Employment

'Since we cannot do anything to stop the loss of employment opportunities, we must learn to live with it,' is a statement taken from a recent paper by Gershuny of Sussex University. Most intellectuals and a growing number of trade union leaders, politicians, economists and businessmen would agree with him. The appropriate-technology movement fundamentally disagrees. We believe that we have a very wide and increasing spectrum of technological options to choose from and there is no technological imperative demanding that only those which minimize human involvement are acceptable. [. . .]

A new approach to technology, which seeks not only to discover the most appropriate quantitative mix of men, money and machines, but also makes the quality of working life a top priority, is the challenge that faces the engineers of the '80s. Much of the potential benefit from past developments has been lost as employment in agriculture and industry has been replaced by employment in mindless, paper-shuffling bureaucracy – both private and public. An alternative system – decentralised, smaller scale, less capital- and energy-intensive, resource economical and more congenial to use and work with – is the best prospect for overcoming many of present social, economic and political problems.

Engineering must serve the people and the needs of the time if it is going to be appropriate technology.

References

1. Schumacher, E. F. (1973) *Small is Beautiful: a study of economics as if people mattered*, Blond & Briggs (Abacus, 1974).
2. Leach, G. *et al.* (1979) *A Low Energy Strategy for the United Kingdom*, Science Reviews.

6.7

Arms Conversion

Transport and General Workers Union

There is no contradiction between trade unions supporting unilateral disarmament and a cut in arms spending while at the same time protecting the jobs of their members.

Conversion of sections of the arms industry to alternative, socially useful, production would:

1. facilitate disarmament, thus removing the greatest threat facing us all – nuclear annihilation;
2. produce greater stability and enhance employment prospects for workers currently employed in the armaments industry;
3. provide a valuable stimulus to economic growth by diverting resources and R & D to civil production;
4. allow urgent social needs to be met.

It is also fair to assume that the majority of workers in the defence industries would prefer that their education, training and skills should be used on useful products.

Yet despite all this there is still a suspicion amongst defence workers that such policies might put their jobs at risk. Such a view is entirely understandable. It is clearly not enough to pass resolutions at conferences and just talk about converting 'swords into ploughshares'. We have to be able to demonstrate that conversion is a viable alternative and that we have a strategy designed to achieve it. And that means we have to define our objectives and carefully consider all the factors which could either hinder or promote the process of conversion.

The objectives of conversion

One of the most important points to emphasise is that arms spending is not simply a pile of money which can easily be diverted to more socially useful forms of production. The kind of resources tied up in arms expenditure are not simply pound notes but specific skills, technical abilities and capital equipment. The resources needed to build a Trident submarine or a Tornado aircraft are not necessarily suited to equipping a school or staffing a

Transport and General Workers Union, *A Better Future – Strategy for Arms Conversion*, TGWU, 1983; extracts from Chapter 3.

new hospital, and to simply call for a transfer of resources between these items of expenditure will understandably offer little comfort to the workers on Trident or Tornado, the majority of whom would be packed off to join four million other unemployed workers. The identification of the resources tied up in armaments production must therefore be our starting point and the task then becomes one of using those resources in a way which will be of maximum benefit to the community whilst protecting the interests of the workers affected. This definition suggests certain criteria which will have to be taken into account:

– the alternative products must use essentially the same skills that the workforce already possesses, although some retraining might be needed;

– the alternative products must be able to be produced in the same workplaces using mainly existing plant and other material resources, although again some modification and new investment will be required;

– the new products must be feasible and must be needed, i.e. they must be products which people will buy or which the government needs;

– the workforce should not have to move;
the whole process of conversion should be fully, and democratically, planned.

Can it be done?

The Transport and General Workers' Union is convinced that conversion is a practical possibility, a view shared by virtually everyone who has examined this question. A report by the United Nations Group of Experts, who looked in detail at the economic and social consequences of disarmament, concluded: 'There would be no insuperable technical difficulties in ensuring the redeployment of the released resources to peaceful uses.'[1] Indeed there is a simple answer to our more cynical questioners and that is that conversion can be done because it has been done.

Of course there are a number of difficulties which would have to be overcome and it is necessary to recognise and confront such problems. One of the greatest obstacles to conversion, for example, will be the massive and largely unaccountable power that resides in what has become known as the military-industrial complex.

The arms companies have built up extensive contacts with military chiefs and the two together are involved in a continual process of lobbying and inducement, designed to make their latest military gadgets seem indispensable to the security of the nation.

The arms contractors have also been insulated from the vagaries of the market and will be reluctant to return to it. With just one customer who pays all their costs and guarantees them a heftier return than they could obtain in virtually any other area of the economy they have led a very cushioned and risk-free existence. Employers will also fear losing control and resent the challenge which workers getting together to discuss alternative products represents. Undoubtedly, too, there will be other obstacles and other parties who are resistant to change. Nevertheless, given the necessary political will to make conversion work, the problems become essentially technical ones which *are* capable of solution. There are several factors which can be identified as helpful to successful conversion.

Table 1   Swords into Ploughshares

SWORDS INTO PLOUGHSHARES

There is no shortage of alternative uses to which the resources released by conversion could be put. New forms of energy, environmental protection, medical equipment, more efficient transport systems, exploitation of the oceans' invaluable resources–these are just some of the areas which could be explored. Already workers in the armaments industry have identified a wide range of socially useful products which they could make instead of military equipment. This diagram shows just a few of those alternatives.

tanks   shipbuilding   airframes   aeroengines   electronics

■ Agriculture

Machinery and equipment

Pumps and pipelines for irrigation systems

Sugar beet crushers

■ Construction

Industrial soundproofing

Machinery and equipment

Pre-design bridges for disaster relief

Pre-fabricated structures

■ Energy

Boilers for power stations ▲

Condensers and evaporators

Disposal of nuclear waste

Flexible power packs ▲

Fuel-cell power plant ▲

Heat exchangers ▲

Heat pumps ▲

Integrated energy systems   ▲

Oil spillage pumps

Oilfield machinery and equipment

Solar-cell systems ▲

Standby power units for computer industry ▲

Submerged oil production systems

Tidal power systems

Wave power systems

Windmills   ▲

■ Environment

Anti-pollution devices

Processing plants   ▲

Recycling machinery and plant

■ Industrial machinery

Advanced machine tools   ▲

Blowers and fans ▲

Fluidised bed boilers

Industrial furnaces and ovens

Machinery for various applications in food production, woodworking, paper industries, etc.

Mechanical power transmission equipment

Precision components ▲

Pumps and compressors

Quality control

■ Medical

Decompression chambers

Electronics for intensive care and
medical analysis ▲

Equipment for the disabled ▲

Kidney machines ▲

Medical mass screening systems

Pacemakers ▲

Sight substituting aids for the blind ▲

Surgical heat exchangers ▲

■ Metalworking

Castings and engravings

Containers ✈

Fabricated metal products

Iron and steel forgings

Machine tool accessories

Machinery for metal cutting and forming

Sheet metal work ✈

Special dies, tools and jigs

■ Oceanics

Submersibles and other equipment for marine
mineral exploration and marine agriculture
✈　▲

Tanks for fish farming

■ Offices and service industries

Automated merchandising machinery

Automated warehousing systems ▲

Commercial laundry machines ✈

Electronic office equipment ▲
Furniture and fittings

Refrigerators and air-conditioners ✈　▲

■ Transport

Airships ▲ ○

Air safety and air traffic control systems ▲

Automated speed/distance warning and
braking systems ▲

Braking systems ▲

Canal gates and heavy duty pumps for canals

Caravans and trailers

Civil helicopters ○ ✈ ▲

Diesel engines for locomotives

Gas turbine engines for ships ○

Hydrofoils ✈

Industrial trucks, including hover trucks

Locomotives

Monorail systems ✈ ▲

Mopeds and motorcycles

Pipelaying and freight barges

Road-rail vehicles ✈ ▲

Rolling stock ✈

Short-to-medium range civil freight and
passenger aircraft ○ ✈ ▲

■ Other

Brewing equipment

Cable-laying equipment

Conveyers

Electronic teaching and library aids ▲

Firefighting equipment

Heavy earth-moving equipment

High-speed motors ▲

Lifts

Micro-processors ▲

Mining machinery and equipment

Telechiric (distant handling) devices for use
in dangerous environments ▲

Occupational Conversion

As we have suggested, it would be quite wrong to assume that workers could be transferred from military production to other forms of production at the drop of a hat. Nevertheless, the problem of occupational conversion should not be exaggerated. Many of those working in the defence industry are supplying goods or services which are essentially the same as those being provided in the civilian sector, for example, producing clothing and food, or carrying out various types of construction or engineering work. In the 1960s, in California where there is a high concentration of military aerospace production, a detailed examination was carried out by the State Department of Employment of 127 production occupations. Twenty-eight of these were described as of basic craft type (electrician, plumber, carpenter, etc.) and were readily matched to occupations in other industries. Ninety-three of the remaining 99 could be immediately matched to one or more non-military production occupations and only six occupations required retraining.

Job skills in armaments engineering are applicable to a wide variety of products and most assembly lines are fairly flexible. Different materials may be involved and work might have to be carried out to different standards but major retraining should not be necessary. Where retraining is required it will tend to be of workers who have a considerable general grounding and expertise making retraining less difficult than might at first be thought. Scientists and technicians may present a larger problem, but here, too, studies have shown that the majority could be employed, fully utilising their skills, after nine months retraining.

Plant and Equipment Conversion

It is often forgotten that all industries are constantly experiencing conversion as they switch from one product to another. Of course they need some new equipment when they do so but overall most plant and machinery is relatively flexible. Many of our defence establishments are in fact ideally suited for conversion to other employment. Our naval dockyards, for example, are almost self-contained units. At Chatham dockyard, where thousands of our members have lost or are about to lose their jobs, they have their own power plant, a saw mill, machine tool department, plants for making ropes and flags, and so on. They have the ability, using essentially the same human skills and equipment, to develop a whole range of oceanic equipment, including vessels and equipment for the extraction of the vast mineral wealth on the sea bed. Other defence establishments are suited for conversion from, say, the manufacture of armoured cars to tractors or irrigation equipment, so badly needed in the third world.

Even in the most advanced technologies there is considerable scope for peaceful application. One recent study has explored what it called 'dual-purpose technologies' which can be used either in military or socially useful production. The technologies examined included nuclear, chemical and biological, space, undersea, environmental control, electronic and computer, and laser. The peaceful applications identified included:

1. *Nuclear:* energy, medical research and treatment, agricultural research, plant breeding, metallurgy
2. *Chemical and biological:* agriculture, nutrition, health, medicine, biomass energy, synthetics
3. *Space:* satellite surveys and sources, satellite communications and navigation systems, civil aircraft
4. *Undersea:* exploration and recovery of sea-bed minerals and oil, fish resources and pollution monitoring

5. *Environmental control:* management of water resources, forests, deserts, local climate, ecological balance
6. *Electronic and computer:* numerous possibilities including remote control in dangerous environments, substituting for tedious jobs, enhancing leisure
7. *Laser:* variety of industrial uses, surgery, energy.

We should also remember that it will be a major task simply decommissioning the weapons we already have. Many of those working in places like Aldermaston and ROF Burghfield, from the research scientists to production workers, could be fully employed for some years in simply dismantling and disposing of nuclear weapons.

Timing

If we reduce arms spending so as to bring Britain's military expenditure as a percentage of GDP into line with the average of our major European allies, a cut of approximately £4,700 million would have to be made. Over a five-year period the number of job changes which would have to take place each year to achieve such a reduction would be about 60,000. Seen in the perspective of hundreds of thousands of 'formal' annual job changes and within the context of a growth economy as envisaged in the TUC/Labour Party Alternative Economic Strategy such a figure is not particularly unmanageable. Moreover, the resources which would be released by such a transfer could be utilised in much more labour-intensive ways, thus encouraging job prospects rather than diminishing them. Experience shows that after 1945 and in the 1950s higher levels of defence cutbacks were accommodated without any overall rise in unemployment.

Finance

There is one very important point which must be stressed. The simple fact is that as it is government expenditure which is being cut the government can afford to be very generous in the transitional finance. Even if all the people engaged in producing equipment for the armed forces were paid their present wages indefinitely for just digging their gardens instead of producing war material, the government balance of revenue and expenditure would not be significantly different.

Reference

1. United Nations (1972) *Disarmament and Development. Report of experts on the economic and social consequeces of disarmament*, New York: UN, p. 24.

6.8

Socially Useful Design

Mike Cooley

There is now growing concern, even alarm, at the multiplier effects of what we call scientific and technological 'progress'. This concern displays a deeper understanding of the contradictions in science and technology than that which was articulated by the anti-science movement of the 1970s whose antecedents lay in the anti-culture movement of the 1960s. It was broadly the premise of those movements that science and technology were evil, totalitarian and devoid of the attributes which could make them amenable to the 'human spirit'. These rather sweeping generalisations are now giving way to a more mature analysis of the contradictions of science and technology. There is a growing realisation that science and technology have embodied within them many of the ideological assumptions of the society which has given rise to them. This has resulted in the questioning of the neutrality of science and technology as they are at present practised in our society. The debate on this issue is likely to be one of major political and philosophical significance. The questioning extends far beyond that of a 'use/abuse model' of science and technology to deeper considerations of the scientific and technological process itself. Science and technology performed within a particular social order reflect the norms and ideology of that order. Science and technology thus ceases to be seen as autonomous but rather as part of an interesting system in which the internalised ideological assumptions help to determine the actual experimental designs and theories of the scientists and technologists themselves.[1]

Arising from this it is held that if we are to have forms of science and technology which will be human-enhancing and liberating, make products that are ecologically desirable, conserve energy and materials in the long term, and help human beings rather than maim them, then we will require forms of science and technology which differ radically from those which predominate at the moment.

Producer and consumer

Given that two of the major components in the motor of the present technological progress are profit maximisation and the consolidation of power (based on the notion of an ever-

M. Cooley, Socially useful design, in R. Langdon and N. Cross (eds.), *Design Policy, Vol. 1: Design and Society*, Design Council, 1984, pp. 51–4.
Mike Cooley is Director of the Greater London Enterprise Board's Technology Division.

increasing rate of production and consumption) it is not surprising that the contradictions manifest themselves so obviously on all sides. Urban dwellers exist in an environment which is choked with traffic and smog, while the very heart of their city declines into slums. Rural dwellers witness the destruction of the countryside through piecemeal industrialisation and urbanisation and the unthinking use of chemicals. Ecologists view with dismay the pollution of our air, rivers and seas. Fossil fuel reserves are depleted at a frantic rate and materials are squandered to manufacture throw-away products for the short-term profit maximisation of a rapacious economic system. The human being as 'producer' is forced to perform grotesque alienated tasks at work to provide products which exploit and dominate the very same human being as 'consumer'. So the dual role of any human being as a producer and consumer is torn apart, with one set against the other.

As a producer the worker, whether by hand or brain, in Western society (and in the so-called socialist countries) is required to undertake highly specialised fragmented alienated tasks performed in the capital-intensive environment of large corporations. With few exceptions these capital-intensive industries are also highly energy-intensive. There is as yet only an embryonic realisation of the social and economic costs of the 'efficient factory' in terms of structural unemployment, frantic energy consumption, loss of job satisfaction and the squandering of some of society's most precious assets, which are the skill, ingenuity, creativity and enthusiasm of ordinary people.

Socially useful design

Socially useful design not merely exposes, criticises and challenges this process, but also presents constructive alternatives. Perhaps one of the most creative and imaginative responses has been that of the Lucas Workers with their corporate plan for socially useful production. Faced with the prospect of growing structural unemployment and reflecting an analysis that structural unemployment would be a growing feature of industrial society, the work-force at Lucas Aerospace proposed 150 socially useful products which they could design and manufacture as an alternative to facing degradation and suffering in the dole queue. The plan has received widespread international support and projects of a similar kind are being instigated in Australia, the United States and throughout Europe. Indeed, recently the French Government suggested that in addition to its measure to reduce the working week, establish early retirement and other features to deal with some of the multiplier effects of growing unemployment, it will also engage in a campaign of socially useful production similar to that advocated by the Lucas workers.[2]

The plan is essentially in two parts. On the one hand it proposes a range of socially useful products, and on the other, new forms of technology which would provide for a human-enhancing and liberating means of making the socially useful products. The product range includes equipment for medical uses: e.g. a vehicle for children with spina bifida and portable kidney machines. Indeed the Lucas workers were alarmed to learn that whilst they were facing structural unemployment 3000 people were dying each year for want of a lightweight portable dialysis machine which uses carbon filters and a small microprocessor for patient monitoring. With this equipment the patient will enjoy the dignity of doing an active job until a more long term solution is found for his or her medical problem.

The design staff and manual workers at Lucas jointly felt there was something seriously wrong about forms of technology which could make product systems as complex as Concorde while in the same society old people were dying of hypothermia. So they have designed, in conjunction with some colleagues at the Energy Research Unit of the Open University, a heat pump using natural gas, which produces a COP (coefficient of perfor-

mance) of 2.8 when the outdoor temperature falls to zero degrees and a COP of 4 in outdoor temperatures of 10 degrees. They have also involved a number of postgraduate
architectural students who were engaged in designing a factory to manufacture the heat
pumps, which would differ radically from factories in the past. It is possible for example
to vary the internal geometry of the factory and thereby alter the nature of the labour process and provide for rotation of job functions.

They have questioned the whole underlying design assumption of the private car as a
throw-away product. In the case of the powerpack they have been greatly influenced by the
work of Professor Thring and his colleagues, and have proposed a hybrid powerpack
which would make the best use of an internal combustion engine and electric motor and
would reduce fuel consumption by some 50% and toxic emissions by 80%. They have
suggested that the powerpack would be so designed that it could readily be repaired and
with suitable maintenance would be capable of running for some 20 years. Initially there
was a response from some critics asserting that this would lead to structural unemployment. It will be argued subsequently that the present forms of science and technology will
in any case give rise to structural unemployment. But these workers had a 'guts feeling' that
if things were designed to last much longer and an infrastructure was set up to repair them,
at least as much work would be created in doing so as was available in the capital-intensive
mass production lines producing the throw-away goods. This has since been supported by
an important report from the Battelle Institute in Geneva which broadly suggests that if
products such as cars or powerpacks were designed to last for 20 years not only would
energy and materials be conserved, but about 65 per cent more work would be created and
the jobs would be the interesting, fiddling, diagnostic type jobs that human beings love
doing.

The Lucas workers also collaborated with Richard Fletcher and his colleagues at the
North East London Polytechnic in producing a unique road/rail vehicle (Figure 1) which
is capable of running through cities as a coach and running on branch railway lines. It
could provide the basis for a cheap, ecologically desirable, integrated transport system. It
would also have many applications in Third World countries since it uses rubber tyres in
both modes and thereby has an incline capability of 1/6. This means that in Third World
countries instead of flattening the mountains and filling the valleys to make them level to
1/80 one could follow the contours of the countryside and enormously reduce civil
engineering costs. The design and development of many of these products was undertaken
by large groups of workers and not limited to a small technocratic elite. The tacit knowledge (in the Polanyi sense of the word) of the wide range of workers was used to the full.
Thus no virtue was made of complexity, and the sense of touch and feel and shape and
form of the workers was utilised to the full and resulted in a process in which it was possible to democratise decision-making processes in design involving masses of people. The
concept of socially useful design extends from products of this kind to the very forefront
of new technology. For example, when the Lucas workers were asked if there were ways
in which it would be possible to protect workers on the North Sea oil pipelines from the
hazards experienced there, they were conscious that their initial reaction was typical of the
existing forms of science and technology: they thought of designing a robotic device which
would eliminate the human being altogether. The more they reflected upon this the more
they realised what a massive amount of fixed capital is necessary to eliminate human intelligence and how incredibly intelligent human beings really are. Indeed, in the labour process the most complicated systems have units of intelligence of about 10^3, whereas a
human being could bring to bear equivalent switching circuits of 10^{14}. A human being
brings in addition the qualities of 'consciousness', 'will', 'ideology', 'humour', 'political
aspirations' – precisely all the things which systems designers regard as uncertainties and

therefore seek to eliminate. Yet on the other hand these are precisely the attributes which make human beings creative and imaginative. Thus the Lucas workers proposed a range of 'telecheric devices' which would react to human intelligence rather than absorb it. Such devices could also be used for coal mining so that the skill of the coal miner could be retained; for fire-fighting in which the skill of the fire-fighter would be retained; and there could be a wide range of medical applications including micro-surgery.

Figure 1   The prototype railbus developed from road-rail vehicle.

New technology

What was perhaps most significant about this product range was that it opened a debate about ways of using new technology which would enhance human intelligence rather than diminish it. Two examples are given here to illustrate how socially useful design could make a fuller and more creative use of new technologies. Technological change may be viewed variously as a process in which human energy and intelligence is replaced by machine intelligence in systems which become capital and energy intensive, or as a process in which systems become active and human beings become more passive and in which there is a shift from the analogical to the digital. Two examples will be cited – one from the field of manual work and one from the field of intellectual work to show how socially useful design would result in quite different systems. These relate to the writer's current research in the Design Discipline at the Open University.

Over the past 200 years turning has been one of the most highly skilled jobs to be found in engineering workshops. Toolroom turning is one of the most highly skilled jobs of all. The historical tendency, certainly since the end of the war, has been to de-skill this function by using NC machines. Noble has suggested that this represents a vivid example of social choice in machine design.[3]

The de-skilling takes place as a result of part programming, the process by which the desired NC motions are converted to finished tapes. Conventional (symbolic) part programming languages require that a part programmer, upon deciding how a something is to be machined, describes the desired tool motion by a series of symbolic commands. These commands are used to define geometric entities, that is points, lines and surfaces which may be given symbolic names.

In practice the part programming languages require the operator to synthesise desired tool motions from a restricted available vocabulary of symbolic commands. However, all this is doing is attempting to build into the machines the intelligence that would have been exercised by a skilled worker in going through the labour process. It is possible, by using computerised equipment in a symbolic way, to link it to the skill of a human being and define the tool motions without a symbolic description. Such a method is called analogic part programming.[4] In this type of part programming tool motion information is conveyed in an analogic form by turning a crank or moving a joystick or some other hand/eye co-ordination task using read-out with a degree of precision appropriate to the machining process.

Using a dynamic visual display of the entire working area of the machine tools including the workpieces, the fixtures, the cutting tool and its position, the skilled craftsmen can directly in-put the desired tool motions to machine the workpiece in the display. Such a system may be described as 'part programming by doing' and would represent a sharp contrast to the main historical tendencies towards symbolic part programming. It would require no knowledge of conventional part programming languages because the necessity to describe symbolically the desired tool motions would be eliminated. This is achieved by providing a system whereby the information regarding the cut is conveyed in a manner closely resembling the conventional process of the skilled machinist. Thus it would be necessary to maintain and enhance the skill and ability of a range of craftspeople who would work in parallel with the system.

Significant research has been carried out in this field. Yet in spite of its obvious advantages it has not been received with any enthusiasm by the large corporations or indeed funding bodies of any description. That this is so would appear to be an entirely 'political' judgement rather than a technological one.

In the field of intellectual work Rosenbrock has questioned the underlying assumptions of the manner in which we are developing computer-aided design systems. He charges firstly that the present techniques fail to exploit the opportunity which interactive computing can offer. The computer and the human mind have different but complementary abilities. If these different abilities can be combined then they represent something more powerful and effective than anything we have ever had before. Rosenbrock objects to the 'automated manual' type of system since it represents as he says 'loss of nerve, a loss of belief in human ability and a further unthinking application of the doctrine of the "division of labour" '.[5]

As in the case of turning described above, Rosenbrock sees two paths open in respect of design. The first is to accept the skill and knowledge of the designer, and to try to give designers improved techniques and improved facilities for exercising this skill and knowledge. Such a system would demand a truly interactive use of computers in a way that allows the very different capabilities of the computer and the human mind to be used to the full. The alternative to this, he suggests, is, 'to subdivide and codify the design process incorporating the knowledge of the existing designers so that it is reduced to a sequence of simple choices'.[6]

This, he points out, would lead to de-skilling such that the job could be done by a person with less training and less experience. Rosenbrock has demonstrated his human enhancing alternative by developing a CAD system with graphic output to develop displays from which the designer can assess stability, state of response, sensitivity to disturbance and other properties of a controlled system. If, having looked at the display, the performance of the system is not satisfactory, the display will suggest how it can be improved. In this respect the displays carry a long tradition of early 'pencil and paper' methods but, of course, they bring with them much greater computing power. So, as with lathes and the skilled turner, so also with the CAD system and the designer, possibilities do exist for a symbiotic relationship between the worker and the equipment. In both cases tacit knowledge and experience is accepted as valid and enhanced and developed.

Socially useful designs of advanced systems would develop these techniques further. This is in glaring contrast to the process in which de-skilling is now so widespread, but it has been suggested in the case of numerically controlled machine tools that the ideal workers are mentally retarded people and a mental age of 12 is advocated.

I have described elsewhere how these Tayloristic techniques have resulted in a situation where factory models are being imposed even on non-alienated work such as that of university professors, whose performance may be assessed by the well-known Frank-Wolfe algorithm. Similar problems are likely to arise in respect of artificial intelligence systems, or heuristic systems as applied to medical diagnosis where again two paths are open but the possibilities which socially useful design make possible would appear to have been ignored.

Indeed, the desire to eliminate uncertainty and human judgement is now so extensive that as Noble pointed out recently: 'Designing for idiots is the highest expression of the engineering art'.[7]

This process of de-skilling all types of work is supercharged by computer scientists who were described by Weizenbaum as 'being like children with a hammer who view the whole world as a nail'.[8] Those technologists and designers who question these processes and challenge the long term viability of the present approach to systems design are written off as being unrealistic or even senile.

Socially useful design provides a way out of this cul-de-sac. Not merely does one criticise the forms of science and technology but one proposes positive human enhancing alternatives. To assert those alternatives and to insist upon them we mean that a designer should cease to be the industrial Eichmann of a large corporation. Above all else designers should become involved with the millions of people who are now experiencing these problems[9] and to demonstrate that the community at large increasingly understands that science and technology are not given. They are made by human beings and if they are not doing what we want then we have a right and a responsibility to change them. The technological future is not out there in the sense that America was out there before Columbus went to discover it. The technological future does not have any predetermined shape or forms or contours. It is waiting to be built by people like us and socially useful design highlights one of the alternative choices.

References

1. Rose and Rose in W. Fuller, (ed.), *The Social Impact of Modern Biology*, London: Routledge and Kegan Paul.

2. J. Palmer, French boost for alternative work pioneers, *The Guardian*, 24 July 1981.
3. D. Noble, *Social Choice in Machine Design: The Case of NC Machine Tools*, Boston: MIT, 1979.
4. D. Gossard and B. Von Turkovitch, Analogic part programming with interactive graphics, in *CIRP*,vol. 27, January 1978.
5. H. H. Rosenbrock, The future of control, in *Automatica*, vol. 13, 1977.
6. Rosenbrock, *ibid*.
7. D. Noble, Introduction in (US edition) M. J. E. Cooley, *Architect orBee*, Boston: South End Press, 1982, p. xvi.
8. J. Weizenbaum, cited in M. J. E. Cooley, The thinking man's dilemma, *The Guardian*, 21 May 1981.
9. M. J. E. Cooley, New technologies – some trade union concerns and possible solutions, Keynote Paper, *EEC FAST Conference*, Selsdon Park, London, 1982.

Index

Index